Applying Pharmacogenomics in Therapeutics

Applying Pharmacogenomics in Therapeutics

Edited by

Xiaodong Feng
Hong-Guang Xie

CRC Press
Taylor & Francis Group
Boca Raton London New York

CRC Press is an imprint of the
Taylor & Francis Group, an **informa** business

CRC Press
Taylor & Francis Group
6000 Broken Sound Parkway NW, Suite 300
Boca Raton, FL 33487-2742

First issued in paperback 2022

ISBN-13: 978-1-466-58267-5 (hbk)
ISBN-13: 978-1-03-234008-1 (pbk)
DOI: 10.1201/b19000

Dedication

We dedicate this book to all of our patients who trust us to touch their lives. To us, therapeutics is personal.

Contents

Preface

Recent advances in high-throughput gene sequencing and other omics biotechnologies have created pharmacogenomics that facilitates the development of personalized medicine and the Precision Medicine Initiative unveiled by the Obama administration in January 2015, the major purpose of which is to improve health and treat diseases. In clinical settings, individuals vary in their response to drugs due to patient heterogeneity. Although DNA or pharmacogenomics is not the whole story about personalized medicine, it is generally accepted as the major determinant of variable drug safety and efficacy as well as cost-effectiveness for some (if not all) drugs. Therefore, widespread use of pharmacogenomics for patient care has become a critical requirement. There is an unprecedented urgency for future clinicians to become trained on how to interpret data from pharmacogenomic testing and to prepare for the upcoming future of healthcare—personalized medicine.

Applying Pharmacogenomics in Therapeutics is intended to educate and train healthcare professionals and students in applying the principles of pharmacogenomics in patient care. It can also serve as an up-to-date reference book covering the broad spectrum of the fundamentals, principles, practice, and the best current thinking, as well as the future potential of pharmacogenomics and personalized medicine. The book has 11 chapters contributed by well-established pharmacologists and scientists from various universities and the genomics industry in the United States and China. The first five chapters describe the principles and practice of pharmacogenomics and its biotechnologies as well as genetic biomarkers in the drug discovery and development, laboratory medicine, and clinical services. Chapters 6 through 9 focus on the use of pharmacogenomics in the treatment of cancers, cardiovascular diseases, neurologic and psychiatric disorders, and pulmonary diseases, and Chapters 10 and 11 address the merging of pharmacogenomics and alternative medicine and the integration of pharmacogenomics into pharmacoeconomics. Each chapter begins with the key concepts, followed by in-depth discussions on case reports or critical evaluation of genetic variants/biomarkers, and concludes with questions for the reader's self-examination.

The editors express their sincere appreciation to each of the chapter authors who have contributed their time and expertise to make this book a comprehensive and valuable resource.

Xiaodong Feng, PhD, PharmD
California Northstate University

Hong-Guang Xie, MD, PhD
Nanjing Medical University

Editors

 **Xiaodong Feng, PhD, PharmD,** earned his PhD in cellular and molecular physiology. During his PharmD training, he specialized in pharmacy education and clinical pharmacy practice. He began his biomedical research career 19 years ago at the Department of Surgery, Stony Brook Wound Center, Stony Brook School of Medicine (East Setauket, New York). Following three years of fellowship, he continued research and teaching as assistant professor in the Department of Dermatology, State University of New York at Stony Brook, for four years. He is currently professor of medical education and associate professor of clinical sciences at California Northstate University College of Medicine and College of Pharmacy (Elk Grove, California). He is also vice president of California Northstate University. Dr. Feng also practices as a clinical pharmacist at Sutter Davis Hospital (Davis, California) and is an oncology pharmacy specialist at Dignity Health Medical Foundation (Rancho Cordova, California). His integrated training, experience, and expertise in biomedical, pharmaceutical, and clinical education, research, and practice has enabled him to incorporate all these essential components of therapeutics into high-quality patient care and pharmaceutical education.

Dr. Feng has over 20 years of clinical and biomedical research experience in cancer, wound healing, and cardiovascular diseases. His current research interests include drug discovery of anti-angiogenesis therapy for macular degeneration and cancer, pharmacovigilance, and application of pharmacogenomics in patient care. Two US patents on strategies of anti-angiogenesis and cancer treatments have recently been issued to Dr. Feng. Personalizing pharmacotherapy using pharmacogenomics has always been his professional passion.

 **Hong-Guang Xie, MD, PhD,** earned his bachelor's degree of medicine (MD equivalent), master's degree of medicine, and doctoral degree of medicine (PhD equivalent) in 1984, 1989, and 1995, respectively, from Central South University Xiangya (formerly Xiang-Yale) School of Medicine (Changsha, China), where he was promoted to lecturer in 1989 and associate professor of pharmacology in 1995. He was appointed associate director of the Pharmacogenetics Research Institute (Changsha, China), as co-founder of the institute in 1997. As a recipient of the Merck International Clinical Pharmacology Fellowship, Dr. Xie joined Vanderbilt University School of Medicine (Nashville, Tennessee) as a research fellow (postdoctoral) in 1997 and went on to become a research instructor (faculty) in 2002. After that, he joined the University of California Washington Center at Washington, DC, as staff (a research fellow) in 2008, and

the United States Food and Drug Administration (FDA) in 2010 as an Oak Ridge Institute for Science and Education (ORISE) fellow. Beginning in December 2011, Dr. Xie became full professor of pharmacology and mentor of PhD students at Nanjing Medical University (Nanjing, China), and was appointed chief of the General Clinical Research Center, Nanjing First Hospital. In 2014, Dr. Xie was awarded the Distinguished Medical Expert of Jiangsu Province, China.

Dr. Xie is active in the research field of clinical pharmacology—pharmacogenomics and personalized medicine, in particular—with at least 130 papers published, many in leading journals worldwide, including *New England Journal of Medicine, Annual Review of Pharmacology and Toxicology, Advanced Drug Delivery Reviews, Pharmacology & Therapeutics, Clinical Pharmacology & Therapeutics, Drug Metabolism Reviews, The Pharmacogenomics Journal, Pharmacogenomics, Pharmacogenetics & Genomics, Journal of Thrombosis & Haemostasis, Drug Metabolism and Disposition,* and *Personalized Medicine*. In addition, he has authored several chapters in books published by Springer, Wiley, ASM Press, and Annual Reviews. Dr. Xie has served on the editorial board of *Pharmacogenomics,* UK (2003–2011); *American Journal of PharmacoGenomics* (currently *Molecular Diagnosis & Therapy*), New Zealand; and *Journal of Geriatric Cardiology*, China. Dr. Xie has also been invited to review about 30 scientific journals in the United States and the United Kingdom as well as other countries. More recently, Dr. Xie was invited to edit the English textbook *Pharmacology* for the medical students nationwide as an associate editor-in-chief, which will be published by the People's Medical Publishing House, Beijing, China.

Contributors

Weiguo Chen, PhD
Department of Medicine
The University of Illinois
Chicago, Illinois

Xiao-Ping Chen, PhD
Department of Clinical Pharmacology
Xiangya Hospital
Central South University
Changsha, China

Megan J. Ehret, PharmD, MS, BCPP
Department of Pharmacy Practice
School of Pharmacy
University of Connecticut
Storrs, Connecticut

Max Feng, B.S. Candidate
Department of Global and Public Health
Cornell University
Ithaca, New York

Xiaodong Feng, PhD, PharmD
California Northstate University
College of Medicine
Elk Grove, California
Dignity Health Medical Foundation
Mercy Cancer Center
Sacramento, California

Gary Yuan Gao, PhD
Department of Biomedical Engineering
School of Medicine
Johns Hopkins University
Baltimore, Maryland

Med Data Quest, Inc.
San Diego, California

Singlera Genomics Inc.
San Diego, California

Grace M. Kuo, PharmD, PhD, MPH, FCCP
UC San Diego Skaggs School of
Pharmacy and Pharmaceutical
Sciences
San Diego, California

Kelly C. Lee, PharmD, MAS, BCPP, FCCP
UC San Diego Skaggs School
of Pharmacy and Pharmaceutical
Sciences
San Diego, California

Mario Listiawan, PharmD
Ronald Reagan Medical Center
University of California Los Angeles
Los Angeles, California

Cong Liu, PhD
Department of Bioengineering
The University of Illinois
Chicago, Illinois

Rui Liu, MD, PhD
Singlera Genomics, Inc.
San Diego, California

Joseph D. Ma, PharmD
UC San Diego Skaggs School
of Pharmacy and Pharmaceutical
Sciences
San Diego, California

Long Ma, PhD
State Key Laboratory of Medical
Genetics and School
of Life Sciences
Central South University
Changsha, China

Jianbo J. Song, PhD, FACMG
Ambry Genetics
Aliso Viejo, California
Kaiser Permanente Northern California
    Regional Genetics Laboratories
San Jose, California

Ruth Vinall, PhD
Department of Pharmaceutical
    and Biomedical Sciences
California Northstate University
    College of Pharmacy
Elk Grove, California

Tibebe Woldemariam, PhD
Department of Biomedical
    and Pharmaceutical Sciences
California Northstate University
    College of Pharmacy
Elk Grove, California

Hong-Guang Xie, MD, PhD
General Clinical Research Center
Nanjing First Hospital
Nanjing Medical University
Nanjing, China

Jason X.-J. Yuan, MD, PhD
Division of Translational
    and Regenerative Medicine
Department of Medicine
College of Medicine
The University of Arizona
Tucson, Arizona

Wei Zhang, PhD (Chapters 3 and 9)
Department of Preventive Medicine
Northwestern University Feinberg
    School of Medicine
Chicago, Illinois

Wei Zhang, PhD (Chapter 10)
Department of Clinical Pharmacology
Xiangya Hospital
Central South University
Changsha, China

Hong-Hao Zhou, MD
Department of Clinical Pharmacology
Xiangya Hospital
Central South University
Changsha, China

1 Concepts in Pharmacogenomics and Personalized Medicine

Kelly C. Lee, Joseph D. Ma, and Grace M. Kuo

CONTENTS

KEY CONCEPTS

- Advances in pharmacogenomics have demonstrated its usefulness in pre-dicting drug efficacy, toxicity, and dosing for various disease states.
- When interpreting pharmacogenomic information, gene/allele of interest, functional effect, clinical relevance, and testing availability and recommen-dations are key areas to consider.
- Pharmacogenomic data contribute to omeprazole pharmacokinetic vari-ability, trastuzumab efficacy, and warfarin dosing. In addition, abacavir, codeine, and carbamazepine toxicity can be explained in part by pharmacogenomics.
- Challenges of pharmacogenomics and clinical implementation include availability of cost-effective, rapid-turnaround pharmacogenomic tests and knowledge and education of healthcare professionals.

INTRODUCTION

Pharmacogenomics is a rapidly expanding field that will have a significant impact on patient care and the role of healthcare professionals in providing personalized care. The clinical implications of utilizing pharmacogenomic information range across maximizing drug efficacy, minimizing drug toxicity, and avoiding unnecessary drug therapy. The ability of clinicians to obtain, interpret, and utilize pharmacogenomic information in a timely and efficient manner will determine their ability to individu-alize pharmacotherapy for their patients.

Pharmacogenomics also has the potential to optimize drug development. Patient populations can be stratified according to their pharmacogenomic profiles and sub-sequently be used to study the effectiveness of an investigational drug for specific patient genotypes and phenotypes. Variations in a gene may impact either the pharmacokinetics (PK) or pharmacodynamics (PD) of a drug, which in turn affects clinical outcomes. PK is the process by which a drug is absorbed, distributed, metabolized, and excreted by the body (ADME); in essence, "what the body does to the drug?" PD is the effect of a drug on the body; in other words, "what the drug does to the body?" These factors can potentially determine the efficacy or toxicity of a drug; for example, a polymorphism may lead to increased drug plasma concentrations, which, in turn, translates to increased drug effectiveness or adverse effects.

Pharmacogenetics is defined as "the study of genetic causes of individual variations in drug response."[1] Pharmacogenomics is defined as "the genome-wide analysis of genetic determinants of drug efficacy and toxicity."[2] Although the terms may be used interchangeably, we use the term *pharmacogenomics* in this chapter.

The purpose of this chapter is to provide an overview of the drugs whose pharmacogenomic applications could demonstrate their place in the prediction of drug efficacy, toxicity, and dosing. We review six important drugs as they relate to each category: omeprazole (efficacy), trastuzumab (efficacy), abacavir (toxicity), codeine (toxicity), carbamazepine (toxicity), and warfarin (dosing). Although these six examples vary in levels of evidence and recommendations for testing in clinical practice, they portray the current state of clinical applications (or lack thereof) in real-world practice. For each drug, we present a clinical case scenario, background of the drug, gene/allele of interest and functional effect, clinical relevance and testing availability, and recommendations. At the end of this chapter, we discuss several challenges in implementing pharmacogenomic testing in clinical practice, including availability of tests, generalizability of scientific evidence, and knowledge and education of healthcare professionals.

DEFINITIONS

A polymorphism is defined as a DNA sequence variant present in >1% of the general population.[3] If a DNA sequence variant is present in <1% of the population, it is defined as a mutation. An allele refers to two different versions, or alternate sequences, of the same gene localized at any position on a chromosome. Within a gene, variations of an individual nucleotide can be considered alleles. Alleles include the wild type: that is, the usual sequence, mutations, and polymorphisms of a given gene. Since humans are diploid organisms, there are two alleles (one from each parent) of each gene. Each person carries a set of two alleles of each gene, and this is defined as the genotype. A phenotype is the set of characteristics or the clinical presentation of an individual, which results from the particular genotype. For example, variations in the *cytochrome P450 (CYP)* genes may result in the contrasting phenotypes of ultrarapid metabolizers (UMs) and poor metabolizers (PMs) of the affected drug. A haplotype is a set of alleles at multiple, neighboring positions that coexist on the same chromosome. These alleles may be in separate locations within a single gene or among different genes. Neighboring alleles (located near one another) are physically tethered and usually inherited as a set that prevents their separation during inheritance. One individual inherits two copies of a haplotype.

There are different types of polymorphisms such as single nucleotide polymorphisms (SNPs), variable number tandem repeats (VNTRs), gene insertion (ins) and deletion (del), copy-number variants (CNVs), and premature stop codons. A SNP is the result of a single base substitution, and novel SNPs continue to be discovered. Some SNPs lie outside the protein-coding regions of the gene and may alter the expression level of a gene, and others lie within the protein-coding regions of the gene. These SNPs may or may not alter protein synthesis and are called nonsynonymous and synonymous polymorphisms, respectively. Enzymes, drug transporters, and/or receptors are examples of proteins that are genetically polymorphic, whose functional levels can be categorized as increased, decreased, or no change in protein activity. Examples exist where a polymorphism is present at a higher frequency in a certain ethnic group.[4] When determining clinical relevance, relationships between the polymorphism, the drug, and the disease should be examined for an individual. Drug dosing, efficacy, toxicity, PK, and/or PD may be affected by the polymorphism. A polymorphism may also influence disease prognosis, susceptibility, or be used as a screening test for certain diseases.[5]

EFFICACY: OMEPRAZOLE

CLINICAL CASE

DS is a 47-year-old Chinese American male who complains of a "burning stomach pain" with "bloating and heartburn." The pain occurs between meals, awakens him at night, and is not relieved with antacid use. He is an otherwise healthy male with no significant past medical history. He denies smoking, alcohol intake, and illicit drug use. Current medications and supplement use include a daily multivitamin. Vital signs are within normal limits with pertinent laboratory values of hemoglobin 13.0 g/dL and hematocrit 43%. He has no signs of bleeding (no bruising, no nose bleeds, and guaiac test negative). The physician is planning to start DS on a two-week triple therapy consisting of omeprazole, amoxicillin, and clarithromycin.

BACKGROUND

Omeprazole is indicated for the treatment of various gastric acid–related disorders (e.g., duodenal ulcer, gastric ulcer, gastroesophageal reflux disease [GERD], and Zollinger–Ellison syndrome). Most patients with GERD and/or *Helicobacter pylori*–related peptic ulcer present with abdominal pain often described as "burning." Omeprazole is used in combination with one or two antibiotics for *H. pylori* eradication.[6–8] It is an irreversible inhibitor of the proton pump (or H^+/K^+-ATPase), thus decreasing acid secretion in the stomach. Omeprazole is primarily metabolized by *CYP2C19* and to some extent by *CYP3A*.[9]

GENE/ALLELE OF INTEREST AND FUNCTIONAL EFFECT

The *CYP2C19* gene is localized at chromosome 10q24.1–q24.3.[10,11] The normal (wild-type) allele is *CYP2C19*1*. At least 34 *CYP2C19* variant alleles have been identified thus far,[12] with *CYP2C19*2* and *CYP2C19*3* being the most extensively studied variant alleles. For *CYP2C19*2*, a splicing defect in exon 5 occurs and results in early termination of protein synthesis.[13] *CYP2C19*3* is a premature stop codon SNP, resulting in a truncated protein. The functional effect of the *CYP2C19*2* and *CYP2C19*3* alleles is a loss-of-function (or null) enzyme activity. Other *CYP2C19* alleles also result in a loss-of-function (*CYP2C19*4–*8*) or reduced (*CYP2C19*9,*10,* and *12*) enzyme activity.[12,14] Increased enzyme activity has been associated with *CYP2C19*17.*[15]

Stratification of subjects by genotype is possible due to observed gene–dose effects for medications metabolized by *CYP2C19*. *CYP2C19*-genotyped homozygous extensive metabolizers (EMs) are individuals who possess two wild-type alleles. *CYP2C19*-genotyped heterozygous EMs are individuals who possess one wild-type and one decreased/null variant allele. *CYP2C19*-genotyped PMs are individuals who possess two decreased/null activity alleles.[16]

CLINICAL RELEVANCE

H. pylori dual/triple therapies that contain omeprazole vary in terms of dosing (Table 1.1). It has been suggested that higher omeprazole doses be considered in homozygous EMs, yet it remains to be implemented in clinical practice.[17] The lack of clinical practice implementation of omeprazole dosing based on the *CYP2C19* genotype is likely due to other nongenetic factors, including the multitude of available dual/triple therapy regimens, low incidence of clinically significant adverse effects, and large therapeutic window of omeprazole.

The majority of clinically relevant data have focused on the effect of *CYP2C19* genotypes on *H. pylori* eradication or cure. Interestingly, there is ample evidence linking omeprazole PK variability to PD variability. Omeprazole area under the concentration–time curve (AUC) is 7- to 14-fold higher in PMs compared with homozygous EMs.[18–20] Omeprazole AUC is correlated with intragastric pH ($r = 0.87$, $p < 0.0001$), of which *CYP2C19* PMs have higher intragastric pH concentrations compared with *CYP2C19* EMs.[20,21] The clinical significance of these findings is that a higher intragastric pH has been shown to increase antibiotic concentrations and improve antibiotic bioavailability and stability.[22–25]

The PK and PD relationship with omeprazole, explained to some extent by *CYP2C19* genetic polymorphisms, has resulted in different *H. pylori* cure rates (Table 1.1). In one study, Japanese patients ($n = 62$) with confirmed *H. pylori* infection were administered dual therapy with omeprazole and amoxicillin for several weeks. *H. pylori* cure rates were 28.6% (8/28 patients), 60% (15/25), and 100% (9/9) in *CYP2C19* homozygous EMs, heterozygous EMs, and PMs, respectively (Table 1.1).[17] These results are consistent with those derived from other omeprazole studies.[6,26–30] In addition, several meta-analyses have concluded that omeprazole efficacy is dependent on *CYP2C19* genotype.[31–33] However, there are conflicting data regarding no difference in *H. pylori* cure rates based on *CYP2C19* genotype for omeprazole (Table 1.1).[8,34–36] In one study, triple therapy with omeprazole, amoxicillin, and clarithromycin was administered for one week in Chinese patients with *H. pylori*–confirmed peptic ulcer disease ($n = 120$). *H. pylori* cure rates

TABLE 1.1

Summary of Dual and Triple Therapy with Omeprazole and Reported *H. pylori* Cure Rates Based on *CYP2C19* Genotype

Drug Regimen[a]	H. pylori Cure Rates (Cured Patients/Total Number of Patients)			Hom EMs vs. PMs	Hom EMs vs. Het EMs	Het EMs vs. PMs	Reference
	Hom EMs[b]	Het EMs[c]	PMs[d]				
Regimen 1	28.6% (8/28)	60% (15/25)	100% (9/9)	p < 0.05	p < 0.001	p < 0.05	17
Regimen 2	73.9% (17/23)	68.8% (33/48)	83.3% (15/18)	NS	NS	NS	35
Regimen 3	33% (3/9)	30% (3/10)	100% (2/2)	NS[e]			26
Regimen 4	43% (3/7)	58.3% (7/12)	100% (2/2)	NS[e]			26
Regimen 5	81% (13/16)	94.7% (18/19)	100% (9/9)	NS[e]			26
Regimen 6	76.2% (16/21)	88.9% (24/27)	90% (9/10)	NS	NS	NS	36
Regimen 7	72.7% (64/88)	92.1% (117/127)	97.8% (45/46)	p < 0.001	p < 0.001	NS	30
Regimen 8	68.9% (31/45)	84.4% (27/32)	91.3% (21/23)	p = 0.023[f]			6
Regimen 9	73.3% (22/30)	86.1% (31/36)	85% (17/20)	NS	NS	NS	34

(Continued)

TABLE 1.1 (Continued)

Summary of Dual and Triple Therapy with Omeprazole and Reported *H. pylori* Cure Rates Based on *CYP2C19* Genotype

Drug Regimen[a]	H. pylori Cure Rates (Cured Patients/Total Number of Patients)			Hom EMs vs. PMs	Hom EMs vs. Het EMs	Het EMs vs. PMs	Reference
	Hom EMs[b]	Het EMs[c]	PMs[d]				
Regimen 10	71.8% (28/39)	80.3% (49/61)	90% (18/20)	NS	NS	NS	8
Regimen 11	69.1% (56/81)	74.4% (93/125)	83.7% (36/43)	NS	NS	NS	28

Note: AMOX, amoxicillin; BID, twice daily; CLAR, clarithromycin; *CYP2C19*, cytochrome P450 2C19; EMs, extensive metabolizers; *H. pylori*, *Helicobacter pylori*; LANS, lansoprazole; NS, non significant; OMP, omeprazole; PMs, poor metabolizers; QD, once daily; QID, four times daily; RABE, rabeprazole; SUC, sucralfate; TID, three times daily.

a Regimen 1 = OMP 20 mg QD × 6 or 8 weeks and AMOX 500 mg QID × 2 weeks; Regimen 2 = OMP 20 mg BID and AMOX 500 mg BID × 2 weeks; Regimen 3 = OMP 40 mg/day and AMOX 2000 mg/day × 1 week; Regimen 4 = OMP 40 mg/day and AMOX 2000 mg/day and SUC 4000 mg/day × 1 week; Regimen 5 = OMP 40 mg/day and AMOX 2000 mg/day and CLAR 800 mg/day × 1 week; Regimen 6 = OMP 20 mg BID and AMOX 500 mg TID and CLAR 200 mg TID × 1 week; Regimen 7 = OMP 20 mg BID or LANS 30 mg BID and AMOX 500 mg TID and CLAR 200 mg TID × 1 week; Regimen 8 = OMP 20 mg BID and AMOX 1000 mg BID and CLAR 500 mg BID × 1 week; Regimen 9 = OMP 20 mg BID and AMOX 750 mg BID and CLAR 400 mg BID × 1 week; Regimen 10 = OMP 20 mg BID and AMOX 1000 mg BID and CLAR 500 mg BID × 1 week; Regimen 11 = OMP 20 mg BID or LANS 30 mg BID or RABE 10 mg BID and AMOX 750 mg TID and CLAR 200 mg BID.

b Homozygous EMs = *CYP2C19*1/*1*.

c Heterozygous EMs = *CYP2C19*1/*2*, *CYP2C19*1/*3*.

d PMs = *CYP2C19*2/*2*, *CYP2C19*2/*3*, *CYP2C19*3/*3*.

e *p*-Value was between homozygous and heterozygous EMs vs. PMs.

f *p*-Value was analyzed by chi-squared applying the linear-by-linear association model to determine if *H. pylori* cure rate was reduced in the patients' order ranked in a trend of extensive metabolizers within each treatment group.

were 71.8% (28/39), 80.3% (49/61), and 90% (18/20) in homozygous EMs, heterozygous EMs, and PMs, respectively, with no difference based on *CYP2C19* genotype ($p = 0.252$; Table 1.1).[8]

TESTING AVAILABILITY AND RECOMMENDATIONS

One genotyping test has been approved by the US Food and Drug Administration (FDA) and is commercially available (AmpliChip CYP450, Roche Diagnostics; www.roche.com) for *CYP2C19* genotyping. This test requires a whole blood sample, whereby DNA is extracted and then tested by a PCR-based microarray that analyzes the presence or absence of *CYP2C19*2* and **3* alleles. There are currently no *CYP2C19* testing recommendations for omeprazole. The clinical pharmacology and drug–drug interaction sections in the prescribing information have been revised regarding *CYP2C19* PMs, but no recommendations for *CYP2C19* testing are documented.[37] Proton-pump inhibitor (PPI) use in patients who are on antiplatelet therapy[38] have prompted professional organizations to publish expert consensus statements. For managing therapy with thienopyridines and PPIs, the ACCF/ACG/AHA states that *CYP2C19* testing has "not yet been established."[39]

EFFICACY: TRASTUZUMAB

CLINICAL CASE

JJ is a 58-year-old female who has just been diagnosed with Stage II breast cancer after first noticing a "hard lump" during a breast self-examination. A mammogram and ultrasound revealed a 3-cm mass. A biopsy was performed, which revealed evidence of invasive ductal carcinoma in the right breast and axillary lymph nodes. A CT scan showed no evidence of metastatic disease. JJ has no first-degree relatives with breast cancer or other family members affected by early breast cancer. The plan for JJ is a lumpectomy with complete axillary node dissection, start dose-dense doxorubicin and cyclophosphamide ("AC regimen") with a taxane, and possibly trastuzumab thereafter. However, tumor marker results regarding estrogen receptor, progesterone receptor, and human epidermal growth factor receptor type 2 (HER2) statuses are pending.

BACKGROUND

Trastuzumab is a humanized monoclonal antibody that binds to the extracellular domain of HER2 and inhibits the proliferation and survival of HER2-dependent tumors.[40,41] The significance of HER2 in cancers was discovered in the 1980s when the rodent homolog *neu* was identified in a rat tumorigenesis model.[42] Trastuzumab is indicated for HER2-overexpressing metastatic breast cancer, as adjuvant treatment of HER2-overexpressing early breast cancer in combination with chemotherapy, and for treatment of HER2-overexpressing metastatic gastric or gastroesophageal junction adenocarcinoma in combination with cisplatin and 5-fluorouracil or capecitabine.

Gene/Allele of Interest and Functional Effect

HER2 is a member of the epidermal growth factor receptor (EGFR) family and plays a significant role in initiating signal transduction pathways that ultimately lead to cellular growth, differentiation, and survival.[41] HER2 is a proto-oncogene that encodes a transmembrane protein, with HER2 overexpression in approximately 20–30% of invasive breast cancers.[41] HER2 overexpression is a significant predictor of overall survival and time to relapse in node-positive patients.[43]

Clinical Relevance

A clinical study of 469 women with HER2-overexpressing metastatic breast cancers who had not been previously treated with chemotherapy were randomized to receive standard chemotherapy alone or standard chemotherapy plus trastuzumab.[44] Standard chemotherapy included anthracycline plus cyclophosphamide or paclitaxel. Treatment response was evaluated at two months, five months, and at three-month intervals thereafter. A longer median time to disease progression was observed with standard chemotherapy plus trastuzumab versus standard chemotherapy alone (7.4 vs. 4.6 months, $p < 0.001$). Trastuzumab plus standard chemotherapy was also associated with other secondary outcomes, including a higher rate of overall response (50% vs. 32%, $p < 0.001$), a longer median duration of response (9.1 vs. 6.1 months, $p < 0.001$), and a longer median time to treatment failure (6.9 vs. 4.5 months, $p < 0.001$).[44] A follow-up study showed a significantly lower rate of death with trastuzumab plus standard chemotherapy versus standard chemotherapy alone after one year.[45]

Additional evidence of trastuzumab efficacy in HER2-positive metastatic breast cancer came from patients with HER2-positive metastatic breast cancer ($n = 186$) who received docetaxel with or without trastuzumab until disease progression.[46] A higher overall response rate was observed in patients who were treated with trastuzumab and docetaxel versus docetaxel alone (61% vs. 34%, $p = 0.0002$). Additionally, compared to docetaxel alone, patients on trastuzumab and docetaxel had a longer overall median survival time (31.2 vs. 22.7 months, $p = 0.03$), longer median time to disease progression (11.7 vs. 6.1 months, $p = 0.0001$), longer median time to treatment failure (9.8 vs. 5.3 months, $p = 0.001$), and longer median duration of response (11.7 vs. 5.7 months, $p = 0.009$).[46]

Efficacy of trastuzumab for HER2-positive and/or -overexpressing breast cancers is also consistent in the adjuvant setting. Romond et al. analyzed the combined results of two trials ($n = 3351$) in women with surgically removed HER2-positive breast cancer.[45] Trastuzumab plus adjuvant paclitaxel reduced mortality by one-third compared to chemotherapy alone ($p = 0.015$).[45] In another study, 3387 women with HER2-positive breast cancer who had completed locoregional therapy and at least four cycles of chemotherapy were randomized to either one year of trastuzumab or observation.[47] The authors reported that trastuzumab treatment for one year after adjuvant chemotherapy significantly improved disease-free survival.[47]

Several clinical studies have also examined the role of trastuzumab-based neoadjuvant chemotherapy in HER2-positive breast cancer.[48,49] In the Neoadjuvant Herceptin (NOAH) study, HER2-positive, locally advanced breast cancer patients who received neoadjuvant chemotherapy with trastuzumab followed by adjuvant trastuzumab had a higher event-free survival at three years (71% vs. 56%, $p < 0.05$) versus patients who were not treated with neoadjuvant trastuzumab.[49] Additionally, patients who received neoadjuvant trastuzumab had a higher pathological complete response in breast tissue (45% vs. 22%, $p < 0.05$).[49] These data are consistent with the results from the GeparQuattro study, where a pathological complete response was observed with an anthracycline–taxane-based neoadjuvant chemotherapy with trastuzumab in locally advanced, HER2-positive breast cancer.[50]

TESTING AVAILABILITY AND RECOMMENDATIONS

Fluorescence in situ hybridization (FISH), immunohistochemistry (IHC), chromogenic in situ hybridization (CISH), a quantitative HER2 total expression, and HER2 homodimer assay are several methods for detecting HER2.[51] Although FISH assesses whether *HER2* gene amplification has occurred, IHC evaluates the level of HER2 protein in invasive breast cancer cells. Combining aspects of FISH and IHC is CISH, which uses permanent staining and ready identification of invasive tissue to selectively stain for the *HER2* gene. This method has proven to be a better predictor of HER2-expression status.[51] Although IHC, FISH, and CISH are indirect measures of the *HER2* gene, the HER mark assay provides a quantitative measurement of HER2 total protein and HER2 homodimer levels.[51,52]

Trastuzumab should only be used in patients with HER2-overexpressing breast cancer. The American Society of Clinical Oncology/College of American Pathologists recommend that HER2 testing be performed in all patients with invasive (early stage or recurrence) breast cancer.[53] The National Comprehensive Cancer Network Task Force Report provides guidelines about IHC and FISH cutoff scores to determine HER2 status.[52]

TOXICITY: ABACAVIR

CLINICAL CASE

A 25-year-old male patient with a six-month history of acquired immunodeficiency syndrome (AIDS) presents to the clinic today with fever and a maculopapular rash on the trunk of his body for the past five days. He has no known drug allergies, and his last viral load was 25,000 copies and CD4 count 110/mm³. His current antiretroviral regimen is abacavir, zidovudine, and efavirenz. He is prescribed with sulfamethoxazole–trimethoprim for *Pneumocystis jiroveci* pneumonia prophylaxis.

BACKGROUND

Abacavir is a nucleoside reverse transcriptase inhibitor used for combination antiretroviral therapy for treating human immunodeficiency virus (HIV) infections. Despite its efficacy in treating HIV infections, abacavir has been associated with treatment-limiting hypersensitivity reaction (HSR). Abacavir-induced HSR occurs in about 5–8% of patients treated with the drug and generally presents within six weeks of drug initiation.[54,55] Symptoms of an abacavir-induced HSR may include skin rash, fever, malaise, gastrointestinal symptoms, and respiratory symptoms. Severe forms of the skin rash may result in Stevens–Johnson syndrome (SJS), toxic epidermal necrolysis (TEN), or systemic lupus erythematosus.[56] If a patient experiences an HSR, abacavir is discontinued and symptoms generally resolve within 72 hours.[57] Restarting abacavir is contraindicated as this may result in a potentially life-threatening reaction.[55]

GENE/ALLELE OF INTEREST AND FUNCTIONAL EFFECT

It appears that the presence of allele *human leukocyte antigen (HLA)-B*5701* confers a high risk of abacavir-induced HSR. Of those with *HLA-B*5701*-negative allele status, less than 1% of individuals are likely to develop HSR. The interaction of abacavir or its metabolite with the major histocompatibility complex (MHC) class I may lead to a CD8+ T-cell-mediated cell death.[58,59] Of those with positive status, more than 70% of patients develop HSR. The prevalence of the *HLA-B*5701* allele is the highest in white populations (5–8%).[56,60–62] In African American, Asian, and Hispanic populations, the prevalence is 0.26–3.6%.[61–64]

CLINICAL RELEVANCE

Dosing and efficacy of abacavir have not been shown to be affected by *HLA-B*5701*. The clinical utility of *HLA-B*5701* testing for abacavir has been demonstrated from several prospective clinical studies. In the PREDICT-1 study, the incidence of confirmed abacavir-associated HSR was 2.7% in the control group versus 0% in the *HLA-B*5701* screened group ($p < 0.001$).[65] In the retrospective, case–control SHAPE study, 47 of 199 white and black patients had suspected HSR cases, with confirmation from a skin patch test. In this study, *HLA-B*5701* screening accurately predicted 100% of abacavir-associated HSR cases. It should be noted that skin patch testing identifies patients who had an immunologically mediated HSR to abacavir. However, it is only useful for confirmation of suspected HSR cases, and it does not work in those without previous exposure to abacavir.[66] In the ARIES study, *HLA-B*5701*-negative patients had less than 1% clinically suspected abacavir-induced HSR, and none had positive skin patch tests at 30 weeks.[67]

TESTING AVAILABILITY AND RECOMMENDATIONS

Abacavir-induced HSR has been associated with the presence of the MHC class I allele *HLA-B*5701*. A screening test for the *HLA-B*5701* allele may allow clinicians to identify patients who are at risk of developing an HSR to abacavir. Due to the severity of the HSR and its potential impact on overall antiretroviral regimens, the *HLA-B*5701* screening test may help minimize potential abacavir toxicity and thus optimize antiretroviral therapy. To test the *HLA-B*5701* allele, a blood or saliva sample specimen is collected. The genetic sequences coding for the *HLA-B*5701* are probed and reported as positive if the allele is present, or negative if the allele is absent.

Among seven international laboratories, specificity and sensitivity of detecting the *HLA-B*5701* allele via PCR sequencing were 100% and 99.4%, respectively.[68] In studies where patients were diagnosed with an abacavir HSR based on symptom presentation, the sensitivity of the *HLA-B*5701* test was 46–78%.[64,69,70] In contrast, that sensitivity was up to 94–100% in patients with an immunologically confirmed (via skin patch testing) abacavir HSR.[65,71,72] There is suggestion that the discrepancy of lower estimates of test sensitivity was the inclusion of non-abacavir-related HSR.[73] However, the specificity of the *HLA-B*5701* test is 90–100%, regardless of whether abacavir HSR is based on symptom presentation or immunologic confirmation.[64,65,69–72] Pooled data from three study populations reported a positive predictive value and a negative predictive value of 82% (95% confidence interval [CI]: 71–90%) and 85% (95% CI: 81–88%), respectively.[64,69,70] This suggests a high genetic penetrance of the *HLA-B*5701* allele in predisposing patients to abacavir HSR.[70]

Several guidelines provide guidance for whether patients should be tested for the *HLA-B*5701* allele prior to initiation. According to the US Department of Health and Human Services,[74] screening for *HLA-B*5701* prior to initiation of abacavir is recommended. If a patient is positive for the *HLA-B*5701* allele, abacavir should not be recommended.[74] The Infectious Diseases Society of America (IDSA) has also recommended *HLA-B*5701* testing prior to initiating abacavir to reduce the risk of an HSR. Patients positive for the *HLA-B*5701* should not be treated with abacavir.[75] In July 2008, the US FDA updated the black box warning in the prescribing information[76] to recommend screening for the *HLA-B*5701* polymorphism prior to starting abacavir treatment. Screening is also recommended prior to reinitiation of abacavir in patients of unknown *HLA-B*5701* status who have previously tolerance abacavir.[77] The benefits of testing for that allele and the potential detection of a potentially life-threatening HSR clearly outweigh the risk of testing.

In at least two pharmacoeconomic studies, testing for the *HLA-B*5701* polymorphism has been shown to be cost-effective according to the high risk of abacavir-induced HSR.[70,78] Some publications have demonstrated the feasibility of implementing such a testing program for abacavir, which involves education and training of staff and continual monitoring.[63,79,80]

TOXICITY: CARBAMAZEPINE

CLINICAL CASE

TL is a 30-year-old Chinese American male with newly diagnosed complex partial seizures who presents to the clinic for evaluation. He has not previously been treated with an antiepileptic drug, and the neurologist is considering starting carbamazepine. TL has no known drug allergies, and his other medical history is pertinent for hypertension and seasonal allergies. His medications include losartan and loratadine.

BACKGROUND

Carbamazepine is an anticonvulsant that is indicated for partial and generalized seizures, trigeminal neuralgia, and bipolar disorder.[30,81] The exact mechanism of action is unknown, but it has been shown to block the voltage-gated sodium channels. It is primarily metabolized by CYP3A4 to produce an active metabolite, carbamazepine-10,11-epoxide.[30] Carbamazepine has been linked to life-threatening idiosyncratic, type B adverse drug reactions (ADRs), compared with type A ADRs which are dose dependent.[82] These reactions can be severe cutaneous reactions ranging from SJS to TEN. The incidence of SJS/TEN is less than two patients per million per year,[83] and the rate of death with these conditions (in the absence of carbamazepine) is about 5% and 35% for SJS and TEN, respectively.[84] Recently, investigators have discovered an association of *HLA-B*1502* allele with the risk for developing SJS and TEN, specifically in Asians who are prescribed carbamazepine. Most recently, other *HLA* alleles (**3101* and **1511*) have been recognized to potentially contribute to an HSR associated with the use of carbamazepine.[81,85,86]

GENE/ALLELE OF INTEREST AND FUNCTIONAL EFFECT

Due to early observations of HSR in families and identical twins, the HLA has been a primary target for analysis of SJS/TEN reactions, which also tend to be familial in pattern.[87] The *HLA-B*1502* allele and its association with SJS/TEN appear to be phenotype specific because that allele is not associated with other HSRs related to carbamazepine, such as mild maculopapular eruptions (MPE) or drug reactions with eosinophilia systemic symptoms (DRESS).[88,89] The relationship between *HLA-B*1502* and SJS/TEN also appears to be drug specific because that allele cannot predict SJS/TEN induced by drugs other than carbamazepine. However, other aromatic anticonvulsants, such as phenytoin, oxcarbazepine, and lamotrigine, may also cause similar HSRs in persons carrying the *HLA-B*1502* allele.[90] In theory, SJS/TEN HSRs may be due to noncovalent binding between carbamazepine and *HLA-B*1502* complex, leading to a CD8+-mediated cell death.[88,91] The high incidence of carbamazepine-induced SJS/TEN in Asians is correlated with the high frequency of *HLA-B*1502* in the same population. The population prevalence of the allele is estimated to be 10–15% in China (Han Chinese), Indonesia, Malaysia, Taiwan, Thailand, the Philippines, and Vietnam; 2–8% in South Asia; and <1% in Japanese, Koreans, African Americans, Europeans, and Hispanics.[92–94]

CLINICAL RELEVANCE

Although a preliminary study has been published regarding the role of microsomal epoxide hydrolase and its prediction of maintenance doses of carbamazepine,[95] there is no definitive study showing that any specific gene/allele can accurately predict doses or clinical efficacy of carbamazepine. The data have confirmed association of the *HLA-B*1502* variant allele with carbamazepine toxicity. It was previously shown that white subjects positive for the *HLA-B*1502* allele are not at risk for carbamazepine-induced HSR.[96] In the landmark study of 44 Han Chinese patients, there was 100% association of the *HLA-B*1502* allele with carbamazepine-induced SJS/TEN.[97] They also reported a follow-up study in which

59 of 60 patients with carbamazepine-induced SJS/TEN had tested positive for the *HLA-B*1502* allele.[89] The one patient who did not test positive for the *HLA-B*1502* allele was tested positive for the *HLA-B*1558* allele. This was compared to 6 out of 144 controls (tolerant to carbamazepine) who were carriers of the *HLA-B*1502* allele (OR = 1357, 95% CI: 193.4–8838.3, $p = 1.6 \times 10^{-41}$).[89] In a separate study, there was a significant difference in SJS/TEN incidences among patients who received carbamazepine, depending on whether they were *HLA-B*1502* carriers (100%) versus noncarriers (14.5%).[93]

HLA-B*1502 was also studied in 4877 Taiwanese patients who were candidates for carbamazepine and had not received testing for the allele.[98] Of them, 7.7% of the patients had tested positive for the allele and was given an alternative medication or a pre-study medication. The 92.3% of the patients who tested negative for the *HLA-B*1502* allele were advised to take carbamazepine. Mild transient rash occurred in 4.3% of subjects, but SJS/TEN did not develop in any patient. This is significantly lower than 0.23% of carbamazepine-induced SJS/TEN from historical incidence ($p < 0.001$).[98]

In addition to Chinese and Taiwanese patients, the *HLA-B*1502* was studied in Thai patients in two studies. A case–control study was conducted in a Thai population in which the odds ratio for developing carbamazepine-induced SJS/TEN was 54.76 (95% CI: 14.62–205.13, $p = 2.89 \times 10^{-12}$) among those who were positive for *HLA-B*1502*.[99] The positive predictive value and negative predictive value of the *HLA-B*1502* allele were 1.92% and 99.96%, respectively. Another study showed a strong association between *HLA-B*1502* and carbamazepine- and phenytoin-induced SJS but not MPE.[100] Of the 81 patients with epilepsy, 31 subjects had antiepileptic drug-induced SJS/MPE. *HLA-B*1502* was associated with carbamazepine-induced SJS and phenytoin-induced SJS ($p = 0.005$ and $p = 0.0005$, respectively).

As discussed before, the *HLA-B*1502* allele may increase the risk of toxicity from other anticonvulsant drugs (lamotrigine, oxcarbazepine, and phenytoin) by contributing to SJS/TEN.[89,101,102] In a case–control study, *HLA-B*1502* was present in 8/26 (30.8%) patients who received phenytoin and developed SJS/TEN (OR = 5.1, 95% CI: 1.8–15.1, $p = 0.0041$) and 3/3 (100%) patients who were tolerant to carbamazepine (OR = 80.7, 95% CI: 3.8–1714.4, $p = 8.4 \times 10^{-4}$). This may be partially due to similarities in aromatic structure between carbamazepine, oxcarbazepine, phenytoin, and lamotrigine.[89] Clinically, there is an estimated 20–30% cross-reactivity probability between these drugs.[103,104]

In addition to *HLA-B*1502*, several other HLA alleles have been recently identified as potential markers for carbamazepine-induced HSRs. The *HLA-A*3101* has been associated with carbamazepine-induced HSRs in Japanese and Europeans.[85,86] In Japanese patients, the prevalence of *HLA-B*1502* is <1% and correlates to the low prevalence of carbamazepine-induced HSRs. Among Japanese, the *HLA-B*1511* has been shown to be a risk factor for carbamazepine-induced SJS/TEN.[81] The *HLA-B*1508*, **1511*, and **1521* as well as **1502* are all members of the HLA-B75 type and have been detected in studies performed in India and Thailand. One study also showed an association of *HLA-B*1518*, *HLA-B*5901*, and *HLA-C*0704* alleles with severe cutaneous ADRs.[81]

A simple HLA typing will indicate whether patients are positive for the *HLA-B*1502*; patients are positive if either one or two alleles of *HLA-B*1502* are present. In 2007, the US FDA recommended that all patients of Asian descent be screened for the *HLA-B*1502* allele before initiating carbamazepine therapy. This recommendation is also reflected in the black box warning in the prescribing information of carbamazepine.[30] Therefore, prior to initiation of this drug in high-risk patients, genotyping is recommended.

It is important to note that patients who have been treated with carbamazepine for an extended period do not need testing of the *HLA-B*1502* allele. The SJS/TEN reactions generally occur within the first two months of treatment, and the risk for developing HSR is considered to be low even among carriers of the *HLA-B*1502* status.[92] However, patients who are negative for the *HLA-B*1502* allele and receive carbamazepine should still be monitored clinically for the development of HSRs since other genetic and nongenetic factors may contribute to these ADRs.

Clinical Pharmacogenetics Implementation Consortium (CPIC) published a guidance for use and dosing of carbamazepine in patients who are carriers and non-carriers of *HLA-B*1502*.[105] They recommend that in carriers of the *HLA-B*1502*, carbamazepine should not be used in those who are new to the drug. In those who have previously used carbamazepine for longer than three months without any adverse effects, the drug can be used with caution.

TOXICITY: CODEINE

CLINICAL CASE

A 55-year-old man is being evaluated for chronic back pain that he has been suffering from for the past five years. He has tried multiple medications, including anti-inflammatory drugs, opiates, and anticonvulsants with little relief. He is considering surgical options for his chronic back pain because he does not want to become "dependent" on his medications, and he has had numerous side effects from his medications.

BACKGROUND

Codeine is a weak opiate agonist that is used to treat mild-to-moderate pain.[106] In order to exert its efficacy, it must be converted to its active metabolite, morphine, via cytochrome P450 2D6 (CYP2D6).[107–109] It is often combined with other analgesics, such as acetaminophen, for mild-to-moderate pain disorders and postsurgical analgesia.

GENE/ALLELE OF INTEREST AND FUNCTIONAL EFFECT

The interindividual variability of CYP2D6 enzyme activity can be explained, in part, by the genetic variation of *CYP2D6*.[110–112] Genetic polymorphisms of *CYP2D6* include phenotyping subgroups of PMs, intermediate metabolizers (IMs), EMs, and UMs.[108,109] The EMs are considered to have normal enzyme activity.[113-116] The median AUC of the morphine metabolite increases from PMs to EMs to UMs.[109] The *CYP2D6*3, *4, *5, *6*, and *7 alleles have been reported to account for the majority of decreased CYP2D6 enzyme activity.[110,112] In addition, the decreased activity of CYP2D6 in IMs has been shown to result from *CYP2D6*9* and *10 alleles.[110] Gene duplication of *CYP2D6* found in UMs is associated with higher plasma concentration and AUCs of morphine than in EMs.[109]

CLINICAL RELEVANCE

The prescribing information for codeine states that the prevalence of CYP2D6 phenotype varies widely and has been estimated at 0.5–1% in Chinese and Japanese; 0.5–1% in Hispanics; 1–10% in whites; 3% in African Americans; and 16–28% in North Africans, Ethiopians, and Arabs. Data are not available for other ethnic groups.[106]

Studies have shown that increased pain threshold is found in EMs but not in PMs, and that morphine in urine samples is not found in PMs.[111,117] The PMs have impaired O-demethylation; therefore, codeine does not have analgesic effects in PMs.[108] Even though PMs may not experience analgesic effects expected from codeine, they can still develop side effects (e.g., sedation, headaches, dizziness, and dry mouth), which may be related to codeine itself rather than its metabolites, including morphine.[118] UMs of CYP2D6 experience greater analgesic effects but have increased risks for toxicity.[109,119,120] For example, UMs have a higher incidence of sedation compared to EMs (91% vs. 50%, $p = 0.069$).[109] Other potential side effects for UMs include euphoria, dizziness, and visual disturbances[119-121] or more severe symptoms, such as extreme sleepiness, confusion, shallow breathing, or respiratory suppression.[106]

A case report by Koren et al. in 2006 described an adverse drug event related to codeine in a breastfeeding mother taking codeine 30 mg and paracetamol 600 mg for postpartum episiotomy pain management.[122,123] Her codeine dosage on day 1 was 60 mg (2 tablets) every 12 hours. However, due to side effects of somnolence and constipation, she lowered the dose by half on day 2 to 14. Unfortunately, her 13-day-old baby died from a morphine overdose with a serum concentration of 70 ng/mL (neonates breastfed by mothers receiving codeine typically have morphine serum concentrations of 0–2.2 ng/mL).[124] The morphine concentration found in her breast milk stored on day 10 was 87 ng/mL (normal range is 1.9–2.5 ng/mL for doses of 60 mg every 6 hours). The mother was genotyped and determined to be an UM of codeine (heterozygous for a *CYP2D6*2A* allele with *CYP2D6*2×2* gene duplication).[122] There is a correlation between increased codeine dosage and ADRs in breastfeeding neonates.[125-127] Subsequently, the US FDA issued a warning, which was included in codeine's prescribing information, stating that "maternal use of codeine can potentially lead to serious adverse reactions, including death, in nursing infants," and that "If a codeine-containing product is selected, the lowest dose should be prescribed for the shortest period of time to achieve the desired clinical effect. Mothers using codeine should be informed about when to seek immediate medical care and how to identify the signs and symptoms of neonatal toxicity, such as drowsiness or sedation, difficulty breastfeeding, breathing difficulties, and decreased tone, in their baby."[106,128]

Inhibitors of *CYP2D6* will interact with codeine and affect its metabolite to display a phenotype similar to that of a PM (i.e., lack of analgesic effect). Therefore, potential adverse drug interactions between codeine and *CYP2D6* inhibitors should be monitored.[106]

TESTING AVAILABILITY AND RECOMMENDATIONS

The AmpliChip® (Roche Diagnostic) DNA microarray, which can detect 33 *CYP2D6* alleles, is an FDA approved test for *CYP2D6* polymorphisms in the United States.[129,130] Additional FDA approved tests will become available that will increase accessibility to patients and providers. Laboratory monitoring is not necessary and no formal recommendation is required prior to initiation of therapy.[123,128]

DOSING: WARFARIN

CLINICAL CASE

KG is a 68-year-old woman who is starting warfarin anticoagulation therapy for a new onset of atrial fibrillation. Her present medical history includes type 2 diabetes, hypertension, and hyperlipidemia. Other oral medications she is taking include metformin 500 mg twice daily with meals, lisinopril 10 mg daily, and atorvastatin 20 mg daily. She does not smoke or drink alcohol. She is advised to change her diet to a heart-healthy one that includes increased salad intake. Her baseline international normalized ratio (INR) is 1.0 and her therapeutic INR goal is 2.5. Her genotype profile reveals *CYP2C9*3/*3* and *VKORC1-1639A/A*.

BACKGROUND

Warfarin is an example of how pharmacogenomic testing can be used to affect medication dosing. It is a widely used oral anticoagulant for the prevention and treatment of thromboembolic diseases (e.g., atrial fibrillation, deep vein thrombosis, pulmonary embolism).[131] Warfarin has a narrow therapeutic window and is associated with a significant risk of bleeding.[132] It also has high interindividual variability in its dosage requirements.[133] Factors accounting for the high interindividual variability come from both nongenetic (e.g., age, disease states, concomitant medications)[134–137] and genetic (e.g., polymorphisms in *CYP2C9* or *VKORC1* variants)[138] sources.

GENE/ALLELE OF INTEREST AND FUNCTIONAL EFFECT

The *CYP2C9* and *VKORC1* polymorphisms are the main genetic factors that have been observed to affect warfarin dosing. They affect either the metabolism of warfarin or the formation of active clotting factors. The *CYP2C9* alleles affect the metabolism of warfarin; patients with the *CYP2C9*1* allele have normal CYP2C9 activity (wild type). Patients with the *CYP2C9*2* or *CYP2C9*3* alleles have decreased CYP2C9 enzyme activity, resulting in impaired warfarin metabolism and increased risk of bleeding.[139,140] The *VKORC1 1173 C>T* or *VKORC1-1639 G>A* polymorphism affects the formation of vitamin K–dependent clotting factors.[141]

Clinical Relevance

The US FDA has approved dosing information based on pharmacogenomic testing for the brand name product of warfarin (Coumadin®).[131] Patients with *CYP2C9*2* or *CYP2C9*3* alleles may have decreased CYP2C9 activities by 50–90%, thus requiring lower warfarin doses.[139,140] Haplotype A from *VKORC1* SNPs is associated with lower warfarin dose requirements, and haplotype B is associated with higher warfarin dose requirements.[141,142] Patients with the *VKORC1* A/A genotype are more likely to require a lower weekly warfarin dose of 32%; those with the B/B genotype are likely to require an increased weekly warfarin dose of 35%.[143]

Warfarin pharmacogenomic testing is helpful in guiding dosing decisions for patients who are either sensitive or resistant to the drug therapy. In a multicenter clinical trial conducted by the International Warfarin Pharmacogenetics Consortium, 4043 patients taking warfarin were randomized to three algorithms—pharmacogenetic, clinical, and fixed dose (35 mg/week) approaches. The study demonstrated that the pharmacogenetic algorithm group showed more accurate dose estimates than the other two dosing algorithms, particularly for patients requiring a total weekly dose of <21 mg or >49 mg.[144] Warfarin pharmacogenomic testing may help patients minimize bleeding risks although the supporting evidence is conflicting.[145] The risk of bleeding in patients with at least one allele variant of *CYP2C9* can increase two- to threefold during the initiation phase of therapy; however, the evidence for the risk of bleeding in patients during long-term therapy is conflicting.[74,133,146,147]

Testing Availability and Recommendations

Pharmacogenomic testing for *CYP2C9* and *VKORC1* may help improve prediction of warfarin target maintenance doses.[143] Not all patients with polymorphisms in *CYP2C9* or *VKORC1* will have a serious bleeding event; those without any polymorphism may still be susceptible to having a bleeding event due to clinical, nongenetic factors. There are several FDA-approved pharmacogenomic tests that detect both *CYP2C9* and *VKORC1* polymorphisms. Pharmacogenomics testing may be cost-effective in helping patients achieve an optimal therapeutic dose and avoid unnecessary bleeding risks.[148]

EDUCATIONAL RESOURCES

Several professional organizations have reviewed, compiled, and endorsed competencies for pharmacogenomic education. In 2001, the National Coalition for Health Professional Education in Genetics (NCHPEG), representing a working group of specialists with experience in genetics and health professions, identified core competencies in genetics for healthcare professionals. In 2007, 18 core competencies were updated to encourage healthcare professionals to "integrate genetics knowledge, skills, and attitudes into routine health care."[149]

The NCHPEG core competencies include baseline competencies requiring each healthcare professional to identify clinical applications of genetics/genomics information, to understand the implications of social and psychological implications, and to know how and when to refer patients to genetics professionals.

Knowledge competencies include obtaining knowledge of terminology; basic principles; disease-associated genetic variations; family history; treatment options; cultural and health beliefs; physical and/or psychosocial benefits; and resource, ethical, legal, and social issues.

Skills competencies include obtaining skills of gathering genetic family history; referring patients for genetics consultation; explaining reasons for and benefits of genetic services, using information technology to obtain credible and current information about genetics; and assuring an appropriate informed-consent process for genetics/genomics tests.

Attitudes competencies include appreciating the sensitivity of genetic information and the need for privacy and confidentiality, and seeking coordination and collaboration with an interdisciplinary team of healthcare professionals.

Several pharmacy professional organizations have adopted the NCHPEG competencies and revised their own specific discipline competencies. For example, the American Association of Colleges of Pharmacy (AACP) Academic Affairs Committee reported in 2002 pharmacist-specific competencies in pharmacogenetics and pharmacogenomics, including competencies related to patient care and patient education.[150] Other pharmacy organizations have emphasized the importance of pharmacogenomics as an important principle in the setting of a patient population.[151] In 2011, the National Human Genome Research Institute (NHGRI) of the National Institutes of Health (NIH) convened a meeting with several pharmacist organizations and other stakeholder groups.[152] Together, they explored the current status of pharmacist genomic education, identified barriers and facilitators to enhanced education, and planned to implement steps to ensure that pharmacists obtain pharmacogenomic competencies.[152] Other healthcare professional organizations have also developed recommendations or programs for genetics/genomics education, including the American Academy of Family Physicians (AAFP) Core Educational Guidelines,[153] the Association of American Medical Colleges (AAMC) Contemporary Issues in Medicine: Genetics Education Report,[154] the physician assistant web-based genetics education program,[37] and the Essential Nursing Competencies and Curricula Guidelines for Genetics and Genomics.[155]

An increasing number of pharmacogenomic educational resources are available via journal articles, books, and online sources. Table 1.2 lists examples of online resources for pharmacogenomic information and educational materials, including resources from the US FDA, Centers for Disease Control and Prevention (CDC), the Pharmacogenomics Knowledge Base (PharmGKB), the Pharmacogenomics Education Program (PharmGenEd), the NHGRI from the NIH, and the Genetics/Genomics Competency Center for Education (G2C2). The FDA lists medications with available pharmacogenomic implications in certain sections of the labeling information. Select drugs with available pharmacogenomic guidelines provided by the CPIC are posted on both the CDC and the PharmGKB websites. The PharmGenEd, using a shared curriculum platform to provide open-access educational materials for students

TABLE 1.2

Pharmacogenomic Online Resources

Resource	Website	Description
US Food and Drug Administration	http://www.fda.gov/Drugs/ ScienceResearch/ ResearchAreas/ Pharmacogenetics/ ucm083378.htm	Table of pharmacogenomic biomarkers in drug labels
Center for Disease Control and Prevention	http://www.cdc.gov/ genomics/gtesting/ guidelines.htm	Lists of pharmacogenomic testing guidelines
Pharmacogenomics Knowledge Base (PharmGKB)	http://www.pharmgkb.org/	Lists of drugs and corresponding pharmacogenomic information and guidelines
Pharmacogenomics Education Program (PharmGenEd)	http://pharmacogenomics. ucsd.edu	Evidence-based lecture videos and slides
National Human Genome Research Institute	http://www.genome.gov/ Education/	Various educational materials about genetics and genomics
Genetics/Genomics Competency Center for Education	http://www.g-2-c-2.org	Provides reviews and organizes educational resources through an interdisciplinary collaborative exchange

and trainers, provides evidence-based lecture videos and slides on pharmacogenomic concepts and clinical applications in various areas.[156,157] The NHGRI website provides resources such as a glossary of genetic terms, fact sheets, the human genome project, and links to genetic education resources. The G2C2 website maps competencies in various genetics/genomics areas and links to available educational resources.

The need for disseminating pharmacogenomic information to healthcare professional students is on the rise. In a 2005 survey conducted by Latif et al., 39% of US Doctor of Pharmacy programs reported offering pharmacogenomics in their curriculum;[158] in comparison, a 2009 survey conducted by Murphy et al. showed that 89.3% of US pharmacy schools included pharmacogenomics in their curriculum.[159] Even though an increasing number of pharmacy schools have incorporated pharmacogenomic content into their curriculum, the depth and clinical applicability may be limited.[159] Furthermore, the majority did not have plans for faculty development in this area.[159] In US and Canadian medical schools, 82% in 2010 reported having incorporated pharmacogenomics into their curriculum; however, only 28% had more than 4 hours of the required didactic pharmacogenomic coursework.[160] In nursing schools, 70% in 2005 reported having had some genetics course content in their curriculum; however, they indicated offering <1 hour or between 1 and 5 hours of genetics education.[161] The future directions of pharmacogenomic education for healthcare professionals will require competency-based educational efforts that will help improve knowledge, skills, and attitudes to effectively translate pharmacogenomic evidence into clinical practice.[152,162]

CHALLENGES OF PHARMACOGENOMICS AND PERSONALIZED MEDICINE

AVAILABILITY OF TESTING

A major challenge to the implementation of pharmacogenomics into clinical practice is pharmacogenomic testing availability.[154,163,164] Conducting pharmacogenomic testing requires specialized genotyping equipment and training of onsite personnel.[154,163,164] Consequently, there may be limited availability of genotyping equipment, lack of suitable training of personnel, and costs associated with such testing. Although centralized laboratories (Labcorp) to perform such testing are available, feasibility information such as the turnaround time for test results or test sensitivity and specificity vary.[165] Practice settings may not have access to testing kits and laboratories in order to conduct testing. In one study, a questionnaire was sent to individuals representing hospitals, laboratories, and universities throughout New Zealand and Australia ($n = 629$) to determine utilization rates of pharmacogenomic testing for drug-metabolizing enzymes.[77] The overall response rate was 81.1% ($n = 510$), with 2% of facilities currently performing clinical genotype testing.[77] Additional evidence includes another study, whereby 20% of respondents from North American medical practices have available warfarin pharmacogenomic testing.[163]

SMALL SAMPLE SIZE

Due to the low prevalence of a specific variant allele in a studied population, numerous pharmacogenomic studies were conducted with small sample sizes. In studies with omeprazole and other PPIs, there was an unequal distribution of homozygous EMs, heterozygous EMs, and PMs, with a smaller number of PMs (Table 1.1). A small sample size in a clinical study is problematic. It is a study design limitation that increases the probability of an error and/or misinterpretation of study results due to lack of statistical power.[166] Ideally, a pharmacogenomic study should have sufficient statistical power of at least 80%, with an equal stratification of subjects across groups. However, this may not be achievable as many variant alleles carry a population frequency of 1–2% and/or the minimum detectable difference used to determine sample size is an unknown value.[4,166,167] Attempts to improve statistical design have been reported with *CYP2C9* and *VKORC1* testing with warfarin. The authors reported that a sample size of 1238 patients is needed to achieve a minimum difference of 5.49% with an 80% statistical power.[166]

KNOWLEDGE AND EDUCATION OF HEALTHCARE PROFESSIONALS

Knowledge deficiencies in genetics and pharmacogenomics exist among healthcare professionals in all disciplines[168] and preclude such individuals from implementing pharmacogenomics and/or personalized medicine into clinical practice. One study evaluated warfarin pharmacogenomic knowledge among pharmacists and nurses/nurse practitioners who provide anticoagulation services.[163] The survey response rate was low (22%), which included five knowledge-based questions. Approximately one-third of respondents correctly answered the knowledge-based questions. Knowledge was poor in areas of the length of time required to perform a test, interpreting test results, and *CYP2C9* and *VKORC1* allele frequencies.[163] In an Internet-based survey of genetics education in psychiatry residency programs, more than 50% of physicians felt that pharmacogenomic training was minimal.[169]

The NCHPEG is a working group of specialists with experience in genetics and health professions, with the intent of developing genomic/pharmacogenomic educational content for healthcare professionals.[149] In 2007, NCHPEG developed 18 core competencies, which include, but are not limited to, understanding basic genetic terminology; identifying genetic variations that facilitate prevention, diagnosis, and treatment options; and identifying available resources to assist those seeking genetic information or services.[149] Such competencies have been adapted in part by professional organizations. For pharmacists, the AACP Academic Affairs Committee reported specific competencies in pharmacogenetics and pharmacogenomics, which were derived from 2001 NCHPEG competencies. Examples include a pharmacist being able to identify patients for whom pharmacogenetic testing is indicated, identify an appropriate pharmacogenomic test for a patient, and be able to provide recommendations based on pharmacogenomic testing results.[150] For physicians, the AAFP Core Educational Guideline and the AAMC Contemporary Issues in Medicine: Genetics Education Report have also provided some considerations in pharmacogenomics/genomics.[170,171]

SUMMARY

The future applicability of pharmacogenomic tests in clinical practice is unknown, and there is both excitement and trepidation by the potential possibilities of these tests among patients and healthcare professionals. With the rapid discoveries of new genes and polymorphisms that affect drug efficacy, toxicity, and dosing, personalized medicine is quickly becoming a reality. There is increasing hope for not only being able to individualize drug treatment but also to treat diseases that were previously considered incurable. As we venture into the next decade of personalized medicine, healthcare professionals in all disciplines need to be knowledgeable in interpreting scientific evidence and appropriately educating patients so that they can make informed decisions about their healthcare.

STUDY QUESTIONS

CASE 1

AS is a 40-year-old Korean woman with a chief complaint of epigastric pain after eating. She denies any blood in her stools and fever. Her past medical history is significant for occasional headaches and seasonal allergies. She currently takes acetaminophen 325 mg every 6 hours as needed. She has no known drug allergies. Her physician orders a urea breath test, which reveals *H. pylori* peptic ulcer disease. She is prescribed a triple therapy consisting of omeprazole, amoxicillin, and clarithromycin.

1. The patient obtains information about a commercially available *CYP2C19* pharmacogenomic test and decides to complete the test. The test reveals that she has a *CYP2C19*3/*3* genotype. Based on the patient's genotype, what is her anticipated *CYP2C19* enzyme activity?
   a. Normal *CYP2C19* enzyme activity
   b. Increased *CYP2C19* enzyme activity
   c. Decreased *CYP2C19* enzyme activity
   d. No *CYP2C19* enzyme activity
2. Based on the patient's *CYP2C19*3/*3* genotype, what is the anticipated effect on omeprazole pharmacokinetics?
   a. Decreased omeprazole area under the concentration–time curve
   b. Increased omeprazole plasma concentrations
   c. Decreased omeprazole volume of distribution
   d. Increased omeprazole clearance

CASE 2

KK is a 29-year-old breastfeeding woman who is receiving codeine after giving birth to a healthy baby boy. After three days, she started to experience symptoms of sleepiness and dizziness. Now, after five days, she noticed that her infant seems sleepy all day, does not want to breastfeed, and may be losing body tone.

3. Codeine is metabolized to morphine by which CYP enzyme?
   a. *CYP2D6*
   b. *CYP2C9*
   c. *CYP2C19*
   d. *CYP3A4*
4. Which metabolizer phenotype most accurately describes a person who does not respond to codeine?
   a. Extensive metabolizer
   b. Intermediate metabolizer
   c. Poor metabolizer
   d. Ultrarapid metabolizer
5. Which ethnic population has the highest incidence of carbamazepine-induced SJS/TEN?
   a. African Americans
   b. Caucasians
   c. Han Chinese
   d. Hispanics

Answer Key
   1. d
   2. b
   3. a
   4. c
   5. c

REFERENCES

1. American Association of Pharmaceutical Scientists. *Pharmacogenomics (PGx)*. Available from: http://www.aaps.org/Pharmacogenomics/ (accessed September 8, 2015).
2. Evans, W.E. and M.V. Relling. Pharmacogenomics: Translating functional genomics into rational therapeutics. *Science*, 1999; 286(5439): 487–91.
3. *Genomics and Its Impact on Science and Society: The Human Genome Project and Beyond*. U.S. Department of Energy Genome Research Programs. Available from: www.genomics.energy.gov (accessed March 9, 2015).
4. Davaalkham, J., et al. Allele and genotype frequencies of cytochrome P450 2B6 gene in a Mongolian population. *Drug Metab Dispos*, 2009; 37(10): 1991–93.
5. Dendukuri, N., et al. Testing for HER2-positive breast cancer: A systematic review and cost-effectiveness analysis. *CMAJ*, 2007; 176(10): 1429–34.
6. Sheu, B.S., et al. Esomeprazole 40 mg twice daily in triple therapy and the efficacy of *Helicobacter pylori* eradication related to CYP2C19 metabolism. *Aliment Pharmacol Ther*, 2005; 21(3): 283–8.
7. Shi, S. and U. Klotz. Proton pump inhibitors: An update of their clinical use and pharmacokinetics. *Eur J Clin Pharmacol*, 2008; 64(10): 935–51.
8. Zhang, L., et al. The effect of cytochrome P2C19 and interleukin-1 polymorphisms on *H. pylori* eradication rate of 1-week triple therapy with omeprazole or rabeprazole, amoxycillin and clarithromycin in Chinese people. *J Clin Pharm Ther*, 2010; 35(6): 713–22.
9. Lawson, E.B., et al. Omeprazole limited sampling strategies to predict area under the concentration-time curve ratios: Implications for cytochrome P450 2C19 and 3A phenotyping. *Eur J Clin Pharmacol*, 2012; 68(4): 407–13.
10. Romkes, M., et al. Cloning and expression of complementary DNAs for multiple members of the human cytochrome P450IIC subfamily. *Biochemistry*, 1991; 30(13): 3247–55.
11. Zaphiropoulos, P.G. RNA molecules containing exons originating from different members of the cytochrome P450 2C gene subfamily (CYP2C) in human epidermis and liver. *Nucleic Acids Res*, 1999; 27(13): 2585–90.
12. *Home Page of the Human Cytochrome P450 (CYP)*. Allele Nomenclature Committee, 2008. Available from: http://www.cypalleles.ki.se/ (accessed March 19, 2014).
13. de Morais, S.M., et al. The major genetic defect responsible for the polymorphism of S-mephenytoin metabolism in humans. *J Biol Chem*, 1994; 269(22): 15419–22.
14. Blaisdell, J., et al. Identification and functional characterization of new potentially defective alleles of human CYP2C19. *Pharmacogenetics*, 2002; 12(9): 703–11.
15. Baldwin, R.M., et al. Increased omeprazole metabolism in carriers of the CYP2C19*17 allele; a pharmacokinetic study in healthy volunteers. *Br J Clin Pharmacol*, 2008; 65(5): 767–74.
16. Kim, M.J., et al. Effect of sex and menstrual cycle phase on cytochrome P450 2C19 activity with omeprazole used as a biomarker. *Clin Pharmacol Ther*, 2002; 72(2): 192–9.
17. Furuta, T., et al. Effect of genetic differences in omeprazole metabolism on cure rates for *Helicobacter pylori* infection and peptic ulcer. *Ann Intern Med*, 1998; 129(12): 1027–30.
18. Furuta, T., et al. Effects of clarithromycin on the metabolism of omeprazole in relation to CYP2C19 genotype status in humans. *Clin Pharmacol Ther*, 1999; 66(3): 265–74.
19. Sakai, T., et al. CYP2C19 genotype and pharmacokinetics of three proton pump inhibitors in healthy subjects. *Pharm Res*, 2001; 18(6): 721–7.
20. Shirai, N., et al. Effects of CYP2C19 genotypic differences in the metabolism of omeprazole and rabeprazole on intragastric pH. *Aliment Pharmacol Ther*, 2001; 15(12): 1929–37.

21. Furuta, T., et al. CYP2C19 genotype status and effect of omeprazole on intragastric pH in humans. *Clin Pharmacol Ther*, 1999; 65(5): 552–61.

22. Goddard, A.F., et al. Effect of omeprazole on the distribution of metronidazole, amoxicillin, and clarithromycin in human gastric juice. *Gastroenterology*, 1996; 111(2): 358–67.

23. Goddard, A.F. and R.C. Spiller. The effect of omeprazole on gastric juice viscosity, pH and bacterial counts. *Aliment Pharmacol Ther*, 1996; 10(1): 105–9.

24. Grayson, M.L., et al. Effect of varying pH on the susceptibility of *Campylobacter pylori* to antimicrobial agents. *Eur J Clin Microbiol Infect Dis*, 1989; 8(10): 888–9.

25. Midolo, P.D., et al. Oxygen concentration influences proton pump inhibitor activity against *Helicobacter pylori* in vitro. *Antimicrob Agents Chemother*, 1996; 40(6): 1531–33.

26. Aoyama, N., et al. Sufficient effect of 1-week omeprazole and amoxicillin dual treatment for *Helicobacter pylori* eradication in cytochrome P450 2C19 poor metabolizers. *J Gastroenterol*, 1999; 34(Suppl 11): 80–3.

27. Sapone, A., et al. The clinical role of cytochrome p450 genotypes in *Helicobacter pylori* management. *Am J Gastroenterol*, 2003; 98(5): 1010–15.

28. Take, S., et al. Interleukin-1beta genetic polymorphism influences the effect of cytochrome P 2C19 genotype on the cure rate of 1-week triple therapy for *Helicobacter pylori* infection. *Am J Gastroenterol*, 2003; 98(11): 2403–8.

29. Tanigawara, Y., et al. CYP2C19 genotype-related efficacy of omeprazole for the treatment of infection caused by *Helicobacter pylori*. *Clin Pharmacol Ther*, 1999; 66(5): 528–34.

30. Furuta, T., et al. Effect of genotypic differences in CYP2C19 on cure rates for *Helicobacter pylori* infection by triple therapy with a proton pump inhibitor, amoxicillin, and clarithromycin. *Clin Pharmacol Ther*, 2001; 69(3): 158–68.

31. Padol, S., et al. The effect of CYP2C19 polymorphisms on *H. pylori* eradication rate in dual and triple first-line PPI therapies: A meta-analysis. *Am J Gastroenterol*, 2006; 101(7): 1467–75.

32. Zhao, F., et al. Effect of CYP2C19 genetic polymorphisms on the efficacy of proton pump inhibitor-based triple therapy for *Helicobacter pylori* eradication: A meta-analysis. *Helicobacter*, 2008; 13(6): 532–41.

33. Tang, H.L., et al. Effects of CYP2C19 loss-of-function variants on the eradication of *H. pylori* infection in patients treated with proton pump inhibitor-based triple therapy regimens: A meta-analysis of randomized clinical trials. *PLoS One*, 2013; 8(4): e62162.

34. Dojo, M., et al. Effects of CYP2C19 gene polymorphism on cure rates for *Helicobacter pylori* infection by triple therapy with proton pump inhibitor (omeprazole or rabeprazole), amoxicillin and clarithromycin in Japan. *Dig Liver Dis*, 2001; 33(8): 671–5.

35. Miyoshi, M., et al. A randomized open trial for comparison of proton pump inhibitors, omeprazole versus rabeprazole, in dual therapy for *Helicobacter pylori* infection in relation to CYP2C19 genetic polymorphism. *J Gastroenterol Hepatol*, 2001; 16(7): 723–8.

36. Inaba, T., et al. Randomized open trial for comparison of proton pump inhibitors in triple therapy for *Helicobacter pylori* infection in relation to CYP2C19 genotype. *J Gastroenterol Hepatol*, 2002; 17(7): 748–53.

37. Food and Drug Administration. *Table of Pharmacogenomic Biomarkers in Drug Labels*. 2011. Available from: http://www.fda.gov/drugs/scienceresearch/researchareas/pharmacogenetics/ucm083378.htm (accessed March 18, 2014).

38. Kwan, J., et al. Effect of proton pump inhibitors on platelet inhibition activity of clopidogrel in Chinese patients with percutaneous coronary intervention. *Vasc Health Risk Manag*, 2011; 7: 399–404.

39. Abraham, N.S., et al. ACCF/ACG/AHA 2010 Expert Consensus Document on the concomitant use of proton pump inhibitors and thienopyridines: A focused update of the ACCF/ACG/AHA 2008 expert consensus document on reducing the gastrointestinal risks of antiplatelet therapy and NSAID use: A report of the American College of Cardiology Foundation Task Force on Expert Consensus Documents. *Circulation*, 2010; 122(24): 2619–33.

40. Rueckert, S., et al. A monoclonal antibody as an effective therapeutic agent in breast cancer: Trastuzumab. *Expert Opin Biol Ther*, 2005; 5(6): 853–66.

41. Hudis, C.A. Trastuzumab—Mechanism of action and use in clinical practice. *N Engl J Med*, 2007; 357(1): 39–51.

42. Shih, C., et al. Transforming genes of carcinomas and neuroblastomas introduced into mouse fibroblasts. *Nature*, 1981; 290(5803): 261–4.

43. Slamon, D.J., et al. Human breast cancer: Correlation of relapse and survival with amplification of the HER-2/neu oncogene. *Science*, 1987; 235(4785): 177–82.

44. Slamon, D.J., et al. Use of chemotherapy plus a monoclonal antibody against HER2 for metastatic breast cancer that overexpresses HER2. *N Engl J Med*, 2001; 344(11): 783–92.

45. Romond, E.H., et al. Trastuzumab plus adjuvant chemotherapy for operable HER2-positive breast cancer. *N Engl J Med*, 2005; 353(16): 1673–84.

46. Marty, M., et al. Randomized phase II trial of the efficacy and safety of trastuzumab combined with docetaxel in patients with human epidermal growth factor receptor 2-positive metastatic breast cancer administered as first-line treatment: The M77001 study group. *J Clin Oncol*, 2005; 23(19): 4265–74.

47. Piccart-Gebhart, M.J., et al. Trastuzumab after adjuvant chemotherapy in HER2-positive breast cancer. *N Engl J Med*, 2005; 353(16): 1659–72.

48. Chang, H.R. Trastuzumab-based neoadjuvant therapy in patients with HER2-positive breast cancer. *Cancer*, 2005; 116(12): 2856–67.

49. Gianni, L., et al. Neoadjuvant chemotherapy with trastuzumab followed by adjuvant trastuzumab versus neoadjuvant chemotherapy alone, in patients with HER2-positive locally advanced breast cancer (the NOAH trial): A randomised controlled superiority trial with a parallel HER2-negative cohort. *Lancet*, 2010; 375(9712): 377–84.

50. Untch, M., et al. Neoadjuvant treatment with trastuzumab in HER2-positive breast cancer: Results from the GeparQuattro study. *J Clin Oncol*, 2010; 28(12): 2024–31.

51. Larson, J.S., et al. Analytical validation of a highly quantitative, sensitive, accurate, and reproducible assay (HERmark) for the measurement of HER2 total protein and HER2 homodimers in FFPE breast cancer tumor specimens. *Patholog Res Int*, 2010; 2010: 814176.

52. Carlson, R.W., et al. HER2 testing in breast cancer: NCCN Task Force report and recommendations. *J Natl Compr Canc Netw*, 2006; 4(Suppl 3): S1–22; quiz S23–4.

53. Wolff, A.C., et al. Recommendations for human epidermal growth factor receptor 2 testing in breast cancer: American Society of Clinical Oncology/College of American Pathologists clinical practice guideline update. *J Clin Oncol*, 2013; 31(31): 3997–4013.

54. Hernandez, J.E., et al. Clinical risk factors for hypersensitivity reactions to abacavir: Retrospective analysis of over 8,000 subjects receiving abacavir in 34 clinical trials. *Programs and abstracts of the 43rd Interscience Conference on Antimicrobial Agents and Chemotherapy*, 2003; 339, September 14–17, 2003; Chicago, Illinois.

55. Hetherington, S., et al. Hypersensitivity reactions during therapy with the nucleoside reverse transcriptase inhibitor abacavir. *Clin Ther*, 2001; 23: 1603–14.

56. Hughes, C.A., et al. Abacavir hypersensitivity reaction: An update. *Ann Pharmacother*, 2008; 42: 387–96.

57. Lucas, A., D. Nolan, and S. Mallal. HLA-B*5701 screening for susceptibility to abacavir hypersensitivity. *J Antimicrob Chemother*, 2007; 59: 591–3.
58. Adam, J., W.J. Pichler, and D. Yerly. Delayed drug hypersensitivity: Models of T-cell stimulation. *Br J Clin Pharmacol*, 2011; 71(5): 701–7.
59. Chessman, D., et al. Human leukocyte antigen class I-restricted activation of CD8+ T cells provides the immunogenetic basis of a systemic drug hypersensitivity. *Immunity*, 2008; 28(6): 822–32.
60. Maiers, M., L. Gragert, and W. Klitz. High-resolution HLA alleles and haplotypes in the United States population. *Hum Immunol*, 2007; 68: 779–88.
61. Orkin, C., et al. Prospective epidemiological study of the prevalence of human leukocyte antigen (HLA)-B*5701 in HIV-1-infected UK subjects. *HIV Med*, 2010; 11: 187–92.
62. Orkin, C., et al. An epidemiologic study to determine the prevalence of the HLA-B*5701 allele among HIV-positive patients in Europe. *Pharmacogenet Genomics*, 2010; 20: 307–14.
63. Faruki, H., et al. HLA-B*5701 clinical testing: Early experience in the United States. *Pharmacogenet Genomics*, 2007; 17: 857–60.
64. Mallal, S., et al. Association between presence of HLA-B*5701, HLA-DR7, and HLA-DQ3 and hypersensitivity to HIV-1 reverse transcriptase inhibitor abacavir. *Lancet*, 2002; 359: 727–32.
65. Mallal, S., et al. HLA-B*5701 screening for hypersensitivity reaction. *N Engl J Med*, 2008; 358: 568–79.
66. Lai-Goldman, M. and H. Faruki. Abacavir hypersensitivity: A model system for pharmacogenetic test adoption. *Genet Med*, 2008; 10(12): 874–8.
67. Young, B., et al. First large, multicenter, open-label study utilizing HLA-B*5701 screening for abacavir hypersensitivity in North America. *AIDS*, 2008; 22: 1673–75.
68. Hammond, E., et al. External quality assessment of HLA-B*5701 reporting: An international multicentre survey. *Antivir Ther*, 2007; 12: 1027–32.
69. Hetherington, S., et al. Genetic variations in HLA-B region and hypersensitivity reactions to abacavir. *Lancet*, 2002; 359: 1121–22.
70. Hughes, D.A., et al. Cost-effectiveness analysis of HLA B*5701 genotyping in preventing abacavir hypersensitivity. *Pharmacogenetics*, 2004; 14: 335–42.
71. Martin, A.M., et al. Predisposition to abacavir hypersensitivity conferred by HLA-B*5701 and a haplotypic Hsp-70-Hom variant. *Proc Natl Acad Sci U S A*, 2004; 101: 4180–5.
72. Saag, M., et al. High sensitivity of human leukocyte antigen-b *5701 as a marker for immunologically confirmed abacavir hypersensitivity in white and black patients. *Clin Infect Dis*, 2008; 46: 1111–18.
73. Hughes, A.R., et al. Pharmacogenetics of hypersensitivity to abacavir: From PGx hypothesis to confirmation to clinical utility. *Pharmacogenomics J*, 2008; 8: 365–74.
74. Margaglione, M., et al. Genetic modulation of oral anticoagulation with warfarin. *Thromb Haemost*, 2000; 84(5): 775–8.
75. Aberg, J.A., et al. Primary care guidelines for the management of persons infected with human immunodeficiency virus: 2009 update by the HIV Medicine Association of the Infectious Diseases Society of America. *Clin Infect Dis*, 2009; 49: 651–81.
76. U.S. Food and Drug Administration. Information for healthcare professionals: Abacavir (marketed as Ziagen) and abacavir-containing medications. Available from: http://www.fda.gov/Drugs/DrugSafety/PostmarketDrugSafetyInformationforPatientsandProviders/ucm123927.htm (accessed March 9, 2015).
77. Gardiner, S.J. and E.J. Begg. Pharmacogenetic testing for drug metabolizing enzymes: Is it happening in practice? *Pharmacogenet Genomics*, 2005; 15(5): 365–9.

78. Schackman, B.R., et al. The cost-effectiveness of HLA-B*5701 genetic screening to guide initial antiretroviral therapy for HIV. *AIDS*, 2008; 22: 2025–33.

79. Lalonde, R.G., et al. Successful implementation of a national HLA-B85701 genetic testing service in Canada. *Tissue Antigens*, 2009; 75: 12–18.

80. Shah, J. Criteria influencing the clinical uptake of pharmacogenomic strategies. *BMJ*, 2004; 328: 1482.

81. Kaniwa, N., et al. HLA-B*1511 is a risk factor for carbamazepine-induced Stevens-Johnson syndrome and toxic epidermal necrolysis in Japanese patients. *Epilepsia*, 2010; 51(12): 2461–5.

82. Gomes, E.R. and P. Demoly. Epidemiology of hypersensitivity drug reactions. *Curr Opin Allergy Clin Immunol*, 2005; 5(4): 309–16.

83. Rzany, B., et al. Histopathological and epidemiological characteristics of patients with erythema exudativum multiforme major, Stevens-Johnson syndrome and toxic epidermal necrolysis. *Br J Dermatol*, 1996; 135(1): 6–11.

84. Roujeau, J.C. The spectrum of Stevens-Johnson syndrome and toxic epidermal necrolysis: A clinical classification. *J Invest Dermatol*, 1994; 102(6): 28S–30S.

85. McCormack, M., et al. HLA-A*3101 and carbamazepine-induced hypersensitivity reactions in Europeans. *N Engl J Med*, 2011; 364(12): 1134–43.

86. Ozeki, T., et al. Genome-wide association study identifies HLA-A*3101 allele as a genetic risk factor for carbamazepine-induced cutaneous adverse drug reactions in Japanese population. *Hum Mol Genet*, 2011; 20(5): 1034–41.

87. Edwards, S.G., et al. Concordance of primary generalised epilepsy and carbamazepine hypersensitivity in monozygotic twins. *Postgrad Med J*, 1999; 75(889): 680–1.

88. Mauri-Hellweg, D., et al. Activation of drug-specific CD4+ and CD8+ T cells in individuals allergic to sulfonamides, phenytoin, and carbamazepine. *J Immunol*, 1995; 155(1): 462–72.

89. Hung, S.I., et al. Genetic susceptibility to carbamazepine-induced cutaneous adverse drug reactions. *Pharmacogenet Genomics*, 2006; 16(4): 297–306.

90. Aihara, M. Pharmacogenetics of cutaneous adverse drug reactions. *J Dermatol*, 2011; 38(3): 246–54.

91. Naisbitt, D.J., et al. Hypersensitivity reactions to carbamazepine: Characterization of the specificity, phenotype, and cytokine profile of drug-specific T cell clones. *Mol Pharmacol*, 2003; 63(3): 732–41.

92. Lee, M.T., et al. Pharmacogenetics of toxic epidermal necrolysis. *Expert Opin Pharmacother*, 2010; 11(13): 2153–62.

93. Lonjou, C., et al. A marker for Stevens-Johnson syndrome …: Ethnicity matters. *Pharmacogenomics J*, 2006; 6(4): 265–8.

94. Schawartz, J. and A. Pollack. Judge invalidates human gene patent. *The New York Times*, 2010. Available from: http://www.nytimes.com/2010/03/30/business/30gene.html (accessed December 2, 2011).

95. Makmor-Bakry, M., et al. Genetic variants in microsomal epoxide hydrolase influence carbamazepine dosing. *Clin Neuropharmacol*, 2009; 32(4): 205–12.

96. Alfirevic, A., et al. HLA-B locus in Caucasian patients with carbamazepine hypersensitivity. *Pharmacogenomics*, 2006; 7(6): 813–18.

97. Chung, W.H., et al. Medical genetics: A marker for Stevens-Johnson syndrome. *Nature*, 2004; 428(6982): 486.

98. Chen, P., et al. Carbamazepine-induced toxic effects and HLA-B*1502 screening in Taiwan. *N Engl J Med*, 2011; 364(12): 1126–33.

99. Tassaneeyakul, W., et al. Association between HLA-B*1502 and carbamazepine-induced severe cutaneous adverse drug reactions in a Thai population. *Epilepsia*, 2010; 51(5): 926–30.

100. Locharernkul, C., et al. Carbamazepine and phenytoin induced Stevens-Johnson syndrome is associated with HLA-B*1502 allele in Thai population. *Epilepsia*, 2008; 49(12): 2087–91.

101. Kuehn, B.M. FDA: Epilepsy drugs may carry skin risks for Asians. *JAMA*, 2008; 300(24): 2845.

102. Yang, C.W., et al. HLA-B*1502-bound peptides: Implications for the pathogenesis of carbamazepine-induced Stevens-Johnson syndrome. *J Allergy Clin Immunol*, 2007; 120(4): 870–7.

103. Hirsch, L.J., et al. Cross-sensitivity of skin rashes with antiepileptic drug use. *Neurology*, 2008; 71(19): 1527–34.

104. Alvestad, S., S. Lydersen, and E. Brodtkorb. Cross-reactivity pattern of rash from current aromatic antiepileptic drugs. *Epilepsy Res*, 2008; 80(2–3): 194–200.

105. Leckband, S.G., et al. Clinical Pharmacogenetics Implementation Consortium guidelines for HLA-B genotype and carbamazepine dosing. *Clin Pharmacol Ther*, 2013; 94(3): 324–8.

106. Codeine sulfate tablets. Prescribing information. 2009. Available from: http://www.accessdata.fda.gov/drugsatfda_docs/label/2009/022402s000lbl.pdf (accessed August 30, 2011).

107. Caraco, Y., et al. Microsomal codeine N-demethylation: Cosegregation with cytochrome P4503A4 activity. *Drug Metab Dispos*, 1996; 24(7): 761–4.

108. Dayer, P., et al. Bioactivation of the narcotic drug codeine in human liver is mediated by the polymorphic monooxygenase catalyzing debrisoquine 4-hydroxylation (cytochrome P-450 dbl/bufI). *Biochem Biophys Res Commun*, 1988; 152(1): 411–16.

109. Kirchheiner, J., et al. Pharmacokinetics of codeine and its metabolite morphine in ultra-rapid metabolizers due to CYP2D6 duplication. *Pharmacogenomics J*, 2007; 7(4): 257–65.

110. Ingelman-Sundberg, M., et al. Influence of cytochrome P450 polymorphisms on drug therapies: Pharmacogenetic, pharmacoepigenetic and clinical aspects. *Pharmacol Ther*, 2007; 116(3): 496–526.

111. Lurcott, G. The effects of the genetic absence and inhibition of CYP2D6 on the metabolism of codeine and its derivatives, hydrocodone and oxycodone. *Anesth Prog*, 1998; 45(4): 154–6.

112. Murphy, M.P., et al. Prospective CYP2D6 genotyping as an exclusion criterion for enrollment of a phase III clinical trial. *Pharmacogenetics*, 2000; 10(7): 583–90.

113. Borges, S., et al. Composite functional genetic and comedication CYP2D6 activity score in predicting tamoxifen drug exposure among breast cancer patients. *J Clin Pharmacol*, 2010; 50(4): 450–8.

114. Gaedigk, A., et al. Optimization of cytochrome P4502D6 (CYP2D6) phenotype assignment using a genotyping algorithm based on allele frequency data. *Pharmacogenetics*, 1999; 9(6): 669–82.

115. Gaedigk, A., et al. The CYP2D6 activity score: Translating genotype information into a qualitative measure of phenotype. *Clin Pharmacol Ther*, 2008; 83(2): 234–42.

116. Zanger, U.M., S. Raimundo, and M. Eichelbaum. Cytochrome P450 2D6: Overview and update on pharmacology, genetics, biochemistry. *Naunyn Schmiedebergs Arch Pharmacol*, 2004; 369(1): 23–37.

117. Sindrup, S.H., et al. Codeine increases pain thresholds to copper vapor laser stimuli in extensive but not poor metabolizers of sparteine. *Clin Pharmacol Ther*, 1990; 48(6): 686–93.

118. Eckhardt, K., et al. Same incidence of adverse drug events after codeine administration irrespective of the genetically determined differences in morphine formation. *Pain*, 1998; 76(1–2): 27–33.

119. Ciszkowski, C., et al. Codeine, ultrarapid-metabolism genotype, and postoperative death. *N Engl J Med*, 2009; 361(8): 827–8.

120. Dalen, P., et al. Quick onset of severe abdominal pain after codeine in an ultrarapid metabolizer of debrisoquine. *Ther Drug Monit*, 1997; 19(5): 543–4.

121. Lotsch, J., et al. Genetic predictors of the clinical response to opioid analgesics: Clinical utility and future perspectives. *Clin Pharmacokinet*, 2004; 43(14): 983–1013.

122. Koren, G., et al. Pharmacogenetics of morphine poisoning in a breastfed neonate of a codeine-prescribed mother. *Lancet*, 2006; 368(9536): 704.

123. Kurtz, M., P. Black Golde, and N. Berlinger. Ethical considerations in CYP2D6 genotype testing for codeine-prescribed breastfeeding mothers. *Clin Pharmacol Ther*, 2010; 88(6): 760–2.

124. Codeine. Breast feeding. Clinical pharmacology. 2011. Available from: http://www.clinicalpharmacology-ip.com/Forms/Monograph/monograph.aspx?cpnum=146&sec=monpreg (accessed August 30, 2011).

125. Ferner, R.E. Did the drug cause death? Codeine and breastfeeding. *Lancet*, 2008; 372(9639): 606–8.

126. Gasche, Y., et al. Codeine intoxication associated with ultrarapid CYP2D6 metabolism. *N Engl J Med*, 2004; 351(27): 2827–31.

127. Madadi, P. and G. Koren. Pharmacogenetic insights into codeine analgesia: Implications to pediatric codeine use. *Pharmacogenomics*, 2008; 9(9): 1267–84.

128. FDA. *FDA Public Health Advisory. Use of Codeine by Some Breastfeeding Mothers May Lead to Life-Threatening Side Effects in Nursing Babies.* 2007. Available from: http://www.fda.gov/Drugs/DrugSafety/PostmarketDrugSafetyInformationforPatientsandProviders/ucm054717.htm (accessed September 8, 2015).

129. de Leon, J. AmpliChip CYP450 test: Personalized medicine has arrived in psychiatry. *Expert Rev Mol Diagn*, 2006; 6(3): 277–86.

130. Roche. *AmpliChip CYP450 Test*. Available from: http://molecular.roche.com/ASSAYS/Pages/AmpliChipCYP450Test.aspx (accessed August 30, 2011).

131. *Coumadin Tablets. Warfarin Sodium Tablets. Prescribing Information.* Bristol-Myers Squibb Company, Princeton, NJ, 2010.

132. Wysowski, D.K., P. Nourjah, and L. Swartz. Bleeding complications with warfarin use: A prevalent adverse effect resulting in regulatory action. *Arch Intern Med*, 2007; 167(13): 1414–19.

133. Higashi, M.K., et al. Association between CYP2C9 genetic variants and anticoagulation-related outcomes during warfarin therapy. *JAMA*, 2002; 287: 1690–8.

134. Demirkan, K., et al. Response to warfarin and other oral anticoagulants: Effects of disease states. *S Med J*, 2000; 93(5): 448–54; quiz 455.

135. Garcia, D., et al. Warfarin maintenance dosing patterns in clinical practice: Implications for safer anticoagulation in the elderly population. *Chest*, 2005; 127(6): 2049–56.

136. Holbrook, A.M., et al. Systematic overview of warfarin and its drug and food interactions. *Arch Intern Med*, 2005; 165(10): 1095–106.

137. Kurnik, D., et al. Complex drug-drug-disease interactions between amiodarone, warfarin, and the thyroid gland. *Medicine (Baltimore)*, 2004; 83(2): 107–13.

138. Takeuchi, F., et al. A genome-wide association study confirms VKORC1, CYP2C9, and CYP4F2 as principal genetic determinants of warfarin dose. *PLoS Genet*, 2009; 5(3): e1000433.

139. Lee, C.R., J.A. Goldstein, and J.A. Pieper. Cytochrome P450 2C9 polymorphisms: A comprehensive review of the in-vitro and human data. *Pharmacogenetics*, 2002; 12(3): 251–63.

140. Schwarz, U.I. Clinical relevance of genetic polymorphisms in the human CYP2C9 gene. *Eur J Clin Invest*, 2003; 33(Suppl 2): 23–30.

141. Rieder, M.J., et al. Effect of VKORC1 haplotypes on transcriptional regulation and warfarin dose. *N Engl J Med*, 2005; 352(22): 2285–93.

142. Gage, B.F., et al. Use of pharmacogenetic and clinical factors to predict the therapeutic dose of warfarin. *Clin Pharmacol Ther*, 2008; 84(3): 326–31.

143. McClain, M.R., et al. A rapid-ACCE review of CYP2C9 and VKORC1 alleles testing to inform warfarin dosing in adults at elevated risk for thrombotic events to avoid serious bleeding. *Genet Med*, 2008; 10(2): 89–98.

144. Klein, T.E., et al. Estimation of the warfarin dose with clinical and pharmacogenetic data. *N Engl J Med*, 2009; 360(8): 753–64.

145. Pirmohamed, M., et al. A randomized trial of genotype-guided dosing of warfarin. *N Engl J Med*, 2013; 369(24): 2294–303.

146. Aithal, G.P., et al. Association of polymorphisms in the cytochrome P450 CYP2C9 with warfarin dose requirement and risk of bleeding complications. *Lancet*, 1999; 353(9154): 717–19.

147. Taube, J., D. Halsall, and T. Baglin. Influence of cytochrome P-450 CYP2C9 polymorphisms on warfarin sensitivity and risk of over-anticoagulation in patients on long-term treatment. *Blood*, 2000; 96(5): 1816–19.

148. Eckman, M.H., et al. Cost-effectiveness of using pharmacogenetic information in warfarin dosing for patients with nonvalvular atrial fibrillation. *Ann Intern Med*, 2009; 150(2): 73–83.

149. National Coalition for Health Professional Education in Genetics. *Core Competencies for All Health Care Professionals*. 2007. Available from: http://www.nchpeg.org/index. php?option=com_content&view=article&id=237&Itemid=84 (accessed March 12, 2014).

150. Johnson, J.A., et al. Pharmacogenomics: A scientific revolution in pharmaceutical sciences and pharmacy practice. Report of the 2001–2002 Academic Affairs Committee. *Am J Pharm Educ*, 2002; 66(Winter Suppl): 12S–15S.

151. Burke, J.M., et al. Clinical pharmacist competencies. *Pharmacotherapy*, 2008; 28(6): 806–15.

152. Ferro, W.G., et al. Pharmacist education in the era of genomic medicine. *J Am Pharm Assoc (2003)*, 2012; 52(5): e113–21.

153. American Academy of Family Physicians (AAFP) Core Educational Guidelines. *Am Fam Physician*, 1999; 60(1): 305–7.

154. Ikediobi, O.N., et al. Addressing the challenges of the clinical application of pharmacogenetic testing. *Clin Pharmacol Ther*, 2009; 86(1): 28–31.

155. Calzone, K.A., et al. Establishing the outcome indicators for the essential nursing competencies and curricula guidelines for genetics and genomics. *J Prof Nurs*, 2011; 27(3): 179–91.

156. Johnson, S.W. and S. Henderson. Genetics of warfarin sensitivity in an emergency department population with thromboembolic. *West J Emerg Med*, 2011; 12(1): 11–16.

157. Kuo, G.M., et al. Institutional Profile: University of California San Diego Pharmacogenomics Education Program (PharmGenEd): Bridging the gap between science and practice. *Pharmacogenomics*, 2011; 12(2): 149–53.

158. Latif, D.A. Pharmacogenetics and pharmacogenomics instruction in schools of pharmacy in the USA: Is it adequate? *Pharmacogenomics*, 2005; 6(4): 317–19.

159. Murphy, J.E., et al. Pharmacogenomics in the curricula of colleges and schools of pharmacy in the United States. *Am J Pharm Educ*, 2010; 74(1): 7.

160. Green, J.S., et al. Pharmacogenomics instruction in US and Canadian medical schools: Implications for personalized medicine. *Pharmacogenomics*, 2010; 11(9): 1331–40.

161. Prows, C., K. Calzon, and J. Jenkins. Genetics content in nursing curriculum. *Proceedings of the National Coalition for Health Professional Education in Genetics*, 2006.

162. Calzone, K.A. and J. Jenkins. Genomics education in nursing in the United States. *Annu Rev Nurs Res*, 2011; 29: 151–72.

163. Kadafour, M., et al. Survey on warfarin pharmacogenetic testing among anticoagulation providers. *Pharmacogenomics*, 2009; 10(11): 1853–60.
164. Marsh, S. and T. van Rooij. Challenges of incorporating pharmacogenomics into clinical practice. *Gastrointest Cancer Res*, 2009; 3(5): 206–7.
165. Lee, K.C., J.D. Ma, and G.M. Kuo. Pharmacogenomics: Bridging the gap between science and practice. *J Am Pharm Assoc (2003)*, 2010; 50(1): e1–14; quiz e15–17.
166. French, B., et al. Statistical design of personalized medicine interventions: The Clarification of Optimal Anticoagulation through Genetics (COAG) trial. *Trials*, 2010; 11: 108.
167. Eiselt, R., et al. Identification and functional characterization of eight CYP3A4 protein variants. *Pharmacogenetics*, 2001; 11(5): 447–58.
168. Dodson, C. Knowledge and attitudes concerning pharmacogenomics among healthcare professionals. *Personal Med*, 2011; 8(4): 421–8.
169. Hoop, J.G., et al. The current state of genetics training in psychiatric residency: Views of 235 U.S. educators and trainees. *Acad Psychiatry*, 2010; 34(2): 109–14.
170. American Academy of Family Physicians. *Core Educational Guidelines. Medical Genetics: Recommended Core Educational Guidelines for Family Practice Residents.* AAFP, 1999.
171. Association of American Medical Colleges. *Contemporary Issues in Medicine: Genetic Education.* AAMC, 2004, Washington DC.

2 Principles of Pharmacogenetic Biotechnology and Testing in Clinical Practice

Rui Liu, Gary Yuan Gao, and Long Ma

CONTENTS

KEY CONCEPTS

- Pharmacogenetics reveals the determinants of genetic variations in drug responses across individuals.
- The rapid development of biotechnologies to precisely identify DNA variations is one of the major driving forces in this field.
- Genetic information is routinely collected in pharmacogenomic analyses by multiple technologies.
- Sanger sequencing is widely used, despite being an early-generation DNA analysis technology.
- Polymerase chain reaction, a widely used DNA amplification method, can be modified to fit broader applications in pharmacogenomic studies.
- Microarray and next-generation sequencing methods are recent biotechnology breakthroughs developed to fulfill the needs for high-throughput sequence analyses.

INTRODUCTION

Pharmacogenetics studies the influence of genetic variations on drug efficacy, absorption, metabolism, elimination, and safety in healthy subjects and patients (Haga and LaPointe 2013; Johnson 2003; Shin et al. 2009). Pharmacogenetics can be used to maximize drug effectiveness and minimize adverse drug reactions with respect to a patient's genetic background, thereby ensuring an optimized drug therapy. In recent years, the development of biotechnology has paved the way for a new era of pharmacogenetics, which undertakes a genome-wide approach to study the effects of multiple genes on drug responses (Cordero and Ashley 2012; Pirmohamed 2001; Squassina et al. 2010). These technologies allow physicians to have a comprehensive view of a patient's unique genetic information. The implementation of a patient's genetic data into clinical decision can facilitate alertness to potential adverse drug reactions, prescription of the most appropriate treatments, prediction of the patient's response to medications, achievement of optimal treatment outcomes, explanation for lack of therapeutic consequences of a treatment, and the search for alternative therapies. Furthermore, a comprehensive understanding of an individual's genotype, phenotype, and environmental factors forms a structured framework that can help optimize the effectiveness of clinical treatments and eventually enable "personalized medicine" (Crews et al. 2012; Levy et al. 2014).

Pharmacogenetics has a history spanning over 80 years. In 1926, hemolytic anemia was first reported in malaria individuals treated with 6-methoxy-8-aminoquinoline, possibly due to different responses to the drug among individuals (Beutler 2008; Cordes 1926). However, the cause of acute hemolysis in these patients, the deficiency in glucose-6-phosphate dehydrogenase (G6PD), was not discovered until three decades later (Alving et al. 1956). A similar case showing that single-gene changes can alter drug effects was discovered from routine injection of succinylcholine, a muscle relaxant during anesthesia. In patients lacking butyrylcholinesterase, the injection caused unexpected adverse drug reactions, such as prolonged paralysis (Kalow 1956). The definitive evidence that genetics affects drug effects was traced back to a study of monozygotic twins in the late 1960s (Vesell and Page 1968). This study revealed the remarkable similarity in phenylbutazone metabolism in identical twins who shared 100% of their genomes in contrast with fraternal twins who shared only 50% of their genomes, suggesting the determining role of genetics in drug metabolism (Motulsky 1957; Motulsky and Qi 2006). Recognizing that individual patients had differential drug responses, Friedrich Vogel coined the term *pharmacogenetics* in 1959 (Vogel 1959). However, until two decades ago, genetics played only a minor role in clinical pharmacology, therapeutic development, and drug prescription, presumably due to the fact that the clinical effects of only a small number of drugs are strongly influenced by a single gene, which obscured our appreciation of the importance of pharmacogenetics in medicine. Since the 1990s, the advance of genomics science, especially the completion of the Human Genome Project (http://www.genome.gov/10001772), has set off a surge of interest in the study of pharmacogenomics. New sequencing technologies have become the driving force for this field. These technologies allow high-throughput analysis of thousands of genetic loci and their effects associated with drug responses.

Genetic variations generally include single-nucleotide polymorphisms (SNPs), genetic insertions and deletions, genomic copy-number variations (CNVs), and chromosomal rearrangements. The most common and inherited sequence variations are SNPs, accounting for approximately 90% of all human genome variations (Carlson et al. 2003; Hinds et al. 2005). To date, over 88 million SNPs have been identified and verified in human genomes (http://www.ncbi.nlm.nih.gov/projects/SNP/). The SNP allele that is more frequent in a population is designated as the major SNP, and the less frequent one is designated as the minor SNP. Accumulating evidence suggests the importance of these SNPs in pharmacogenomics (Carlson et al. 2004; Ramsey et al. 2012). Genome-wide association studies (GWASs) take advantage of the enormous number of SNPs that frequently occur in human populations to locate the susceptible genomic loci associated with disease phenotypes or drug responses (Gabriel et al. 2002; Roses 2000). The underlying genetics is that SNPs in close proximity, usually within 50 kb distance of a gene, are typically inherited together along generations. These closely linked SNPs are viewed as haplotype blocks, and the level of concomitant inheritance is called linkage disequilibrium (LD). The human genome consists of regions of low and high LD. Therefore, an unknown disease-causing gene that has a high LD with known SNPs can be identified based on the strong association of these SNPs with the disease trait among several generations or specific patient populations.

One well-established genotype–phenotype relationship is the gene *CYP2D6* that encodes the drug-metabolizing enzyme CYP2D6 (also known as debrisoquine hydroxylase), which was discovered after the adverse deficiency of the use of debrisoquine was known (Mahgoub et al. 1977). The metabolism of a large number of drugs relies heavily on the highly polymorphic nature of more than 100 SNPs in *CYP2D6*. One example is the analgesic codeine, which is converted to morphine at different pharmacokinetic efficiencies depending on genetic variations in *CYP2D6* (Kirchheiner et al. 2007; Lotsch et al. 2009). For example, the US FDA recently released a statement addressing a serious concern regarding the connection between children who have the *CYP2D6* UM (ultrarapid metabolizer) variation and fatal reactions to codeine following tonsillectomy and/or adenoidectomy (http://www.fda. gov/Drugs/DrugSafety/ucm313631.htm).

Studies indicate that the human genome contains 60 families and subfamilies of *CYP* genes, in which *CYP2C9*, *CYP2C19*, *CYP2D6*, *CYP3A4*, and *CYP3A5* are commonly tested in clinical practice. These genes are involved in metabolic pathways of approximately 80–90% of currently available prescription drugs (Ingelman-Sundberg 2004).

Pharmacogenetic studies further indicate that ethnicity-specific SNPs affect drug effectiveness in different human populations (Kalow 1982). In the study of barbiturate metabolism, it was found that only East Asian students exhibited enzymatic deficiency among a mixed student population (Kalow et al. 1979). The difference might be caused by genomic variations across two populations that potentially lead to different protein structures or activities.

Drug efficacy is influenced not only by SNP variations in drug-metabolizing genes but also by polymorphisms in genes that encode drug transporters, targets, receptors, and effectors. For example, the *KRAS* gene, which is involved in the cell membrane–bound small GTPase signaling pathway, is the target for the drug Vectibix® or Erbitux® in patients with metastatic colorectal cancer. However, these drugs become ineffective if the *KRAS* gene contains point mutations at several positions in certain colorectal tumor tissues (Stintzing 2014).

CNV is another form of genomic variation that involves relatively large genome regions being deleted or duplicated on certain chromosome(s) (Iafrate et al. 2004; Sebat et al. 2004). CNVs usually alter the copy numbers of multiple genes and therefore might cause more severe consequences on gene expression, regulation, and function than SNPs. An individual usually carries 4 million bases of CNVs (1 in every 800 bp) (Kidd et al. 2008). Besides stable and heritable CNVs, *de novo* CNVs are also present, which are confirmed by studies of identical twins. *De novo* CNVs may arise through diverse mechanisms at various stages of development. Like SNPs, some CNVs have been associated with disease susceptibility or drug efficacy. For example, excessive expression or extra copies of the *HER2* gene could lead to a very aggressive breast cancer in patients. A monoclonal antibody drug, Herceptin® (trastuzumab), could effectively treat these cancer patients with *HER2* overexpression. To date, many companion diagnostic tests have been approved by the US FDA for various cancer treatments (http://www.fda.gov/). These tests reveal specific genomic mutations in cancer patients and can greatly increase the success rate of drug treatments by identifying and matching a patient's genotype with the target(s) of a given cancer drug.

Epigenetics is the study of biological changes that are not caused by changes in DNA sequences. DNA methylation and histone modification are two major epigenetic events and can control the on- and off-switch of gene expression. The pattern of DNA methylation changes in development, aging, and certain diseases (Jones and Baylin 2002; Singal and Ginder 1999). Like gene mutations or deletions, DNA methylation frequently silences gene expression and could lead to aberrant function of normal tumor suppressor(s). A better understanding of epigenetic mechanisms underlying diseases has allowed therapeutic applications of DNA methylation inhibitors, such as azacitidine (5-azacytidine; Vidaza®, Pharmion Corp., Boulder, Colorado) and decitabine (Dacogen™, SuperGen, Inc., Dublin, California, and MGI Pharma, Inc., Minneapolis, Minnesota). These drugs provide new and effective options for patients.

Hence, a pharmacogenetic understanding of genomic variations (e.g., point mutations and CNVs) and epigenetic changes in a patient will play an important role in the age of personalized medicine.

MAJOR BIOTECHNOLOGIES IN PHARMACOGENETICS

Numerous biomedical technologies have advanced our knowledge of pharmacogenetics, among which DNA sequencing and analysis methods are the major driving forces. The chain-termination method (Sanger sequencing) (Sanger et al. 1977) and the less frequently used chemical sequencing method (Maxam–Gilbert sequencing) (Maxam and Gilbert 1977) are first-generation DNA sequencing technologies. The principle of Sanger sequencing is that DNA polymerase selectively incorporates chain-terminating dideoxynucleotides during in vitro DNA replication, which results in synthesized DNA fragments of variable lengths that match the positions of the bases substituted by the corresponding dideoxynucleotides (Sanger et al. 1977). These DNA fragments can be separated by electrophoresis to reveal the sequence of the DNA template. Sanger sequencing is the underpinning sequencing technology for the first human genome (Lander et al. 2001) and is still the primary sequencing method in most basic and clinical genetics laboratories. In this chapter, we briefly review three other major DNA analysis technologies widely used today.

POLYMERASE CHAIN REACTION

An extremely important technique in molecular biology is the polymerase chain reaction (PCR) proposed in 1983 by Kary Mullis. PCR initiated a new era of highly efficient gene analysis and manipulation (Saiki et al. 1988). This method utilizes the basic principle of DNA replication and cycles this process in a test tube, which allows the generation of millions of copies of a particular DNA sequence. An ample supply of DNA copies allows easy detection and manipulation of genetic information encoded in the DNA. The basic setup of a PCR requires the following: (1) a template that contains the DNA region of interest; (2) two amplification primers that are complementary to the 3′-ends of the double-stranded DNA template; (3) deoxynucleoside triphosphates (dNTP), the building blocks used to synthesize new DNA strands; (4) a thermostable DNA polymerase that synthesizes new DNA strands complementary to the template strands; (5) buffer solution with suitable pH, salt concentrations, and magnesium or manganese ions (Figure 2.1a). Typically, a PCR consists of 20–40 cycles of heating and cooling steps. In each cycle, a DNA template is denatured, annealed to primers, and synthesized into two new copies. Specifically, the denaturation step occurs at 94–98°C for 10–30 seconds, which disrupts the hydrogen bonds between complementary bases and unwinds the double-stranded templates into single strands. At the annealing step, the reaction is cooled down to the annealing temperature to allow hybridization of the primers to the single-stranded DNA template. Then the temperature is increased to 68–72°C (the optimal temperature for thermostable DNA polymerase) to allow new DNA strands to be synthesized. At this stage, DNA polymerase adds dNTPs to the annealed primers to assemble a nascent DNA strand complementary to the template strand in the 5′ to 3′ direction. In principle, DNA polymerase doubles the starting DNA strands at each extension step, leading to an exponential amplification of a given DNA region.

PCR has been adopted into many variations to fit a variety of applications. For example, allele-specific PCR is designed to detect allele-specific variations, such as single nucleotide changes. Prior knowledge of the variations in a DNA sequence is required to design primers specific for such SNPs. The amplification of a specific allele is achieved by stringently setting the temperatures for the annealing and elongation steps. The presence of a mismatch between the template DNA and the complementary primer would cause failed annealing and subsequent PCR reaction under the stringent conditions, whereas only a perfect match will lead to an allele-specific amplification. By identifying allele-specific DNA fragments, SNP information can be easily detected. Similarly, two pairs of primers can be used in one PCR reaction (a tetra-primer set)

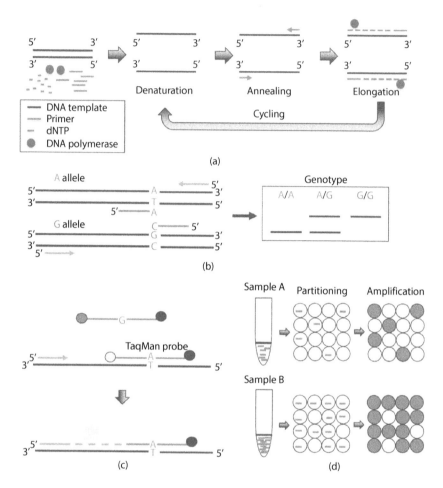

FIGURE 2.1 **(See color insert.)** Illustration of polymerase chain reaction (PCR)–related methods. (a) A PCR cycle generally includes denaturation, annealing, and elongation steps. (b) A representative design for a tetra-primer set, which integrates allele-specific nucleotides into the 3′-ends of primers. (c) A modified graph showing the mechanism of SNP genotyping by TaqMan assay (Life Technologies) using allele-specific fluorophore-conjugated probes. The perfectly matched TaqMan probe that anneals to the template will be degraded by DNA polymerase and release the fluorescence (yellow), while the fluorophore (blue) of the mismatched TaqMan probe that does not anneal will remain quenched. (d) Strategy of digital PCR to quantitatively estimate the starting DNA templates. Sample A has fewer starting templates, which are shown in fewer positive digital PCR reactions. For sample B, more starting templates are shown in more positive digital PCR reactions.

to target different allele-specific SNPs (You et al. 2008) (Figure 2.1b). These primer pairs are designed such that one single mismatch at the 3′-end will make the primers nonfunctional and terminate the amplification. As a result, the tetra-primer can amplify only the specific allele present in the PCR reaction but not the alternative allele with a different SNP. To distinguish the SNPs, PCR regions spanning the two different alleles are designed in distinct lengths to allow separation of amplified DNA fragments by gel electrophoresis or melting curve analysis (Figure 2.1b).

Real-time PCR is developed to amplify and simultaneously quantify target DNA molecules. One approach is to include a double-stranded DNA-binding dye in the PCR reaction. The binding of the dye to double-stranded DNA causes fluorescence that can be measured, therefore generating a real-time quantification of the amount of amplified DNA molecules in the reaction (Ponchel et al. 2003; Zipper et al. 2004). An alternative strategy is the use of fluorophore-conjugated probes that anneal to different alleles, as exemplified by the TaqMan® assay (ThermoFisher Scientific; www.lifetechnologies.com) (Figure 2.1c). In this assay, allele-specific oligonucleotide (or oligo) probes are labeled with different fluorophores (e.g., 6-carboxyfluorescein or tetrachlorofluorescein) at the 5'-ends and with a quencher molecule (e.g., tetra-methylrhodamine) at the 3'-ends. The quencher molecules quench the fluorescence of the fluorophores by fluorescence resonance energy transfer, which is most effective when the fluorophore and the quencher are in close proximity in the same probe. The allele-specific probes are included in the PCR reaction with a specific set of primers that amplify the region complementary to the probes. In the annealing phase, the probes and primers hybridize to their target. In the extension phase, the 5'–3' exonuclease activity of the DNA polymerase (Holland et al. 1991) degrades the perfectly matched, annealed probes. The degraded probes are released into the solution as single nucleotide, separating the fluorophore from the quencher, which results in an increase in fluorescence. In contrast, mismatched probes are not annealed to their targets and will not be degraded by the DNA polymerase, resulting in fluorescence still quenched. The difference in fluorescence can be monitored in a quantitative PCR thermal cycler, in which fluorescence released from degraded nucleotides of perfectly matched probes indicate how much the target SNP is amplified. The TaqMan assay can be multiplexed by combining the detection of up to seven SNPs in one reaction. However, since each SNP requires a distinct probe, the TaqMan assay is limited by how close the SNPs locate from each other on the DNA template. Generally, TaqMan is limited to applications that involve a small number of SNPs since optimal probes and reaction conditions must be designed for each SNP.

More recently, digital PCR has been developed and routinely used for clonal amplification of samples in next-generation sequencing (NGS). The digital PCR procedure was originally aimed to precisely quantify the input template rather than the final PCR product. The very first clinical application of digital PCR was to measure the absolute lowest number of leukemic cells in a leukemia patient with a goal to monitor residue disease and detect recurrence in patients as early as possible (Sykes et al. 1992). Evolution of this technology has allowed for a broad use in studying variations in gene sequences—such as CNVs and point mutations. The key difference between digital PCR and traditional PCR lies in the methods of treating the DNA templates. Digital PCR separates each single starting DNA molecule into distinct partitioned reactors and carries out one single reaction within each partition individually (Kalinina et al. 1997). The localization of individual DNA molecules in separate partitions provides an estimation of the starting molecule number by assuming that the population follows the Poisson distribution. In other words, each partition is assumed to contain either zero or one starting template for PCR.

After amplification, the PCR-positive (1 template) and PCR-negative (0 template) partitions are counted to provide an absolute quantification of the starting DNA copies in digital form (Figure 2.1d).

How to efficiently capture or isolate individual molecules is the major determinant for performing a high-quality digital PCR. The first commercial system for digital PCR was introduced by Fluidigm (www.fluidigm.com) in 2006 based on integrated fluidic circuits (chips) composed of chambers and valves to partition samples (Heyries et al. 2011). Droplet Digital PCR technology is another method of partitioning innovated by QuantaLife (now part of Bio-Rad [www.biorad.com]). This method separates a DNA template into 20,000 nanoliter-sized droplets for individual PCR reactions and provides digital counting of each target (Hindson et al. 2013). RainDance Technologies further reduced the size of the droplets, leading to the generation of up to 10 million picoliter-sized droplets per assay (http://raindancetech.com/digital-pcr-tech/).

Given that digital PCR technology could detect DNA molecules with additional sensitivity, accuracy, and precision, it has many applications. Digital PCR can be used in the detection and quantification of rare genetic sequences, CNVs, single-cell genetic variations, and rare pathogens. Moreover, sample preparation in many NGS platforms, including Roche/454, ABI/SOLiD, and Life Technologies/Ion Torrent, is enabled by single-step digital PCR as a key factor to reduce the time and cost (Sandberg et al. 2011; Williams et al. 2006).

Microarray Technology

Although PCR is a widely used and easily applied method for analyzing genetic variations, the number of primers that can be mixed together and annealed to different target sequences is limited. To fulfill the need for throughput increase, the technology of microarray, a multiplex lab-on-a-chip, was demonstrated by Schena et al. (1995). Since then, several companies, including Affymetrix, Agilent, Applied Microarrays, NimbleGen, and Illumina, have greatly facilitated the expansion of the microarray technology. Microarray technology fixes designed single-stranded DNA probes on a solid substrate, for example, a glass slide or silicon thin-file cell, and processes with the biological specimen for hybridization or capture of the probes' targets (complementary DNA or RNA) (Figure 2.2). It relies on the base-pairing principle that nucleic acids bind to their complementary strands (A to T and G to C) to differentiate targets with different sequences. The target samples are usually labeled with fluorescence or other chemiluminescent molecules, and hybridization to the probes can be detected by specialized equipment. This allows high-throughput screening of target samples with miniaturized, multiplexed, and parallel processing and detection. The throughput has increased dramatically in past decades from roughly 400 bacterial gene targets manually handled in a 1982 study (Augenlicht and Kobrin 1982) to approximately 2 million genome-wide human SNPs coupled with automatic scanning and image processing recently (Genome-Wide Human SNP Array 6.0, Affymetrix [www.affymetrix.com/catalog]). These changes allow DNA microarrays to be used for broad applications that include the measurement of gene expression levels, detection of SNPs, and targeted resequencing on a genome-wide scale.

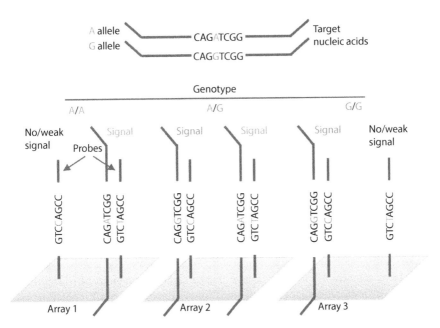

FIGURE 2.2   (See color insert.) Simplified illustration of an SNP microarray. This simplified array has only two probes, each corresponding to a different SNP allele (A or G). The affinities of A or G targets to allele-specific probes are different, which result in differential hybridization strength. Targets are often labeled with fluorophores or chemiluminescence. After nonbound targets are washed off the microarray, the scanning of fluorescence signal emitted by bound targets on the probe spots can reveal SNP information of the targets. In fact, thousands to millions of probes can be printed on designated spots of each microarray, with each probe representing a unique sequence variation.

Microarray fabrication differs in the types and number of probes, the technology for attaching probes to the test point, and the solid surface used for probe fixation. The number of probes can range from hundreds to millions, depending on the experimental designs and scientific questions to be addressed. A widely used fabrication method is to spot oligos on the solid surface of the array using robotic arm–controlled pins or needles, which capture presynthesized DNA probes by dipping into solutions of the probes. The array can be customized for specialized research purposes. The use of such miniaturized microarrays for gene expression profiling was first reported by Schena et al. (1995), and the first analysis of a complete eukaryotic genome was published by Lashkari et al. (1997).

An alternative approach to manufacturing a microarray is to directly synthesize probes on solid surfaces. Probes as long as 60 bases (e.g., Agilent design) or as short as 25 bases (e.g., Affymetrix design) can be generated in this manner. Longer probes are more specific to individual target genes, whereas shorter ones may be spotted in higher density across the array and are economic for manufacturing. The biotechnology company Roche NimbleGen Systems (www.nimblegen.com) also developed a new method called maskless array synthesis, which combines sequence flexibility with the large-scale probes (Nuwaysir et al. 2002). In standard microarrays, the probes are attached to a solid surface by a covalent bond via epoxy-silane, amino-silane, lysine, polyacrylamide, or other chemical matrix. The solid surface can be a glass, plastic, or silicon biochip (Affymetrix) or microscopic beads (Illumina).

The biological principle behind microarrays is the property of complementation between the probe and the nucleic acid target. More complementary base pairs in the target sequence mean stronger hydrogen binding between the probe and the target, while the presence of mismatches reduces this binding. Thus, only strongly paired targets will remain hybridized to their probes after several rounds of washing from a mild to stringent condition. The total strength of signals generated from fluorescence-labeled targets is determined by the amount of targets bound to the probes on a given spot.

Since an array can contain tens of thousands of different microscopic probes, a microarray experiment can accomplish many genetic tests in parallel and therefore dramatically expand the scope of investigation. The Affymetrix Genome-Wide Human SNP Array series serves as a good representative for the application of microarray in detection of whole-genome SNPs and CNVs. The latest version (6.0) of this array features 1.8 million genetic markers and has demonstrated impressive performance in detecting genetic variations (www.affymetrix.com). Therefore, such microarrays and similar ones have enabled GWASs with a larger sample size in the initial screen and replication phases, and significantly increased the overall genetic power of these studies.

NGS Technology

Microarray-based technology has been remarkably successful at high-throughput detection of genetic variations and expression profiles. However, both sensitivity and specificity are limited with microarrays. More importantly, microarrays are restricted to known genetic annotations with little ability to detect novel genetic variations.

The demand for sequencing technologies that are capable of delivering faster, less expensive, and massive genomic information has led to the invention of NGS technologies. NGS technologies can generate millions or billions of sequences (Church 2006; Schuster 2008) at a much faster speed and at an extremely low cost compared to the standard Sanger sequencing method, which underlies the decoding of the first human genome (Lander et al. 2001) that costs about US$3 billion (http://www.genome.gov/11006943). The first example of NGS was the massively parallel signature sequencing technology developed over a decade ago (Brenner et al. 2000). The polony sequencing method (Shendure et al. 2005) developed in the laboratory of George M. Church was a more applicable, early NGS system. This method combines emulsion PCR (a type of digital PCR), automated microscope system, and ligation-based sequencing chemistry (sequence by ligation) and was used to sequence a full genome of the *Escherichia coli* bacteria at an accuracy of >99.99% and a cost approximately one-ninth of that of the Sanger method (Shendure et al. 2005). The same strategy was used in a meta-genomic study that sequenced the whole genomes of single bacterial cells and provided critical tools for systematic characterization of genome diversity in the biosphere (Zhang et al. 2006).

The major feature of NGS platforms is massively parallel sequencing, in which millions of DNA fragments from a single sample are sequenced in parallel at a microscopic level. With NGS, genetic information of an entire human genome can now be revealed in one day. To date, several NGS platforms have been developed and modified to provide low-cost, high-throughput sequencing. Among them, the Life Technologies Ion Torrent and the Illumina (Solexa) sequencing are the two most commonly used platforms in research and clinical laboratories. Different models of these NGS platforms have been developed to meet the needs of research and clinical diagnostics.

Although Life Technologies Ion Torrent and Illumina have distinct sequencing technologies, they use similar methodologies for preparing sequencing libraries. The library construction consists of a series of steps that include DNA fragmentation, end repair, adaptor ligation, and PCR amplification (Harakalova et al. 2011) (Figure 2.3). In principle, the final sequencing library should cover the complete genomic view of every single starting template.

Once constructed, libraries are clonally amplified in preparation for sequencing. The Ion Torrent method utilizes emulsion PCR to amplify single template fragments onto microbeads, whereas the Illumina method utilizes bridge amplification to form template clusters on a flow cell (Berglund et al. 2011).

Both platforms make use of the sequencing-by-synthesis approach to sequence the amplified libraries, by which a new DNA strand is synthesized complementary to a strand of sequencing libraries. Through cycles of flashing with the nucleotides and washing off unbound ones in a sequential order, sequencing occurs when a certain nucleotide is incorporated into the extending strand. The incorporation of each single nucleotide into the newly synthesized strand is detected either by the Ion Torrent semiconductor sequencer based on the induced pH change or by the Illumina sequencer upon the released fluorescence (Figure 2.3). By these methods, millions of DNA fragments are sequenced in parallel. Once sequencing is complete, terabytes of genomic data will be analyzed. Sequence analysis can reveal almost endless information on genomic variations that include SNPs, the insertion or deletion of bases, and the detection of novel genes. Analysis can also include identification of both somatic and germline mutations that may contribute to the diagnosis of diseases or genetic conditions (Gogol-Doring and Chen 2012).

The applications of NGS have been enormous, allowing for rapid advances in many fields of biological sciences and clinic practice. Various NGS assays have been developed for genetic analyses, including whole-genome sequencing, whole-exome sequencing, and focused assays that target only a handful of genes. Parallel to this, different assays provide another set of tools for analyzing epigenetic modifications of DNA (e.g., ChIP-seq, bisulfite sequencing, DNase-I hypersensitivity site sequencing, formaldehyde-assisted isolation of regulatory elements sequencing, and more). These methods together with NGS are routinely used for analyzing gene expression–related processes or gene expression (RNA-sequencing) itself.

Because of relative cost-effectiveness and ease of accessibility, NGS has been used in a broad spectrum of companion diagnostics, although more evidence needs to be accumulated to fully evaluate this technology for clinical applications.

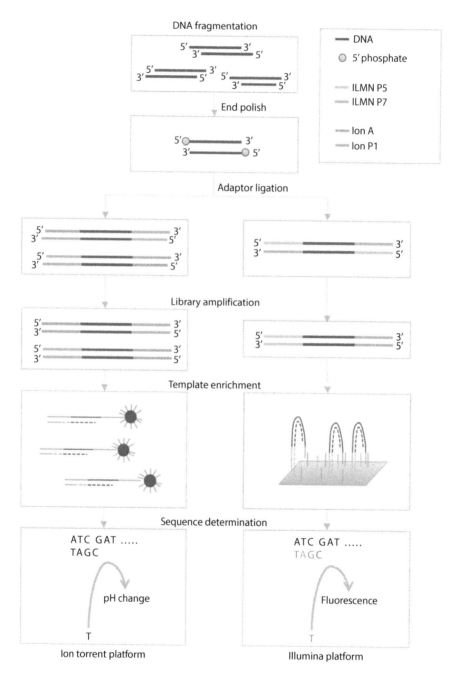

FIGURE 2.3 Overview of next-generation sequencing workflow. Fragmented DNAs are end-polished, ligated with platform-specific adapters (ILMN P5, ILMN P7, Ion A, Ion P1, etc.), and amplified to generate sequencing libraries. Different sequencing platforms (Ion Torrent or Illumina) use different enrichment methods (beads-based or flow cell–based) and sequencing strategies (pH or fluorescence changes during strand synthesis) to reveal sequence information.

Nevertheless, it is foreseeable that in the near future, NGS will be an indispensable approach to probing genetic changes relevant to numerous diseases.

CONCLUDING REMARKS

Today, traditional nucleic acid analysis methods, such as the Sanger sequencing method, and more advanced technologies, such as NGS, coexist to serve different needs in biomedical analyses. These new technologies have expanded the use of pharmacogenomics in numerous medical areas, including but not limited to pain management, cardiology, oncology, pathology, and psychiatry. In forensic pathology, these technologies may be used to determine the cause of nontraumatic deaths (Karch 2007). Pharmacogenomic tests are also used to identify cancer patients who are most likely to respond to certain cancer drugs (Lee and McLeod 2011). For psychiatric patients, pharmacogenomic tests will provide tools for better drug selection and more informative side effect amelioration (Gardner et al. 2014). Similarly, pharmacogenomics will improve the treatment of cardiovascular disorders, where a patient's possible response to drugs, including warfarin, clopidogrel, β blockers, and statins, may be determined by identifying the genotypes of genes involved in the efficacy and metabolism of the affected drugs (Turner and Pirmohamed 2014). In the near future, technological advances will likely transform pharmacogenetics into an indispensable tool for pharmacists and clinicians to evaluate, guide, and monitor various treatments for a broad variety of diseases.

STUDY QUESTIONS

1. Which of the following is not a common genetic variation?
   a. DNA methylation
   b. Nucleotide insertions and deletions
   c. Copy-number variations
   d. Chromosomal rearrangements
2. Which pair of patients will most likely have similar responses to a drug?
   a. Two strangers
   b. A married couple
   c. Father and son
   d. Identical twins
3. Sanger sequencing is also called:
   a. Chemical sequencing method
   b. Chain-termination method
   c. Shotgun sequencing method
   d. TaqMan method
4. A polymerase chain reaction requires the following ingredient:
   a. ATP
   b. RNA
   c. RNA polymerase
   d. DNA polymerase

5. Microarray uses the following biological principle to detect genomic variations:
   a. Base-paring principle of nucleic acids
   b. Gene transcription
   c. Protein modification
   d. DNA replication
6. Next-generation sequencing can be used for:
   a. Sequencing human genomes
   b. Detecting genetic variations in a population
   c. Identifying disease-causing genes
   d. All of the above

Answer Key
   1. a
   2. d
   3. b
   4. d
   5. a
   6. d

REFERENCES

Alving, A. S., P. E. Carson, C. L. Flanagan, and C. E. Ickes. 1956. Enzymatic deficiency in primaquine-sensitive erythrocytes. *Science* 124: 484–485.

Augenlicht, L. H., and D. Kobrin. 1982. Cloning and screening of sequences expressed in a mouse colon tumor. *Cancer Res* 42: 1088–1093.

Berglund, E. C., A. Kiialainen, and A. C. Syvanen. 2011. Next-generation sequencing technologies and applications for human genetic history and forensics. *Investig Genet* 2: 23.

Beutler, E. 2008. Glucose-6-phosphate dehydrogenase deficiency: A historical perspective. *Blood* 111: 16–24.

Brenner, S., M. Johnson, J. Bridgham, G. Golda, D. H. Lloyd, et al. 2000. Gene expression analysis by massively parallel signature sequencing (MPSS) on microbead arrays. *Nat Biotechnol* 18: 630–634.

Carlson, C. S., M. A. Eberle, L. Kruglyak, and D. A. Nickerson. 2004. Mapping complex disease loci in whole-genome association studies. *Nature* 429: 446–452.

Carlson, C. S., M. A. Eberle, M. J. Rieder, J. D. Smith, L. Kruglyak, D. A. Nickerson. 2003. Additional SNPs and linkage-disequilibrium analyses are necessary for whole-genome association studies in humans. *Nat Genet* 33: 518–521.

Church, G. M. 2006. Genomes for all. *Sci Am* 294: 46–54.

Cordero, P., and E. A. Ashley. 2012. Whole-genome sequencing in personalized therapeutics. *Clin Pharmacol Ther* 91: 1001–1009.

Cordes, W. 1926. Experiences with plasmochin in malaria. In: *Anonymous. 15th Annual Report*. Boston, MA: United Fruit Co, pp. 66–71.

Crews, K. R., J. K. Hicks, C. H. Pui, M. V. Relling, and W. E. Evans. 2012. Pharmacogenomics and individualized medicine: Translating science into practice. *Clin Pharmacol Ther* 92: 467–475.

Gabriel, S. B., S. F. Schaffner, H. Nguyen, J. M. Moore, J. Roy, et al. 2002. The structure of haplotype blocks in the human genome. *Science* 296: 2225–2229.

Gardner, K. R., F. X. Brennan, R. Scott, and J. Lombard. 2014. The potential utility of pharmacogenetic testing in psychiatry. *Psychiatry J* 2014: 730956.

Gogol-Doring, A., and W. Chen. 2012. An overview of the analysis of next generation sequencing data. *Methods Mol Biol* 802: 249–257.

Haga, S. B., and N. M. LaPointe. 2013. The potential impact of pharmacogenetic testing on medication adherence. *Pharmacogenomics J* 13: 481–483.

Harakalova, M., M. Mokry, B. Hrdlickova, I. Renkens, K. Duran, et al. 2011. Multiplexed array-based and in-solution genomic enrichment for flexible and cost-effective targeted next-generation sequencing. *Nat Protoc* 6: 1870–1886.

Heyries, K. A., C. Tropini, M. Vaninsberghe, C. Doolin, O. I. Petriv, et al. 2011. Megapixel digital PCR. *Nat Methods* 8: 649–651.

Hinds, D. A., L. L. Stuve, G. B. Nilsen, E. Halperin, E. Eskin, et al. 2005. Whole-genome patterns of common DNA variation in three human populations. *Science* 307: 1072–1079.

Hindson, C. M., J. R. Chevillet, H. A. Briggs, E. N. Gallichotte, I. K. Ruf, et al. 2013. Absolute quantification by droplet digital PCR versus analog real-time PCR. *Nat Methods* 10: 1003–1005.

Holland, P. M., R. D. Abramson, R. Watson, and D. H. Gelfand. 1991. Detection of specific polymerase chain reaction product by utilizing the 5′–3′ exonuclease activity of *Thermus aquaticus* DNA polymerase. *Proc Natl Acad Sci U S A* 88: 7276–7280.

Iafrate, A. J., L. Feuk, M. N. Rivera, M. L. Listewnik, P. K. Donahoe, et al. 2004. Detection of large-scale variation in the human genome. *Nat Genet* 36: 949–951.

Ingelman-Sundberg, M. 2004. Pharmacogenetics of cytochrome P450 and its applications in drug therapy: The past, present and future. *Trends Pharmacol Sci* 25: 193–200.

Johnson, J. A. 2003. Pharmacogenetics: Potential for individualized drug therapy through genetics. *Trends Genet* 19: 660–666.

Jones, P. A., and S. B. Baylin. 2002. The fundamental role of epigenetic events in cancer. *Nat Rev Genet* 3: 415–428.

Kalinina, O., I. Lebedeva, J. Brown, and J. Silver. 1997. Nanoliter scale PCR with TaqMan detection. *Nucleic Acids Res* 25: 1999–2004.

Kalow, W. 1956. Familial incidence of low pseudocholinesterase level. *Lancet* 268: 576–577.

Kalow, W. 1982. Ethnic differences in drug metabolism. *Clin Pharmacokinet* 7: 373–400.

Kalow, W., B. K. Tang, D. Kadar, L. Endrenyi, and F. Y. Chan. 1979. A method for studying drug metabolism in populations: Racial differences in amobarbital metabolism. *Clin Pharmacol Ther* 26: 766–776.

Karch, S. B. 2007. Changing times: DNA resequencing and the "nearly normal autopsy." *J Forensic Leg Med* 14: 389–397.

Kidd, J. M., G. M. Cooper, W. F. Donahue, H. S. Hayden, N. Sampas, et al. 2008. Mapping and sequencing of structural variation from eight human genomes. *Nature* 453: 56–64.

Kirchheiner, J., H. Schmidt, M. Tzvetkov, J. T. Keulen, J. Lotsch, et al. 2007. Pharmacokinetics of codeine and its metabolite morphine in ultra-rapid metabolizers due to CYP2D6 duplication. *Pharmacogenomics J* 7: 257–265.

Lander, E. S., L. M. Linton, B. Birren, C. Nusbaum, M. C. Zody, et al. 2001. Initial sequencing and analysis of the human genome. *Nature* 409: 860–921.

Lashkari, D. A., J. L. DeRisi, J. H. McCusker, A. F. Namath, C. Gentile, et al. 1997. Yeast microarrays for genome wide parallel genetic and gene expression analysis. *Proc Natl Acad Sci U S A* 94: 13057–13062.

Lee, S. Y., and H. L. McLeod. 2011. Pharmacogenetic tests in cancer chemotherapy: What physicians should know for clinical application. *J Pathol* 223: 15–27.

Levy, K. D., B. S. Decker, J. S. Carpenter, D. A. Flockhart, P. R. Dexter, et al. 2014. Prerequisites to implementing a pharmacogenomics program in a large health-care system. *Clin Pharmacol Ther* 96: 307–309.

Lotsch, J., M. Rohrbacher, H. Schmidt, A. Doehring, J. Brockmoller, et al. 2009. Can extremely low or high morphine formation from codeine be predicted prior to therapy initiation? *Pain* 144: 119–124.

Mahgoub, A., J. R. Idle, L. G. Dring, R. Lancaster, and R. L. Smith. 1977. Polymorphic hydroxylation of Debrisoquine in man. *Lancet* 2: 584–586.

Maxam, A. M., and W. Gilbert. 1977. A new method for sequencing DNA. *Proc Natl Acad Sci U S A* 74: 560–564.

Motulsky, A. G. 1957. Drug reactions enzymes, and biochemical genetics. *J Am Med Assoc* 165: 835–837.

Motulsky, A. G., and M. Qi. 2006. Pharmacogenetics, pharmacogenomics and ecogenetics. *J Zhejiang Univ Sci B* 7: 169–170.

Nuwaysir, E. F., W. Huang, T. J. Albert, J. Singh, K. Nuwaysir, et al. 2002. Gene expression analysis using oligonucleotide arrays produced by maskless photolithography. *Genome Res* 12: 1749–1755.

Pirmohamed, M. 2001. Pharmacogenetics and pharmacogenomics. *Br J Clin Pharmacol* 52: 345–347.

Ponchel, F., C. Toomes, K. Bransfield, F. T. Leong, S. H. Douglas, et al. 2003. Real-time PCR based on SYBR-Green I fluorescence: An alternative to the TaqMan assay for a relative quantification of gene rearrangements, gene amplifications and micro gene deletions. *BMC Biotechnol* 3: 18.

Ramsey, L. B., G. H. Bruun, W. Yang, L. R. Trevino, S. Vattathil, et al. 2012. Rare versus common variants in pharmacogenetics: SLCO1B1 variation and methotrexate disposition. *Genome Res* 22: 1–8.

Roses, A. D. 2000. Pharmacogenetics and the practice of medicine. *Nature* 405: 857–865.

Saiki, R. K., D. H. Gelfand, S. Stoffel, S. J. Scharf, R. Higuchi, et al. 1988. Primer-directed enzymatic amplification of DNA with a thermostable DNA polymerase. *Science* 239: 487–491.

Sandberg, J., B. Werne, M. Dessing, and J. Lundeberg. 2011. Rapid flow-sorting to simultaneously resolve multiplex massively parallel sequencing products. *Sci Rep* 1: 108.

Sanger, F., S. Nicklen, and A. R. Coulson. 1977. DNA sequencing with chain-terminating inhibitors. *Proc Natl Acad Sci U S A* 74: 5463–5467.

Schena, M., D. Shalon, R. W. Davis, and P. O. Brown. 1995. Quantitative monitoring of gene expression patterns with a complementary DNA microarray. *Science* 270: 467–470.

Schuster, S. C. 2008. Next-generation sequencing transforms today's biology. *Nat Methods* 5: 16–18.

Sebat, J., B. Lakshmi, J. Troge, J. Alexander, J. Young, et al. 2004. Large-scale copy number polymorphism in the human genome. *Science* 305: 525–528.

Shendure, J., G. J. Porreca, N. B. Reppas, X. Lin, J. P. McCutcheon, et al. 2005. Accurate multiplex polony sequencing of an evolved bacterial genome. *Science* 309: 1728–1732.

Shin, J., S. R. Kayser, and T. Y. Langaee. 2009. Pharmacogenetics: From discovery to patient care. *Am J Health Syst Pharm* 66: 625–637.

Singal, R., and G. D. Ginder. 1999. DNA methylation. *Blood* 93: 4059–4070.

Squassina, A., M. Manchia, V. G. Manolopoulos, M. Artac, C. Lappa-Manakou, et al. 2010. Realities and expectations of pharmacogenomics and personalized medicine: Impact of translating genetic knowledge into clinical practice. *Pharmacogenomics* 11: 1149–1167.

Stintzing, S. 2014. Management of colorectal cancer. *F1000Prime Rep* 6: 108.

Sykes, P. J., S. H. Neoh, M. J. Brisco, E. Hughes, J. Condon, et al. 1992. Quantitation of targets for PCR by use of limiting dilution. *Biotechniques* 13: 444–449.

Turner, R. M., and M. Pirmohamed. 2014. Cardiovascular pharmacogenomics: Expectations and practical benefits. *Clin Pharmacol Ther* 95: 281–293.

Vesell, E. S., and J. G. Page. 1968. Genetic control of drug levels in man: Phenylbutazone. *Science* 159: 1479–1480.

Vogel, F. 1959. Moderne problem der humangenetik. *Ergeb Inn Med U Kinderheilk* 12: 52–125.

Williams, R., S. G. Peisajovich, O. J. Miller, S. Magdassi, D. S. Tawfik, et al. 2006. Amplification of complex gene libraries by emulsion PCR. *Nat Methods* 3: 545–550.

You, F. M., N. Huo, Y. Q. Gu, M. C. Luo, Y. Ma, et al. 2008. BatchPrimer3: A high throughput web application for PCR and sequencing primer design. *BMC Bioinformatics* 9: 253.

Zhang, K., A. C. Martiny, N. B. Reppas, K. W. Barry, J. Malek, et al. 2006. Sequencing genomes from single cells by polymerase cloning. *Nat Biotechnol* 24: 680–686.

Zipper, H., H. Brunner, J. Bernhagen, and F. Vitzthum. 2004. Investigations on DNA intercalation and surface binding by SYBR Green I, its structure determination and methodological implications. *Nucleic Acids Res* 32: e103.

3 Essential Pharmacogenomic Biomarkers in Clinical Practice

Cong Liu, Weiguo Chen, and Wei Zhang

CONTENTS

KEY CONCEPTS

- Individual response to therapeutics is likely a complex trait influenced by both genetic and nongenetic factors.
- Pharmacogenomics aims to elucidate the relationships between therapeutic phenotypes and various molecular targets, such as gene expression, genetic variants, and epigenetic markers.
- Clinical implications of pharmacogenomic biomarkers hold the promise of realizing personalized medicine by identifying patients who may benefit most from a particular drug as well as those who may perform the worst and with severe adverse side effects.
- Pharmacogenomic biomarkers, particularly genetic variants associated with therapeutic phenotypes, have begun to be replicated and applied in several common diseases, including cardiovascular diseases, cancers, and psychiatric disorders.
- Future pharmacogenomic discovery will integrate genetic variants and other critical molecular targets, such as epigenetic biomarkers, for a more comprehensive understanding of drug response.

INTRODUCTION

Clinical responses to therapeutic treatments may vary significantly among individual patients. For a particular drug, individual responses may range from beneficial effects to no response to severe side effects, including even fatal adverse drug reactions (ADRs). Side effects from therapeutic treatments represent leading causes for hospital admissions and mortality. It is estimated that ADRs account for over 2.2 million hospitalization cases and more than 100,000 deaths annually in the United States alone.[1]

In particular, for drugs with a narrow therapeutic window (i.e., a narrow range of doses between efficacy and side effects), identifying patients who may benefit from treatment and/or predicting those who may exhibit severe ADRs prior to treatment will substantially improve clinical practice, acting by providing precision and personalized care for patients. For example, in oncology, anticancer chemotherapeutics often have a narrow range of drug dose requirements between therapeutic response and resistance, as well as between tumor response and cellular toxicities. Chemotherapy-induced toxicities may affect vital organs such as heart (cardiotoxicity),[2] kidney (nephrotoxicity),[3] nervous system (neurotoxicity),[4] and liver (hepatotoxicity)[5] and may cause severe ADRs including fatalities. Another well-established example of a drug with a narrow therapeutic index is warfarin, a commonly prescribed anticoagulant for the prevention of thrombosis and thromboembolism, featuring large interindividual variability in dose requirements, which, if not properly monitored and managed, may increase either thrombosis risk if doses are too low, or serious bleeding risk if doses are too high.[6]

Concurrent with our great improvement of knowledge in human genetic variations (linkage disequilibrium patterns between genetic variants) since the launch of the Human Genome Project[7,8] and the advances in molecular profiling technologies (genotyping using microarrays and sequencing) during the last 15 years, the development of pharmacogenetic and pharmacogenomic research has begun to pave the road to precision and personalized medicine by investigating the relationships between human genetic variations and drug response variation in patients. Significant progress has been made in elucidating the genetic basis of drug response phenotypes. In particular, pharmacogenomic loci or biomarkers with clinical implications have been identified for a variety of therapeutic treatments.

In this chapter, the general background of drug response as a complex trait as well as human genetic variations as the foundation of pharmacogenomic discovery will be introduced. The commonly used research strategy for identifying pharmacogenomic biomarkers will be reviewed to provide an overview of the investigative approaches utilized in pharmacogenomic discovery. The focus of this chapter is some of the essential pharmacogenomic biomarkers with currently the strongest evidence and known clinical implementations for drugs used to treat common diseases. We also describe some potential pharmacogenomic biomarkers in development that will likely have high clinical impact.

DRUG RESPONSE IS A COMPLEX TRAIT

An individual patient's response to the drug is a complex phenotype that can be influenced by a variety of genetic and nongenetic factors (diet, life style, and environment) (Figure 3.1). Nongenetic factors may contribute to drug response variability due to drug–drug or drug–diet interactions. Notably, concomitant administration of statins with dietary compounds was found to alter statin pharmacokinetics or pharmacodynamics, thus increasing the risk of statin-induced ADRs (myopathy or rhabdomyolysis) or reducing their pharmacological action.[9] Mechanistically, grapefruit juice components may inhibit CYP3A4 (cytochrome P450, family 3, subfamily A, polypeptide 4), reducing the presystemic or first-pass metabolism of drugs, such as simvastatin, lovastatin, and atorvastatin.[9]

In contrast, drug response has been demonstrated to be an inheritable phenotype, suggesting that an individual's genetic make-up may contribute substantially to drug response variability. For example, using linkage analysis based on a large pedigree of human lymphoblastoid cell lines (LCLs) derived from individuals of European ancestry, the heritability for cisplatin, a platinum-containing chemotherapeutic agent used to treat various cancers,[10] was estimated to be approximately 47%.[11] Therefore, sensitivity to the cytotoxic effects of cisplatin is likely under appreciable genetic influence. In addition, there is evidence that drug response phenotypes can be due to multiple genomic loci or regions (polygenic traits), each of which may only contribute a small proportion of the total variability. Shukla et al. identified 11 genomic regions on 6 chromosomes that may be significantly associated with the susceptibility to cisplatin-induced cytotoxicity using an LCL model and linkage analysis.[12]

Given the complexity of potential contributions from both genetic and nongenetic factors, therefore, it is often not a straightforward decision to determine whether a patient will respond well to the drug or not. Elucidating the relationships between genetic factors and drug response variability is crucial for a comprehensive

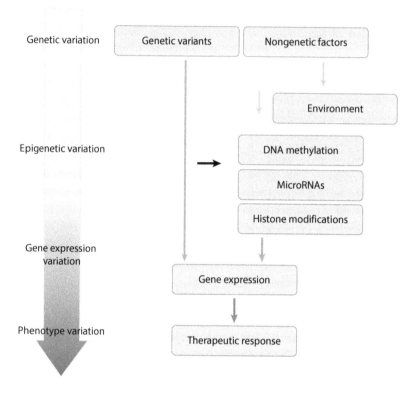

FIGURE 3.1 Drug response is a complex trait. Individual drug response can be influenced by various genetic and nongenetic factors, such as genetic variants, epigenetic factors (e.g., DNA methylation, histone modifications), and environment. Gene expression, a complex phenotype itself, may be intermediate to drug response phenotypes.

understanding of the mechanisms underlying drug response phenotypes as well as providing a basis for implementing precision and personalized medicine in clinical practice. Pharmacogenomics aims to systematically and comprehensively investigate the genetic architectures of drug response variability by leveraging the great advances of our scientific knowledge in human genetic variations and the technological advances, including various microarray- and sequencing-based profiling platforms[13,14] (Figure 3.2).

HUMAN GENETIC VARIATIONS UNDERLIE PHARMACOGENOMICS

A primary task of pharmacogenomic studies is to identify genetic loci that are responsible for the clinically observed variability of drug response phenotypes. Genetic variations within and between human populations, in the forms of single nucleotide polymorphisms (SNPs) and other structural variants such as copy-number variants (CNVs), are the basis for pharmacogenomic studies. There are known to be

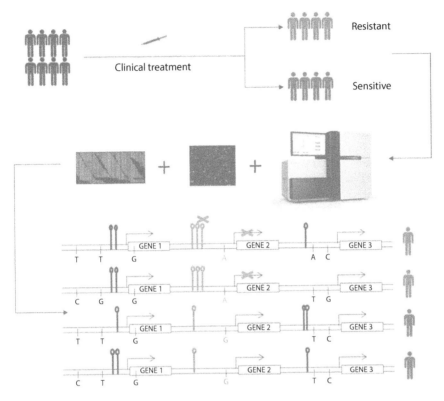

FIGURE 3.2 Overview of pharmacogenomic discovery. For the aim of precision and personalized medicine, pharmacogenomic biomarkers can be used to separate patients based on a particular therapeutic phenotype (resistant or sensitive to a drug). Various high-throughput technologies, such as microarray- and sequencing-based platforms, are now available to profile molecular targets to be tested for genes and/or genomic loci associated with a therapeutic phenotype.

extensive genetic variations among individuals from global populations. For example, the 1000 Genomes Project (phase I integrated release) reported >17 million SNPs by sequencing 1092 individuals from 11 global populations from Asia, Africa, the Americas, and Europe.[15] The majority (approximately 90%) of genetic variations is found within major global populations (Asians, Europeans, and Africans), and only an additional 5–15% of variations is found between any two populations.[16] In contrast to other global populations, African individuals typically have a higher level of genetic diversity than non-African populations.[17,18]

In particular, it has been estimated that a 99.9% genetic identity exists between any two randomly chosen individuals. The 0.1% differences in an individual's genetic make-up (corresponding to approximately 3 million base-pair differences between any two randomly picked individuals) likely play an important role in defining complex traits and phenotypes, such as adult height, skin color, body mass index, as well as risks for common diseases.[19] Importantly, gene expression, an intermediate phenotype reflecting cellular functions, is a complex trait itself that may be affected by both genetic and nongenetic factors (Figure 3.1), integrating SNP genotypic data with quantitative gene expression levels using the HapMap Project[17,18] samples, a widely used human genetics model, have demonstrated that common genetic variants, known as eQTL (expression quantitative trait loci) contributed substantially to gene expression variation within and between populations.[20–24] Given that gene expression is such a fundamental phenotype that may be intermediate to other cellular and whole-body phenotypes, it is likely that genetic control of complex traits (drug response) could be through eQTL that regulate gene expression in relevant pathways. For the variability of drug response, though the extent of genetic contribution is still controversial, at least some of the individual variations in drug response phenotypes may be attributed to the 0.1% genetic differences between individuals.

APPROACHES TO IDENTIFYING PHARMACOGENOMIC BIOMARKERS

In general, the research approaches used in pharmacogenomic research are similar to those used in other complex trait studies, though investigation of drug response may present certain unique challenges in study design and data analysis. An important limitation for pharmacogenomic research is that it can be both unethical and difficult to identify healthy controls in a conventional case–control association study, because many drugs (cytotoxic anticancer drugs) may cause serious harms to healthy individuals. Therefore, in vitro systems, such as cell lines, are often used in pharmacogenomic discovery. For example, for cytotoxicities induced by anticancer agents, the LCLs samples from the HapMap Project[17,18] have been developed as a cell-based model.[25,26] Significant progress in pharmacogenomic discovery for chemotherapies of various mechanisms, such as cisplatin,[27] etoposide,[28] and daunorubicin,[29] has been made using this model by integrating the extensive genotypic and gene expression data available for these samples.[30]

There are two general research frameworks in pharmacogenomic discovery: the candidate-gene approach and the whole-genome approach (Figure 3.3). Studies applying the candidate-gene approach focus on selected genes or pathways with known functions and/or relationships with a drug response phenotype, such as genes known to be involved in pharmacokinetics and pharmacodynamics. In contrast, studies applying the whole-genome approach aim to identify pharmacogenomic loci associated with drug response through an unbiased, genome-wide scan. A combination of these two frameworks is often necessary for elucidating the genetic contribution to a particular drug response phenotype under investigation.

Statistically, various analytical approaches and methods have been utilized to identify genetic loci associated with complex traits. Genome-wide association studies (GWASs) have been used to map genetic loci associated with more than 500 complex traits including common human traits (adult height and skin color) and risks for complex diseases, such as cancers, cardiovascular diseases, and psychiatric disorders.[19] Specifically, genetic association studies, which take advantage of the characteristics of human genetic variations (linkage between a genotyped variant and an nontyped

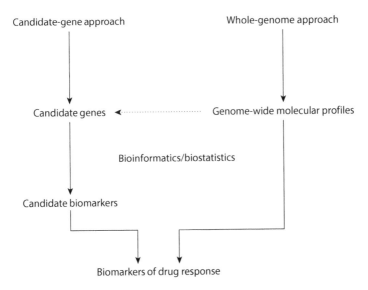

FIGURE 3.3   General approaches to detecting biomarkers of drug response. Either a knowledge-driven candidate gene or an unbiased, whole-genome approach may be used to detect biomarkers (genetic variants) of drug response phenotypes.

causal variant), have been extensively utilized to identify pharmacogenomic biomarkers or primarily genetic variants associated with drug response phenotypes. Some essential pharmacogenomic biomarkers identified from previous association studies are introduced in the following sections.

PHARMACOGENOMIC BIOMARKERS WITH CLINICAL IMPLICATIONS

The Pharmacogenomics Knowledgebase (PharmGKB) (http://www.pharmgkb. org/)[31] guidelines were followed to obtain a list of essential pharmacogenomic biomarkers for drugs used to treat common, complex diseases with the strongest level of evidence for association and clinical implications. Specifically, we focus on essential pharmacogenomic biomarkers that contain a genetic variant–drug combination in a CPIC (Clinical Pharmacogenetics Implementation Consortium)[32] or medical society–endorsed pharmacogenomic guideline as well as those implemented at a PGRN (Pharmacogenomics Research Network)[33] site or another major health system. In addition, essential pharmacogenomic biomarkers must contain a variant–drug combination where the preponderance of evidence shows an association, and must be replicated in more than one cohort with significant p-values, and preferably, with a strong effect size. Table 3.1 shows some essential pharmacogenomic biomarkers by disease type. Given the important roles of drug-metabolizing enzymes, many of the current pharmacogenomic biomarkers with clinical implementation are genetic variants located in these enzymes: for example, cytochrome P450 (*CYP*) genes.

TABLE 3.1

Examples of Pharmacogenomic Biomarkers with Clinical Implementation and Strong Evidence

Condition	Drug	Biomarker	Host Gene
Cardiovascular diseases	Warfarin	CYP2C9*1, *2, and *3[39–42] rs7294, rs9923231[39]	CYP2C9 VKORC1
	Clopidogrel	CYP2C19*1, *2, *3, *4,*5, *6, *8, and *17[48]	CYP2C19
	Simvastatin	rs4149056[57]	SLCO1B1
Cancers	Cisplatin	rs2228001[64]	XPC
	Erlotinib, gefitinib	rs121434568[67,68]	EGFR
	Methotrexate	rs1801133	MTHFR
Psychiatric disorders	Amitriptyline, nortriptyline	CYP2C19*1, *17, CYP2C19 *2, *3[73–75]	CYP2C19
		CYP2D6*1, *10, *1XN, *2, *2XN, *3, *4, *5, and *6[73–75]	CYP2D6
	Citalopram	rs1954787[81]	GRIK4
Viral infections	Abacavir	HLA-B*5701[87]	HLA-B
Cystic fibrosis	Ivacaftor	rs113993960, rs75527207[91–93]	CFTR
Pain	Codeine	CYP2D6*10[99]	CYP2D6
Hyperuricemia	Allopurinol	HLA-B*5801[102–106]	HLA-B

CARDIOVASCULAR DISEASES

Warfarin

Warfarin is an oral anticoagulant that is frequently used to prevent thrombosis and reduce the risk for thromboembolism. Warfarin features a narrow therapeutic index and a wide interindividual variation in dose requirements. Overdose of warfarin increases the risk for severe bleeding. Dosing of warfarin is complicated as it is known to interact with many commonly used drugs and some food components.[34] Though patient-specific clinical characteristics such as age, body size, race, and concurrent diseases may explain some of its dose requirement variability, genetic variants have been demonstrated to contribute a significant proportion of the variability in warfarin dose.[35] Several studies have derived dosing algorithms that integrate both genetic and nongenetic factors to predict an individual patient's warfarin dose.[36,37] In particular, genetic variants in CYP2C9 (encoding cytochrome P450, family 2, subfamily C, polypeptide 9) and VKORC1 (encoding vitamin K epoxide reductase complex, subunit 1) have been found to affect warfarin dosing, accounting for up to 15% and 25% of the variation in dose requirements, respectively.[38] A common noncoding variant rs7294 located in the 5′-UTR (untranslated region) of VKORC1 is associated with an increased sensitivity to warfarin.[36,39] In contrast, CYP2C9*2 and CYP2C9*3, two common variants in CYP2C9, are associated with reduced enzyme activity of CYP2C9. Compared to patients with normal enzyme activity, carriers of CYP2C9*2 or *3 alleles are more sensitive to warfarin with a greater risk of severe bleeding, thus requiring lower starting doses.[40–42] Notably, there are significant population differences in the frequencies of the *CYP2C9* alleles with *2 allele more common in Caucasian than Asian or African populations, while the *3 allele is less common in all ethnic groups and is extremely rare in African populations.[43–45] Other alleles such as *CYP2C9*5, *6, and *8 may contribute to the variability in African American patient response to warfarin.[46] Therefore, a patient's *CYP2C9* and *VKORC1* genotypes may be used clinically to help determine the optimal starting dose of warfarin. The United States Food and Drug Administration (FDA)–approved warfarin drug label now provides a dosing table that is derived from multiple clinical studies, which recommends that warfarin initial doses vary between the patients based on their combinations of *CYP2C9* and *VKORC1* genotypes.

Clopidogrel

Clopidogrel, an antiplatelet prodrug, can specifically inhibit platelet activation and aggregation through blockade of the adenosine diphosphate (ADP) receptor P2Y12 expressed on platelets after its metabolic activation in the liver. Clopidogrel is a commonly prescribed drug used to prevent thrombotic events after myocardial infarction, ischemic stroke, and coronary stent placement. Serious ADRs have been associated with clopidogrel, including severe neutropenia, hemorrhage, and thrombotic thrombocytopenic purpura.[47] *CYP2C19* (cytochrome P450, family 2, subfamily C, polypeptide 19) is one of the principal enzymes involved in the metabolism of clopidogrel. Pharmacogenetic studies have demonstrated the importance of *CYP2C19* genotypes in clopidogrel treatment. For example, carriers of *CY2C19* loss-of-function alleles, for example, *CYP2C19*2, were found to have significantly lower levels of the active metabolite of clopidogrel, diminished platelet inhibition, and a higher rate of major adverse cardiovascular events (MACEs) than did noncarriers of these alleles.[48,49] Among patients treated with clopidogrel for percutaneous coronary intervention, carriage of even one reduced-function *CYP2C19* allele was found to be associated with a significantly increased risk of MACEs, particularly stent thrombosis.[50] The *CYP2C19*2 variant was also found to be a major determinant of prognosis in young patients who were receiving clopidogrel treatment after myocardial infarction.[51] The US FDA recommends that *CYP2C19* genotyping be considered prior to prescribing clopidogrel, although the association between the *CYP2C19* genotypes and cardiovascular events is still controversial.[52,53]

Simvastatin

Simvastatin is used to control elevated cholesterol in blood, or hypercholesterolemia and the prevention of cardiovascular diseases.[54] Common side effects of simvastatin include abdominal pain, diarrhea, indigestion, and a general feeling of weakness. In rare cases, myopathy may occur in association with statin treatments, especially when the statins are administered at higher doses (80 mg of simvastatin daily vs. standard doses of 20–40 mg daily) and in combination with certain other drugs (cyclosporine).[55] A genome-wide pharmacogenomic scan has been performed to evaluate the genetic contribution to simvastatin-induced myopathy in patients with prior myocardial infarction in a clinical trial to determine whether a daily dose of 80 mg of simvastatin safely produces greater benefits than does a daily dose of 20 mg of simvastatin.[56,57] A strong association with a nonsynonymous polymorphism, SNP rs4149056, located in *SLCO1B1* (encoding solute carrier organic anion transporter family, member 1B1), which regulates the hepatic statin uptake transporter, has been linked to simvastatin transport.[57,58] In particular, more than 60% of these myopathy cases were attributed to the C allele of this SNP.[57]

CANCERS

Cisplatin

Cisplatin is a platinum-based chemotherapeutic agent used in various types of cancers. Common side effects of cisplatin include nephrotoxicity, ototoxicity, myelotoxicity, and nausea.[59–62] The main dose-limiting side effect with cisplatin chemotherapy is peripheral neurotoxicity (i.e., nerve damage).[63] Nucleotide excision repair gene polymorphisms, especially a missense variant, SNP rs2228001, located in *XPC* (encoding xeroderma pigmentosum, complementation group C), were found to be potentially predictive for toxicities induced by cisplatin chemoradiotherapy in bladder cancer[64] and osteosarcoma patients.[65]

Erlotinib and Gefitinib

Erlotinib and gefitinib, two drugs of the EGFR (epidermal growth factor receptor) tyrosine kinase inhibitors (TKIs), are targeted therapies used to treat non–small cell lung cancer, pancreatic cancer, and several other cancers.[66] EGFR overexpression may lead to inappropriate activation of the anti-apoptotic Ras signaling cascade, eventually leading to uncontrolled cell proliferation. The tolerability profiles of erlotinib and gefitinib are generally better than other cytotoxic agents, because they are selective chemotherapeutic agents. *EGFR* mutations have been found to affect EGFR TKI efficacy.[67,68] A missense variant, SNP rs121434568 in *EGFR* coding regions, is associated with the efficacy of EGFR TKIs as well as the toxicity and ADRs induced by erlotinib.

Methotrexate

Methotrexate is an antimetabolite and antifolate drug. It is a chemotherapy either used alone or in combination with other anticancer agents. It is effective for the treatment of a number of cancers, such as breast, head and neck, leukemia, lymphoma, lung, osteosarcoma, bladder, and trophoblastic neoplasms as well as autoimmune diseases, such as rheumatoid arthritis. The most common ADRs include ulcerative stomatitis; low white blood cell count and thus predisposition to infections; nausea; abdominal pain; fatigue; fever; dizziness; acute pneumonitis; and, rarely, pulmonary fibrosis. Central nervous system reactions (myelopathy and leukoencephalopathy) to methotrexate have been reported, especially when given via the intrathecal route. It may also cause a variety of cutaneous side effects, particularly when administered at high doses.[69] The C677T polymorphism, SNP rs1801133, located in the *MTHFR* (encoding methylenetetrahydrofolate reductase) gene, was found to be associated with the toxicity of methotrexate in a Spanish rheumatoid arthritis population, though other meta-analyses still suggested some controversies in terms of this association as a reliable predictor of methotrexate response in rheumatoid arthritis patients.

PSYCHIATRIC DISORDERS

Amitriptyline

Amitriptyline is a tricyclic antidepressant that has been considered one of the reference compounds for depression therapy.[70] Amitriptyline is also used systemically for the management of neuropathic pain.[71] Though amitriptyline is an efficacious antidepressant drug, it is associated with a number of side effects, such as clumsiness, drowsiness, muscle aches, and sleepiness.[72] Polymorphisms in *CYP2C19* and *CYP2D6* have been associated with its efficacy and the risk for side effects in amitriptyline therapy.[73–75] For example, it has been confirmed that the increased activity of the *CYP2C19*17* allele is associated with the increased metabolism of amitriptyline.[76] In contrast, the *CYP2D6*4* allele, the main polymorphism resulting in reduced enzyme activity in Caucasians, has been associated with the increased toxicity (risk of developing symptoms of hyponatremia or low blood sodium concentrations)[77] of antidepressants including amitriptyline.[78]

Nortriptyline

Nortriptyline, a second-generation tricyclic antidepressant, is an active metabolite of amitriptyline. It is used in the treatment of major depression and childhood nocturnal enuresis. Major side effects of nortriptyline include sedation, hypotension, and anticholinergic effects. Nortriptyline is metabolized by the hepatic enzyme *CYP2D6*. Approximately 7–10% of Caucasian individuals are poor metabolizers with lower *CYP2D6* enzyme activity, and might experience more adverse effects: therefore, a lower dosage may be required in these individuals. Similar to amitriptyline, *CYP2D6* polymorphisms, such as the *CYP2D6*4* allele, has been associated with the increased toxicity of nortriptyline.

Citalopram

Citalopram, a drug of the selective serotonin reuptake inhibitor (SSRI) class that affects neurotransmitters,[79] has been approved to treat major depression, given that an imbalance of neurotransmitters is known to be the cause of depression. Although SSRIs have better overall safety and tolerability than older antidepressants, adverse effects, such as sexual dysfunction, weight gain, sleep disturbance, and suicidal thinking, may be associated with long-term SSRI therapy.[80] Genetic variation in a GRIK4 (kainic acid–type glutamate receptor) has been demonstrated to be reproducibly associated with response to the antidepressant citalopram, suggesting that the glutamate system plays an important role in modulating response to SSRIs.[81] In addition, since citalopram is metabolized in the liver mostly by *CYP2C19*, as well as by *CYP3A4* and *CYP2D6*, genetic variants in these P450 genes are likely to influence the response to this drug.

OTHER DISORDERS

Abacavir for Viral Infection

Abacavir, a nucleoside analog reverse transcriptase inhibitor, is an antiretroviral agent used together with other medicines to treat human immunodeficiency virus (HIV) infection.[82] Abacavir has been well tolerated with main side effects of hypersensitivity,[83] a treatment-limiting and potentially life-threatening ADR. Side effects of abacavir therapy could be severe and, in rare cases, fatal. Though controversial, the use of abacavir has been associated with an increased risk of cardiovascular diseases in some cohort studies.[84,85] Genetic testing can indicate whether an individual would be hypersensitive to abacavir therapy. Two retrospective pharmacogenomic studies were conducted to identify HIV-1 patients at increased risk for abacavir hypersensitivity, indicating a strong statistical association between the major histocompatibility complex or human leukocyte antigen (HLA) allele, *HLA-B*5701*, and clinically diagnosed abacavir hypersensitivity between racial populations.[86,87] Screening patients for *HLA-B*5701* prior to the initial administration of abacavir represents a clinical tool to further decrease the risk of hypersensitivity reactions as well as unnecessary discontinuation of abacavir.[83]

Ivacaftor for Cystic Fibrosis

Ivacaftor is a treatment for cystic fibrosis aiming to improve the function of the mutant CFTR (cystic fibrosis transmembrane conductance regulator).[88] Ivacaftor was associated with relative improvements (within subject) in CFTR and lung function.[89,90] A phase II clinical trial showed that the safety profile of ivacaftor was comparable to that of the placebo in subjects with cystic fibrosis who are homozygous for F508del (SNP rs113993960), the most prevalent disease-causing *CFTR* mutation.[91] Polymorphisms in the *CFTR* gene have also been associated with ivacaftor efficacy. Ivacaftor was shown to offer an effective and well-tolerated treatment for the clinical management of cystic fibrosis patients with the G551D mutation (SNP rs75527207).[92,93]

Codeine for Pain

Codeine is a prodrug that belongs to the opiate class used to treat mild-to-moderate pain. Codeine's function stems from its conversion to morphine and morphine-6-glucuronide, a strong opioid agonist, by *CYP2D6* in the liver. Common adverse effects of opiates may include nausea, constipation, abdominal pain, respiratory depression, urinary retention, sedation, itching, and addiction.[94] The efficacy and safety of codeine as an analgesic have been associated with *CYP2D6* polymorphisms.[95,96] In particular, the association between *CYP2D6* metabolizer phenotype and the bioconversion of morphine from codeine has been well defined.[97] Codeine was shown to have little therapeutic effect in patients who were *CYP2D6* poor metabolizers (with little or no *CYP2D6* activity), while the risk of morphine toxicity was higher in ultrarapid metabolizers (with greater-than-normal *CYP2D6* activity).[98] For example, the *CYP2D6*10* allele, which is the most common allele with reduced *CYP2D6* activity, plays an important role in the pharmacokinetics of the O-demethylated metabolites of codeine including morphine, the most active metabolite, after oral administration.[99] Therefore, clinical implementation of genotyping of *CYP2D6* may help prevent diminished pain relief and/or severe opioid side effects.

Allopurinol for Hyperuricemia

Allopurinol, a purine analog, is primarily used to treat hyperuricemia or excess uric acid in blood, and its complications, including chronic gout.[100] Allopurinol can cause serious ADRs, such as Stevens–Johnson syndrome (a rare, serious disorder of skin and mucous membranes) and toxic epidermal necrolysis (a dermatologic disorder characterized by symptoms such as widespread erythema and necrosis) that are associated with significant mortality and morbidity.[101] The *HLA-B*5801* allele was shown to be strongly associated with allopurinol-induced ADRs in several Asian and non-Asian populations.[102–106] *HLA-B*5801* allele screening is currently recommended to be considered in patients who will be treated with allopurinol.[107]

PHARMACOGENOMIC BIOMARKERS UNDER DEVELOPMENT

Although significant progress has been made in detecting clinically implementable pharmacogenomic biomarkers, there still remain grand challenges to optimize precision and personalized medicine for patients with various diseases and disorders, given the complex nature of pathogenesis and drug response. With the advances in high-throughput profiling technologies, particularly those based on the next-generation sequencing (NGS), numerous trials and studies are ongoing with the ultimate goals of detecting more sensitive and specific biomarkers of drug response phenotypes. Some of these biomarkers under development are introduced with a focus on novel biomarkers beyond traditional drug-metabolizing genes for certain common, complex diseases (Table 3.2).

CARDIOVASCULAR DISEASES

Besides these well-established *CYP* genes, previous studies on cardiovascular diseases, especially recent genome-wide scans enabled by the advanced high-throughput profiling technologies, have implicated several strong candidate genes that may be responsible for the interindividual variation in drug response phenotypes. Of particular interest are genes that may determine therapeutic responses to statin treatments, given that statins are widely used to reduce the risk of major cardiovascular events. Among the top candidate genes for statin response are *ABCB1* (encoding ATP-binding cassette, subfamily B, member 1), *CETP* (encoding cholesterol ester transfer protein), *HMGCR* (encoding 3-hydroxy-3-methylglutaryl-coA reductase), and

TABLE 3.2
Some Novel Pharmacogenomic Biomarkers under Development

Disorder	Drug	Biomarker	Host Gene
Cardiovascular diseases	Statins	rs1128503, rs2032582, rs1045642[108]	*ABCB1*
		rs5883, rs9930761[109]	*CETP*
		rs5908, rs12916[110]	*HMGCR*
		rs4149056[111]	*SLCO1B1*
Chronic myeloid leukemia	Imatinib, nilotinib	Philadelphia chromosome[113–115]	*BCR-ABL*
Acute myeloid leukemia	Ruxolitinib	V617F[117]	*JAK2*
Obsessive–compulsive disorder	Selective serotonin reuptake inhibitors	rs17162912[118] rs9957281[118]	*DISP1* *DLGAP1*

SLCO1B1 (encoding solute carrier organic anion transporter family, member 1B1). For example, three SNPs, rs1128503 (synonymous), rs2032582 (missense), and rs1045642 (synonymous) located in *ABCB1*, have been demonstrated to influence directly on statin pharmacokinetics and indirectly on statin pharmacodynamics, considering the important role of ABCB1 in transporting statins and metabolites;[108] SNPs rs5883 (synonymous) and rs9930761 (intronic) in *CETP* gene, the protein product of which shuttles cholesterol esters from high-density lipoprotein (HDL) particles to low-density lipoproteins (LDL), have been associated with HDL levels, risk for coronary artery disease, and response to statin therapy.[109] SNPs rs5908 (missense) and rs12916 (3'-UTR) in *HMGCR* gene, which encodes an enzyme cata-lyzing a rate-limiting step in cholesterol biosynthesis, were found to be significantly associated with attenuated LDL cholesterol reduction as well as racial differences in statin therapy;[110] and SNP rs4149056 (missense) in *SLCO1B1* gene, which encodes a transporter that transports certain hormones, toxins, and drugs into the liver for removal from the body, was found to be associated with simvastatin-induced myopa-thy in patients, while the effect of rs4149056 on statin pharmacokinetics was shown to vary with statin type.[111] In addition, a multiple-loci interaction model that com-bines polymorphisms from different candidate genes was shown to be a better pre-dictor for LDL cholesterol-lowering effects.[112]

CANCERS

During the past decade, cancer research has witnessed extensive utilization of novel high-throughput profiling technologies. Large cancer-focused projects, such as the Cancer Genome Atlas (TCGA) Project sponsored by the National Institutes of Health, have begun to significantly enhance our current knowledge of cancer patho-genesis and responses to therapeutics. Pharmacogenomic biomarker discovery for anticancer drugs has benefited from these technological advances, with novel bio-markers being identified for better prediction of sensitivity to cancer therapies. We showcase here some novel biomarkers recently proposed or developed for anticancer drug responses, particularly using hematological malignancies as an example.

For example, Ph+/Ph− status for the Philadelphia chromosome has been proposed to be a "chromosomal" biomarker that may predict survival response to TKIs, includ-ing imatinib and nilotinib, used to treat chronic myeloid leukemia (CML). Previous studies suggested that treatment with these TKIs significantly improves survival compared with survival rates in the pre-TKI era in patients with Ph+ CML.[113–115] With the availability of these TKIs, a molecular biomarker, *BCR-ABL* gene fusion, which represents the product of Ph+ status, could enable us to measure the molecular response to the TKIs, thus helping determine treatment efficacy and guide further clinical decision. A recent study showed that assessment of *BCR-ABL1* transcript levels at three months was the only requirement for predicting outcome for patients with CML treated with TKIs.[116] However, this molecular biomarker has not yet been validated sufficiently to be cleared by the US FDA for clinical use.

Another example of recently implicated novel biomarkers for anticancer drug response is the mutation status of *JAK2* gene (which encodes Janus kinase 2) in treating acute myeloid leukemia (AML). Previous studies showed that although

JAK2 mutations (*JAK2*V617F) are not commonly expressed in AML, a small subset of patients may harbor these mutations, including those with myeloproliferative neoplasms that evolved into AML. Those patients with these *JAK2* mutations were found to respond to ruxolitinib, a JAK1/2 inhibitor.[117]

PSYCHIATRIC DISORDERS

Pharmacogenetic and pharmacogenomic studies have also benefited psychiatric disorders management by identifying novel biomarkers that are associated with responders and/or patients who may have severe side effects, such as suicidal thinking. We showcase here some examples of biomarkers based on recent studies, including GWASs on drug response. For example, SSRIs are currently the most commonly administered first-line treatment for depression and obsessive–compulsive disorder (OCD); however, more than 30% of patients do not respond to SSRIs. Therefore, novel drugs and biomarkers are needed for individualizing treatment in patients with depression. A recent whole-genome association study identified novel loci associated with altered antidepressant response to SSRIs in 1773 OCD cases from a family-based GWAS,[118] including SNPs rs17162912 (an intergenic variation near *DISP1* gene on 1q41-q42, a microdeletion region related to mental retardation) and rs9957281 (located within a newly identified OCD susceptibility gene *DLGAP1*, a member of the neuronal postsynaptic density complex). In addition, 13 other significant variants were also identified in 6 genes, namely *FAM225B* (encoding family with sequence similarity 225, member B, nonprotein coding), *CDYL2* (encoding chromodomain protein, Y-like 2), *PAFAH1B1* (encoding platelet-activating factor acetylhydrolase 1b, regulatory subunit 1), *TACC1* (encoding transforming, acidic coiled-coil containing protein 1), *CSMD1* (encoding CUB and Sushi multiple domains 1), and *GAS2* (encoding growth arrest-specific 2). Therefore, these newly identified drug response loci can be tested in the development of personalized care of OCD patients treated with SSRIs. In addition, these results would also provide new targets for developing other novel drugs for treating nonresponders.

THOUGHTS FOR FURTHER CONSIDERATION

The majority of currently known pharmacogenomic loci have not yet become pharmacogenomic biomarkers implemented in clinical practice. Large-scale clinical trials in the future on these biomarkers will provide essential knowledge on the usefulness of pharmacogenetics and pharmacogenomics in alleviating severe ADRs and enhancing therapeutic efficacy in patients. Notably, given the ethnic disparities in drug treatments, expanding future pharmacogenomic discovery studies to target ethnicity- or population-specific biomarkers for therapeutic phenotypes may help provide better healthcare to different populations, an important component of precision and personalized medicine. More recently, the relationships across genetic variants, epigenetic traits (DNA methylation, histone modifications, microRNAs) as well as gene expression phenotypes have also begun to be elucidated in human genetics models. For example, the complex relationships across common genetic variants (QTL), gene expression, and epigenetic systems such as microRNA[119] and cytosine modification (primarily methylation at CpG dinucleotides) levels[120–123] have been investigated using the HapMap LCL samples. Interestingly, in a recent pharmacogenomic study, integrating CpG methylation has been shown to significantly improve our understanding of the cytotoxicities induced by clofarabine, a purine nucleoside analog used in the treatment of hematologic malignancies and as induction therapy for stem cell transplantation, compared to genetic variants alone,[124] indicating the potential of epigenetic biomarkers of drug response phenotypes. These advances in elucidating the complexity of human genome and gene regulation, together with advances in high-throughput profiling technologies (NGS-based approaches), suggest the promise of the next wave of pharmacogenomic discovery that aims to integrate genetic variants with other molecular targets (epigenetic biomarkers) in detecting pharmacogenomic biomarkers with clinical implications.

STUDY QUESTIONS

1. What is the major aim of pharmacogenetic and pharmacogenomic studies?
2. What is the clinical goal of applying pharmacogenomic biomarkers in patients?
3. What is the major difference between pharmacogenetic and pharmacogenomic studies?
4. Give an example for the current pharmacogenomic biomarkers with clinical practice.
5. Besides genetic variants, what other genomic features and molecular targets can be integrated into the next wave of pharmacogenomic discovery?

Answers
1. Pharmacogenetic and pharmacogenomic studies aim to elucidate the relationships between genetic variations and therapeutic phenotypes.
2. Clinical applications of pharmacogenomic biomarkers in patients are used to identify patients who may benefit most from a particular drug as well as those who may perform the worst, with severe adverse side effects.
3. Pharmacogenetic studies usually focus on well-defined candidate genes and/or pathways. In contrast, pharmacogenomic studies are intended to be unbiased, genome-wide scans for pharmacogenomic discovery.
4. Currently, pharmacogenomic biomarkers are being implemented in clinical practice in patients with several common, complex diseases, such as cardiovascular diseases, cancers, and psychiatric disorders. For example, extensive pharmacogenetic and pharmacogenomic studies have been carried out for the phenotype of warfarin dosing, implicating genetic variants in *VKORC1* and *CYP2C9* for determining interindividual variability of dose requirements.
5. Given their critical roles in regulating gene expression, which is fundamental to complex traits including drug response, epigenetic biomarkers, such as CpG methylation; histone modifications; and microRNAs, can be a novel class of pharmacogenomic biomarkers.

REFERENCES

1. Lazarou J, Pomeranz BH, Corey PN. Incidence of adverse drug reactions in hospitalized patients: A meta-analysis of prospective studies. *JAMA* 1998;279:1200–5.
2. Bonita R, Pradhan R. Cardiovascular toxicities of cancer chemotherapy. *Seminars in Oncology* 2013;40:156–67.
3. Perazella MA, Moeckel GW. Nephrotoxicity from chemotherapeutic agents: Clinical manifestations, pathobiology, and prevention/therapy. *Seminars in Nephrology* 2010;30:570–81.
4. Grisold W, Cavaletti G, Windebank AJ. Peripheral neuropathies from chemotherapeutics and targeted agents: Diagnosis, treatment, and prevention. *Neuro-Oncology* 2012;14(Suppl 4):iv45–54.
5. Thatishetty AV, Agresti N, O'Brien CB. Chemotherapy-induced hepatotoxicity. *Clinics in Liver Disease* 2013;17:671–86, ix–x.
6. Hirsh J, Fuster V, Ansell J, Halperin JL. American Heart Association/American College of Cardiology Foundation guide to warfarin therapy. *Journal of the American College of Cardiology* 2003;41:1633–52.
7. Lander ES, Linton LM, Birren B, et al. Initial sequencing and analysis of the human genome. *Nature* 2001;409:860–921.
8. Venter JC, Adams MD, Myers EW, et al. The sequence of the human genome. *Science* 2001;291:1304–51.
9. Vaquero MP, Sanchez Muniz FJ, Jimenez Redondo S, Prats Olivan P, Higueras FJ, Bastida S. Major diet-drug interactions affecting the kinetic characteristics and hypolipidaemic properties of statins. *Nutricion Hospitalaria* 2010;25:193–206.
10. Rosenberg B, VanCamp L, Trosko JE, Mansour VH. Platinum compounds: A new class of potent antitumour agents. *Nature* 1969;222:385–6.
11. Dolan ME, Newbold KG, Nagasubramanian R, et al. Heritability and linkage analysis of sensitivity to cisplatin-induced cytotoxicity. *Cancer Research* 2004;64:4353–6.
12. Shukla SJ, Duan S, Badner JA, Wu X, Dolan ME. Susceptibility loci involved in cisplatin-induced cytotoxicity and apoptosis. *Pharmacogenetics and Genomics* 2008;18:253–62.
13. Mardis ER. The impact of next-generation sequencing technology on genetics. *Trends in Genetics* 2008;24:133–41.
14. Mardis ER. Next-generation DNA sequencing methods. *Annual Review of Genomics and Human Genetics* 2008;9:387–402.
15. Abecasis GR, Altshuler D, Auton A, et al. A map of human genome variation from population-scale sequencing. *Nature* 2010;467:1061–73.
16. Watkins WS, Rogers AR, Ostler CT, et al. Genetic variation among world populations: Inferences from 100 Alu insertion polymorphisms. *Genome Research* 2003;13:1607–18.
17. HapMap. The International HapMap Project. *Nature* 2003;426:789–96.
18. HapMap. A haplotype map of the human genome. *Nature* 2005;437:1299–320.
19. Hindorff LA, Sethupathy P, Junkins HA, et al. Potential etiologic and functional implications of genome-wide association loci for human diseases and traits. *Proceedings of the National Academy of Sciences of the United States of America* 2009;106:9362–7.
20. Spielman RS, Bastone LA, Burdick JT, Morley M, Ewens WJ, Cheung VG. Common genetic variants account for differences in gene expression among ethnic groups. *Nature Genetics* 2007;39:226–31.
21. Stranger BE, Nica AC, Forrest MS, et al. Population genomics of human gene expression. *Nature Genetics* 2007;39:1217–24.
22. Duan S, Huang RS, Zhang W, et al. Genetic architecture of transcript-level variation in humans. *American Journal of Human Genetics* 2008;82:1101–13.
23. Morley M, Molony CM, Weber TM, et al. Genetic analysis of genome-wide variation in human gene expression. *Nature* 2004;430:743–7.

24. Zhang W, Duan S, Kistner EO, et al. Evaluation of genetic variation contributing to differences in gene expression between populations. *American Journal of Human Genetics* 2008;82:631–40.

25. Zhang W, Dolan ME. Use of cell lines in the investigation of pharmacogenetic loci. *Current Pharmaceutical Design* 2009;15:3782–95.

26. Welsh M, Mangravite L, Medina MW, et al. Pharmacogenomic discovery using cell-based models. *Pharmacological Reviews* 2009;61:413–29.

27. Huang RS, Duan S, Shukla SJ, et al. Identification of genetic variants contributing to cisplatin-induced cytotoxicity by use of a genomewide approach. *American Journal of Human Genetics* 2007;81:427–37.

28. Huang RS, Duan S, Bleibel WK, et al. A genome-wide approach to identify genetic variants that contribute to etoposide-induced cytotoxicity. *Proceedings of the National Academy of Sciences of the United States of America* 2007;104:9758–63.

29. Huang RS, Duan S, Kistner EO, et al. Genetic variants contributing to daunorubicin-induced cytotoxicity. *Cancer Research* 2008;68:3161–8.

30. Zhang W, Zheng Y, Hou L. Pharmacogenomic discovery delineating the genetic basis of drug response. *Current Genetic Medicine Reports* 2013;1:143–9.

31. Whirl-Carrillo M, McDonagh EM, Hebert JM, et al. Pharmacogenomics knowledge for personalized medicine. *Clinical Pharmacology and Therapeutics* 2012;92:414–7.

32. Relling MV, Klein TE. CPIC: Clinical Pharmacogenetics Implementation Consortium of the Pharmacogenomics Research Network. *Clinical Pharmacology and Therapeutics* 2011;89:464–7.

33. Giacomini KM, Brett CM, Altman RB, et al. The pharmacogenetics research network: From SNP discovery to clinical drug response. *Clinical Pharmacology and Therapeutics* 2007;81:328–45.

34. Holbrook AM, Pereira JA, Labiris R, et al. Systematic overview of warfarin and its drug and food interactions. *Archives of Internal Medicine* 2005;165:1095–106.

35. Kamali F, Wynne H. Pharmacogenetics of warfarin. *Annual Review of Medicine* 2010;61:63–75.

36. Klein TE, Altman RB, Eriksson N, et al. Estimation of the warfarin dose with clinical and pharmacogenetic data. *The New England Journal of Medicine* 2009;360:753–64.

37. Gage BF, Eby C, Johnson JA, et al. Use of pharmacogenetic and clinical factors to predict the therapeutic dose of warfarin. *Clinical Pharmacology and Therapeutics* 2008;84:326–31.

38. Jonas DE, McLeod HL. Genetic and clinical factors relating to warfarin dosing. *Trends in Pharmacological Sciences* 2009;30:375–86.

39. Puehringer H, Loreth RM, Klose G, et al. VKORC1 -1639G>A and CYP2C9*3 are the major genetic predictors of phenprocoumon dose requirement. *European Journal of Clinical Pharmacology* 2010;66:591–8.

40. Higashi MK, Veenstra DL, Kondo LM, et al. Association between CYP2C9 genetic variants and anticoagulation-related outcomes during warfarin therapy. *JAMA* 2002;287:1690–8.

41. Aithal GP, Day CP, Kesteven PJ, Daly AK. Association of polymorphisms in the cytochrome P450 CYP2C9 with warfarin dose requirement and risk of bleeding complications. *The Lancet* 1999;353:717–9.

42. Lindh JD, Holm L, Andersson ML, Rane A. Influence of CYP2C9 genotype on warfarin dose requirements—A systematic review and meta-analysis. *European Journal of Clinical Pharmacology* 2009;65:365–75.

43. Sistonen J, Fuselli S, Palo JU, Chauhan N, Padh H, Sajantila A. Pharmacogenetic variation at CYP2C9, CYP2C19, and CYP2D6 at global and microgeographic scales. *Pharmacogenetics and Genomics* 2009;19:170–9.

44. Solus JF, Arietta BJ, Harris JR, et al. Genetic variation in eleven phase I drug metabolism genes in an ethnically diverse population. *Pharmacogenomics* 2004;5:895–931.
45. Lee CR, Goldstein JA, Pieper JA. Cytochrome P450 2C9 polymorphisms: A comprehensive review of the in-vitro and human data. *Pharmacogenetics* 2002;12:251–63.
46. Johnson JA, Gong L, Whirl-Carrillo M, et al. Clinical Pharmacogenetics Implementation Consortium guidelines for CYP2C9 and VKORC1 genotypes and warfarin dosing. *Clinical Pharmacology and Therapeutics* 2011;90:625–9.
47. Zakarija A, Bandarenko N, Pandey DK, et al. Clopidogrel-associated TTP: An update of pharmacovigilance efforts conducted by independent researchers, pharmaceutical suppliers, and the Food and Drug Administration. *Stroke; A Journal of Cerebral Circulation* 2004;35:533–7.
48. Mega JL, Close SL, Wiviott SD, et al. Cytochrome p-450 polymorphisms and response to clopidogrel. *The New England Journal of Medicine* 2009;360:354–62.
49. Simon T, Verstuyft C, Mary-Krause M, et al. Genetic determinants of response to clopidogrel and cardiovascular events. *The New England Journal of Medicine* 2009;360:363–75.
50. Mega JL, Simon T, Collet JP, et al. Reduced-function CYP2C19 genotype and risk of adverse clinical outcomes among patients treated with clopidogrel predominantly for PCI: A meta-analysis. *JAMA* 2010;304:1821–30.
51. Collet JP, Hulot JS, Pena A, et al. Cytochrome P450 2C19 polymorphism in young patients treated with clopidogrel after myocardial infarction: A cohort study. *The Lancet* 2009;373:309–17.
52. Holmes MV, Perel P, Shah T, Hingorani AD, Casas JP. CYP2C19 genotype, clopidogrel metabolism, platelet function, and cardiovascular events: A systematic review and meta-analysis. *JAMA* 2011;306:2704–14.
53. Bauer T, Bouman HJ, van Werkum JW, Ford NF, ten Berg JM, Taubert D. Impact of CYP2C19 variant genotypes on clinical efficacy of antiplatelet treatment with clopidogrel: Systematic review and meta-analysis. *BMJ (Clinical Research Edition)* 2011;343:d4588.
54. Steinberg H, Anderson MS, Musliner T, Hanson ME, Engel SS. Management of dyslipidemia and hyperglycemia with a fixed-dose combination of sitagliptin and simvastatin. *Vascular Health and Risk Management* 2013;9:273–82.
55. Armitage J. The safety of statins in clinical practice. *The Lancet* 2007;370:1781–90.
56. Bowman L, Armitage J, Bulbulia R, Parish S, Collins R. Study of the effectiveness of additional reductions in cholesterol and homocysteine (SEARCH): Characteristics of a randomized trial among 12064 myocardial infarction survivors. *American Heart Journal* 2007;154:815–23, 23 e1–6.
57. Link E, Parish S, Armitage J, et al. SLCO1B1 variants and statin-induced myopathy—A genomewide study. *The New England Journal of Medicine* 2008;359:789–99.
58. Daly AK. Using genome-wide association studies to identify genes important in serious adverse drug reactions. *Annual Review of Pharmacology and Toxicology* 2012; 52:21–35.
59. Loehrer PJ, Einhorn LH. Drugs five years later. Cisplatin. *Annals of Internal Medicine* 1984;100:704–13.
60. Yasui N, Adachi N, Kato M, et al. Cisplatin-induced hearing loss: The need for a long-term evaluating system. *Journal of Pediatric Hematology/Oncology* 2014;36:e241–5.
61. Waissbluth S, Daniel SJ. Cisplatin-induced ototoxicity: Transporters playing a role in cisplatin toxicity. *Hearing Research* 2013;299:37–45.
62. Windsor RE, Strauss SJ, Kallis C, Wood NE, Whelan JS. Germline genetic polymorphisms may influence chemotherapy response and disease outcome in osteosarcoma: A pilot study. *Cancer* 2012;118:1856–67.

63. Donzelli E, Carfi M, Miloso M, et al. Neurotoxicity of platinum compounds: Comparison of the effects of cisplatin and oxaliplatin on the human neuroblastoma cell line SH-SY5Y. *Journal of Neuro-Oncology* 2004;67:65–73.

64. Sakano S, Hinoda Y, Sasaki M, et al. Nucleotide excision repair gene polymorphisms may predict acute toxicity in patients treated with chemoradiotherapy for bladder cancer. *Pharmacogenomics* 2010;11:1377–87.

65. Caronia D, Patino-Garcia A, Milne RL, et al. Common variations in ERCC2 are associated with response to cisplatin chemotherapy and clinical outcome in osteosarcoma patients. *The Pharmacogenomics Journal* 2009;9:347–53.

66. D'Arcangelo M, Cappuzzo F. Erlotinib in the first-line treatment of non-small-cell lung cancer. *Expert Review of Anticancer Therapy* 2013;13:523–33.

67. Cho SH, Park LC, Ji JH, et al. Efficacy of EGFR tyrosine kinase inhibitors for non-adenocarcinoma NSCLC patients with EGFR mutation. *Cancer Chemotherapy and Pharmacology* 2012;70:315–20.

68. Han SW, Kim TY, Lee KH, et al. Clinical predictors versus epidermal growth factor receptor mutation in gefitinib-treated non-small-cell lung cancer patients. *Lung Cancer (Amsterdam, Netherlands)* 2006;54:201–7.

69. Scheinfeld N. Three cases of toxic skin eruptions associated with methotrexate and a compilation of methotrexate-induced skin eruptions. *Dermatology Online Journal* 2006;12:15.

70. Guaiana G, Barbui C, Hotopf M. Amitriptyline for depression. *The Cochrane Database of Systematic Reviews* 2007;3:CD004186.

71. Estebe JP, Myers RR. Amitriptyline neurotoxicity: Dose-related pathology after topical application to rat sciatic nerve. *Anesthesiology* 2004;100:1519–25.

72. Leucht C, Huhn M, Leucht S. Amitriptyline versus placebo for major depressive disorder. *The Cochrane Database of Systematic Reviews* 2012;12:CD009138.

73. Steimer W, Zopf K, von Amelunxen S, et al. Amitriptyline or not, that is the question: Pharmacogenetic testing of CYP2D6 and CYP2C19 identifies patients with low or high risk for side effects in amitriptyline therapy. *Clinical chemistry* 2005;51:376–85.

74. Koski A, Sistonen J, Ojanpera I, Gergov M, Vuori E, Sajantila A. CYP2D6 and CYP2C19 genotypes and amitriptyline metabolite ratios in a series of medicolegal autopsies. *Forensic Science International* 2006;158:177–83.

75. Shimoda K, Someya T, Yokono A, et al. The impact of CYP2C19 and CYP2D6 genotypes on metabolism of amitriptyline in Japanese psychiatric patients. *Journal of Clinical Psychopharmacology* 2002;22:371–8.

76. de Vos A, van der Weide J, Loovers HM. Association between CYP2C19*17 and metabolism of amitriptyline, citalopram and clomipramine in Dutch hospitalized patients. *The Pharmacogenomics Journal* 2011;11:359–67.

77. Kwadijk-de Gijsel S, Bijl MJ, Visser LE, et al. Variation in the CYP2D6 gene is associated with a lower serum sodium concentration in patients on antidepressants. *British Journal of Clinical Pharmacology* 2009;68:221–5.

78. Bijl MJ, Visser LE, Hofman A, et al. Influence of the CYP2D6*4 polymorphism on dose, switching and discontinuation of antidepressants. *British Journal of Clinical Pharmacology* 2008;65:558–64.

79. Ravna AW, Sylte I, Dahl SG. Molecular mechanism of citalopram and cocaine interactions with neurotransmitter transporters. *The Journal of Pharmacology and Experimental Therapeutics* 2003;307:34–41.

80. Schmidt HM, Hagen M, Kriston L, Soares-Weiser K, Maayan N, Berner MM. Management of sexual dysfunction due to antipsychotic drug therapy. *The Cochrane Database of Systematic Reviews* 2012;11:CD003546.

81. Paddock S, Laje G, Charney D, et al. Association of GRIK4 with outcome of antidepressant treatment in the STAR*D cohort. *The American Journal of Psychiatry* 2007;164:1181–8.

82. Rizzardini G, Zucchi P. Abacavir and lamivudine for the treatment of human immuno-deficiency virus. *Expert Opinion on Pharmacotherapy* 2011;12:2129–38.
83. Hughes CA, Foisy MM, Dewhurst N, et al. Abacavir hypersensitivity reaction: An update. *The Annals of Pharmacotherapy* 2008;42:387–96.
84. Cruciani M, Zanichelli V, Serpelloni G, et al. Abacavir use and cardiovascular disease events: A meta-analysis of published and unpublished data. *AIDS (London, England)* 2011;25:1993–2004.
85. Costagliola D, Lang S, Mary-Krause M, Boccara F. Abacavir and cardiovascular risk: Reviewing the evidence. *Current HIV/AIDS Reports* 2010;7:127–33.
86. Hughes AR, Spreen WR, Mosteller M, et al. Pharmacogenetics of hypersensi-tivity to abacavir: From PGx hypothesis to confirmation to clinical utility. *The Pharmacogenomics Journal* 2008;8:365–74.
87. Saag M, Balu R, Phillips E, et al. High sensitivity of human leukocyte antigen-b*5701 as a marker for immunologically confirmed abacavir hypersensitivity in white and black patients. *Clinical Infectious Diseases* 2008;46:1111–8.
88. McPhail GL, Clancy JP. Ivacaftor: The first therapy acting on the primary cause of cystic fibrosis. *Drugs Today (Barc)* 2013;49:253–60.
89. Accurso FJ, Rowe SM, Clancy JP, et al. Effect of VX-770 in persons with cystic fibro-sis and the G551D-CFTR mutation. *The New England Journal of Medicine* 2010; 363:1991–2003.
90. Davies JC, Wainwright CE, Canny GJ, et al. Efficacy and safety of ivacaftor in patients aged 6 to 11 years with cystic fibrosis with a G551D mutation. *American Journal of Respiratory and Critical Care Medicine* 2011;187:1219–25.
91. Flume PA, Liou TG, Borowitz DS, et al. Ivacaftor in subjects with cystic fibrosis who are homozygous for the F508del-CFTR mutation. *Chest* 2012;142:718–24.
92. Sermet-Gaudelus I. Ivacaftor treatment in patients with cystic fibrosis and the G551D-CFTR mutation. *European Respiratory Review* 2013;22:66–71.
93. Harrison MJ, Murphy DM, Plant BJ. Ivacaftor in a G551D homozygote with cystic fibrosis. *The New England Journal of Medicine* 2013;369:1280–2.
94. Labianca R, Sarzi-Puttini P, Zuccaro SM, Cherubino P, Vellucci R, Fornasari D. Adverse effects associated with non-opioid and opioid treatment in patients with chronic pain. *Clinical Drug Investigation* 2012;32(Suppl 1):53–63.
95. Crews KR, Gaedigk A, Dunnenberger HM, et al. Clinical Pharmacogenetics Implementation Consortium (CPIC) guidelines for codeine therapy in the context of cytochrome P450 2D6 (CYP2D6) genotype. *Clinical Pharmacology and Therapeutics* 2012;91:321–6.
96. Zhou SF. Polymorphism of human cytochrome P450 2D6 and its clinical significance: Part I. *Clinical Pharmacokinetics* 2009;48:689–723.
97. Lotsch J, Rohrbacher M, Schmidt H, Doehring A, Brockmoller J, Geisslinger G. Can extremely low or high morphine formation from codeine be predicted prior to therapy initiation? *Pain* 2009;144:119–24.
98. VanderVaart S, Berger H, Sistonen J, et al. CYP2D6 polymorphisms and codeine anal-gesia in postpartum pain management: A pilot study. *Therapeutic Drug Monitoring* 2011;33:425–32.
99. Wu X, Yuan L, Zuo J, Lv J, Guo T. The impact of CYP2D6 polymorphisms on the phar-macokinetics of codeine and its metabolites in Mongolian Chinese subjects. *European Journal of Clinical Pharmacology* 2014;70:57–63.
100. Pacher P, Nivorozhkin A, Szabo C. Therapeutic effects of xanthine oxidase inhibi-tors: Renaissance half a century after the discovery of allopurinol. *Pharmacological Reviews* 2006;58:87–114.
101. Lee HY, Ariyasinghe JT, Thirumoorthy T. Allopurinol hypersensitivity syndrome: A preventable severe cutaneous adverse reaction? *Singapore Medical Journal* 2008; 49:384–7.

102. Niihara H, Kaneko S, Ito T, et al. HLA-B*58:01 strongly associates with allopurinol-induced adverse drug reactions in a Japanese sample population. *Journal of Dermatological Science* 2013;71:150–2.

103. Tassaneeyakul W, Jantararoungtong T, Chen P, et al. Strong association between HLA-B*5801 and allopurinol-induced Stevens-Johnson syndrome and toxic epidermal necrolysis in a Thai population. *Pharmacogenetics and Genomics* 2009;19:704–9.

104. Somkrua R, Eickman EE, Saokaew S, Lohitnavy M, Chaiyakunapruk N. Association of HLA-B*5801 allele and allopurinol-induced Stevens Johnson syndrome and toxic epidermal necrolysis: A systematic review and meta-analysis. *BMC Medical Genetics* 2011;12:118.

105. Goncalo M, Coutinho I, Teixeira V, et al. HLA-B*58:01 is a risk factor for allopurinol-induced DRESS and Stevens-Johnson syndrome/toxic epidermal necrolysis in a Portuguese population. *The British Journal of Dermatology* 2013;169:660–5.

106. Cao ZH, Wei ZY, Zhu QY, et al. HLA-B*58:01 allele is associated with augmented risk for both mild and severe cutaneous adverse reactions induced by allopurinol in Han Chinese. *Pharmacogenomics* 2012;13:1193–201.

107. Lee MH, Stocker SL, Williams KM, Day RO. HLA-B*5801 should be used to screen for risk of Stevens-Johnson syndrome in family members of Han Chinese patients commencing allopurinol therapy. *The Journal of Rheumatology* 2013;40:96–7.

108. Wang D, Johnson AD, Papp AC, Kroetz DL, Sadee W. Multidrug resistance polypeptide 1 (MDR1, ABCB1) variant 3435C>T affects mRNA stability. *Pharmacogenetics and Genomics* 2005;15:693–704.

109. Papp AC, Pinsonneault JK, Wang D, et al. Cholesteryl ester transfer protein (CETP) polymorphisms affect mRNA splicing, HDL levels, and sex-dependent cardiovascular risk. *PLoS One* 2012;7:e31930.

110. Krauss RM, Mangravite LM, Smith JD, et al. Variation in the 3-hydroxyl-3-methylglutaryl coenzyme a reductase gene is associated with racial differences in low-density lipoprotein cholesterol response to simvastatin treatment. *Circulation* 2008;117:1537–44.

111. Stewart A. SLCO1B1 polymorphisms and statin-induced myopathy. *PLoS Currents* 2013;5, available at ecurrents.eogt.d21e7f0c58463571bb0d9d3a19b82203.

112. Poduri A, Khullar M, Bahl A, Sehrawat BS, Sharma Y, Talwar KK. Common variants of HMGCR, CETP, APOAI, ABCB1, CYP3A4, and CYP7A1 genes as predictors of lipid-lowering response to atorvastatin therapy. *DNA and Cell Biology* 2010;29:629–37.

113. Kantarjian HM, Talpaz M, O'Brien S, et al. Imatinib mesylate for Philadelphia chromosome-positive, chronic-phase myeloid leukemia after failure of interferon-alpha: Follow-up results. *Clin Cancer Res* 2002;8:2177–87.

114. Saglio G, Kim DW, Issaragrisil S, et al. Nilotinib versus imatinib for newly diagnosed chronic myeloid leukemia. *The New England Journal of Medicine* 2010;362:2251–9.

115. Kantarjian H, O'Brien S, Jabbour E, et al. Improved survival in chronic myeloid leukemia since the introduction of imatinib therapy: A single-institution historical experience. *Blood* 2012;119:1981–7.

116. Marin D, Ibrahim AR, Lucas C, et al. Assessment of BCR-ABL1 transcript levels at 3 months is the only requirement for predicting outcome for patients with chronic myeloid leukemia treated with tyrosine kinase inhibitors. *Journal of Clinical Oncology* 2012;30:232–8.

117. Ganetsky A. Ruxolitinib: A new treatment option for myelofibrosis. *Pharmacotherapy* 2013;33:84–92.

118. Qin HD, Wang Y, Grados MA, et al. Whole genome association study identifies novel antidepressant response loci for the treatment of obsessive-compulsive disorder with selective serotonin re-uptake inhibitors. *ASHG Annual Meeting*, Boston, MA, October 22–26, 2013.

119. Huang RS, Gamazon ER, Ziliak D, et al. Population differences in microRNA expression and biological implications. *RNA Biology* 2011;8:692–701.
120. Bell JT, Pai AA, Pickrell JK, et al. DNA methylation patterns associate with genetic and gene expression variation in HapMap cell lines. *Genome Biology* 2011;12:R10.
121. Fraser HB, Lam LL, Neumann SM, Kobor MS. Population-specificity of human DNA methylation. *Genome Biology* 2012;13:R8.
122. Moen EL, Zhang X, Mu W, et al. Genome-wide variation of cytosine modifications between European and African populations and the implications for complex traits. *Genetics* 2013;194:987–96.
123. Zhang X, Moen EL, Liu C, et al. Linking the genetic architecture of cytosine modifications with human complex traits. *Human Molecular Genetics* 2014;23:5893–905.
124. Eadon MT, Wheeler HE, Stark AL, et al. Genetic and epigenetic variants contributing to clofarabine cytotoxicity. *Human Molecular Genetics* 2014;22:4007–20.

4 Applying Pharmacogenomics in Drug Discovery and Development

Ruth Vinall

CONTENTS

KEY CONCEPTS

- Identification of genetic alterations that are causative in the disease process allows for the development of targeted therapies that are likely to have higher efficacy and lower toxicity compared to nontargeted therapies.
- Polymorphisms that alter binding of the drug to a target, and drug absorption; distribution; metabolism; and excretion, as well as transport, can affect both drug efficacy and toxicity.
- *In silico*, cell line, animal model studies can be used to assess "druggability" of targets and to predict clinical efficacy and toxicity in patients.
- Pharmacogenomic studies can be used to identify new indications for existing drugs.
- Inclusion of patients from a diverse patient population in pharmacogenomic analyses is important to maximize benefit and minimize toxicity for all patients.
- Pharmacogenomic analysis of biospecimens from patients recruited in phase I, II, and III clinical trials can help guide subsequent patient selection and drug usage.

INTRODUCTION

Over 90% of the new drug applications (NDAs) fail to receive approval from the US Food and Drug Administration (FDA) due to lack of efficacy or unacceptable levels of toxicity.[1-3] This high failure rate is widely believed to be one of the main reasons why drug development costs are so high and why there has been a steady decline in the number of NDAs in recent years.[4,5] More importantly, this high failure rate implies that many clinical trial participants receive treatment that is not of benefit to them and that may, in fact, cause more harm than standard of care treatment. This chapter will discuss how pharmacogenomics can be used in drug discovery and development to maximize the likelihood that a new drug will have high efficacy and low toxicity in patients, bringing the new drug to market quickly and safely, and also benefiting all stakeholders.

USE OF PHARMACOGENOMICS TO IDENTIFY DRUG TARGETS

It is well known that some genetic alterations can drive disease initiation and progression as well as resistance to drug therapies. Targeting these genetic alterations through the design and usage of drugs that "hit" the genetic alteration (target) can help slow, halt, or even reverse the disease process as well as minimize the likelihood of drug-related toxicity.[6] Many new anticancer drugs are targeted therapies. For example, trastuzumab (Herceptin®), a drug used to treat metastatic breast cancer, negates the effects of genetic alterations (typically gene amplification) that mediate HER2/Neu receptor overexpression by binding to the HER2/Neu receptor and preventing activation of downstream pathways that drive cell proliferation.[7] Treatment regimens that include trastuzumab have been shown to significantly improve overall survival rates in patients who harbor this genetic alteration.[8] Genomic analyses can be used to identify genetic alterations that are associated with a particular disease and thereby serve as a starting point for drug target identification and subsequent drug development. There are two main genomic-based approaches that can be used to identify genetic alterations that are associated with a disease: target gene analysis and genomic profiling. For target gene analysis, potential drug targets are identified based on our existing understanding of disease pathophysiology. This approach focuses on identifying genetic alterations that occur in components of specific molecular pathways known to be responsible for disease initiation and/or progression. An advantage of this approach is that the number of analyses needed are limited, thus reducing time and cost. Also because it is based on the existing understanding of disease process, it is more likely to identify a genetic alteration that is causative in the disease process. A disadvantage is that the likelihood of identifying a potential target is largely reduced if the underlying pathophysiology for a disease is not well understood or if prior studies have failed to identify good targets based on existing knowledge of the disease process. In contrast, genomic profiling allows for drug targets to be identified in an unbiased way; the presence of genetic alterations is assessed in all known genes, not only those that have been associated with a particular disease and/or molecular pathway.[9] The genomic profiling approach therefore not only has greater potential to identify drug targets, but also increases the likelihood of identifying novel targets. It can also enhance our understanding of disease pathophysiology through identification of previously unknown mediators. Evidence supports this: the genomic profiling approach has led to major advances in drug target identification for Crohn disease and rheumatoid arthritis by allowing for the identification and delineation of the inflammatory and autoimmune pathways that drive these diseases.[10,11] Disadvantages of genomic profiling include increased cost of analysis compared to target gene analysis and, due to generation of very large data sets, the need for subsequent complex biostatistical and bioinformatic analyses.[9] In addition, generation of false positives is of concern due to the large number of genes being surveyed.

Similar technologies can be employed for both target gene and genomic profiling approaches, and these technologies can be used to assess either DNA or RNA samples. The former are much easier to collect and work with; however, the latter are often more relevant to drug target discovery because they provide information regarding genes that are being actively expressed and are therefore more likely to be causative in the disease process. Quantitative real-time polymerase chain reaction (qRT-PCR) and microarray can be used to survey the presence of and/or expression levels of genetic alterations in both small and large formats (between one gene and thousands of genes assessed). Sequencing can also be used. Again, anywhere from a single gene to an entire genome may be sequenced. The benefit of sequencing versus using qRT-PCR or microarray is that many different types of genetic alterations can be detected simultaneously, including single nucleotide polymorphisms (SNPs), copy-number variants (CNVs), and structural variants (SVs), and that sensitivity of detection is higher.[12,13] Disadvantages are increased costs and the technical challenges posed by a generation of such large and complex data sets. Genome-wide association studies (GWASs) and next-generation sequencing (NGS) have been widely employed for genomic profiling analyses. GWASs typically use microarray technology and screen samples for the presence of common genetic variants (genetic alterations, most often SNPs, that occur within >1% of the population). While the entire genome is not interrogated by GWASs, thousands of representative SNPs (Tag SNPs) are assessed, and this strategy has enabled the identification of multiple novel potential drug targets.[14] For example, GWASs identified the complement factor H (CFH) allele as being associated with age-related macular degeneration (AMD).[15] As a result of these studies, several complementary component inhibitors are being developed to treat AMD patients.[16,17] The rapidly decreasing cost of sequencing analyses has allowed NGS to play an important role in drug target discovery. For example, NGS analyses identified *DHODH*, a gene that encodes an enzyme needed for *de novo* pyrimidine biosynthesis, as playing a role in Miller syndrome, a rare Mendelian disorder whose cause is poorly understood.[18] An overview of the genomic profiling methodologies that can be employed to identify different genetic alterations is shown in Figure 4.1.[19]

Ideally, patient specimens (biospecimens) should be used for genomic-based identification of drug targets; however, in some cases, panels of immortalized cell lines that are derived from patient samples are used for screening purposes. Either way, samples that originate from patients with, versus without, the disease of interest need to be assessed. A problem with the cell line approach is that cell lines

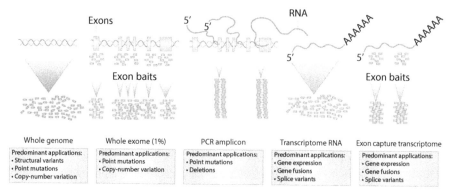

FIGURE 4.1   Genomic profiling allows for the unbiased identification of genetic alterations within patient samples. As outlined in the diagram below, different methodologies can be used to detect different types of genetic alterations. Analysis of RNA samples is often desirable for drug discovery studies because the genetic alterations that are identified are more likely to be causative in the disease process and are therefore more likely to make good drug targets. (Adapted from Simon R, Roychowdhury S, *Nat Rev Drug Discov*, 12, 358–69, 2013.)

can acquire genetic alterations while in culture that may drive disease but are not relevant to patients. In addition, in many cases an insufficient number of cell lines that are representative of the patient population exists. An advantage is that the cell line–based approach is inexpensive and less time consuming compared to patient specimen collection and analysis. Although the use of patient specimens faces challenges, it is more likely to identify relevant drug targets. One of the biggest challenges is obtaining a large enough sample set that is representative of the patient population for the analyses. The size of biorepositories is often limited by the cost associated with their maintenance and the need for patients to consent to donation of biospecimens.[20,21] In addition, patients from minority populations are less likely to donate biospecimens to biorepositories, meaning that the sample set may not be representative of that patient population.[22–24] Having a large sample set is usually important for drug target discovery because the vast majority of diseases have more than one cause, and it is highly likely that not all patients with the disease will harbor a particular genetic alteration; a large sample set is therefore necessary to identify genetic alterations that are associated with disease. Working with a large sample set also provides a good estimate of the prevalence of the genetic alteration in the patient population, and if a large number of genes are being surveyed, working with a large sample set helps reduce the risk of generating false positives.

Histological and/or cytological analyses have been successfully used to identify genetic alterations and have led to the development of several "blockbuster" drugs. However, it is challenging to use these methodologies for large-scale screenings of samples, and for the most part histological and/or cytological studies can only be used to identify major alterations such as chromosomal rearrangements. The development of imatinib (Gleevec; www.gleevec.com) was made possible by cytological analyses; cytological analyses identified a chromosomal abnormality, referred to as the Philadelphia chromosome, in patients with chronic myelogenous leukemia (CML).[25]

The chromosomal rearrangement (a reciprocal translocation) that creates the Philadelphia chromosome in CML patients results in the fusion of two genes, Bcr and Abl, and this causes Abl to be expressed at much higher levels than usual. Imatinib targets the Abl protein and is effective in 89% of CML patients as a first-line therapy if they harbor this genetic alteration.[26] Genomic analyses have the ability to identify many different types of genetic alterations compared to histological and/or cytological analyses, and can be used for large-scale screenings; it is widely believed that genomic analyses have the potential to dramatically increase our ability to develop new and effective drugs.

TARGET VALIDATION

Once an association between a disease and a genetic alteration (the potential drug target) has been identified, a next step is to determine whether or not the genetic alteration is causative in the disease process. The majority of genetic alterations—in particular, SNPs—are not causal in the disease process, or only make a minimal contribution, and therefore the majority of genetic alterations that are identified as being associated with a disease are highly unlikely to be useful drug targets (although they may still be useful biomarkers).[27] A significant amount of money is often needed to conduct these studies, and it can take years to generate enough evidence to warrant subsequent development of a suitable drug for the target.

Typically target validation experiments include both cell line and animal experiments. If cell lines were used to identify the genetic alteration, patient samples may also be assessed at this point to ensure that the potential target is clinically relevant. Cell lines and animal models can be genetically engineered to express a particular genetic alteration ("knock-in" experiments). The impact of the genetic alteration on disease pathogenesis can then be assessed to establish whether a cause-and-effect relationship exists. Alternatively, anti-sense technologies can be employed. These technologies allow for "knock-out" of a particular genetic alteration that is present within a cell line or animal model, and can be used to determine whether elimination of the genetic alteration can provide "rescue" and prevent disease initiation and/or progression in cell lines and animal models with endogenous expression of the genetic alteration.

Cell line studies typically focus on the impact of "knocking-in" or "knocking-out" the genetic alteration of signaling pathways that are known to be associated with the potential target, and on physiological processes that pertain to a particular cell type and the disease of interest. For example, if a genetic alteration in a component of the mTOR signaling pathway (a pathway that can impact cell proliferation and survival) is found to be associated with prostate cancer, an investigator would assess the expression and activity levels of multiple components of the mTOR pathway to help establish whether the genetic alteration does impact cell signaling. They would also assess the impact of "knocking-in" or "knocking-out" the genetic alteration on cell proliferation and survival, physiological processes relevant to cancer cells. In addition to establishing the cause-and-effect relationship, cell line experiments are relatively fast and cheap to perform, whereas animal experiments are also necessary as they take into account systemic effects similar to patients. As with cell line studies, the impact of "knocking-in" or "knocking-out" the genetic alteration on associated signaling pathways and on physiological processes can be assessed, but more importantly, the impact on disease initiation and/or progression can be measured. Disadvantages of animal studies include time and cost, as well as ethical concerns. If the cause-and-effect relationship is established, the cell lines and animal models generated for target validation experiments can subsequently be used to test drug efficacy and toxicity in preclinical studies of potential drug candidates. It is important that data from target validation studies can also provide insights into whether targeting a genetic alteration is likely to cause significant toxicity. For example, toxicity is more likely if the genetic alteration is found to have a broad tissue distribution. In addition, target validation studies often improve our understanding of disease etiology because the focus is placed on understanding the impact of the genetic alteration on related pathways. They can even result in the identification of additional drug and more suitable drug targets.[28,29]

Cell lines and animal models were instrumental in validating the involvement of Bcr-Abl, the genetic alteration that results from a chromosomal translocation in hematopoietic progenitor cells, in driving CML and for the development and testing of drugs to treat this disease.[30-33] Some of the cell lines used for these studies were generated from CML patient samples that harbored the *Bcr-Abl* gene fusion, and others were generated by genetically engineering cells to include this genetic alteration.[31,33] Phenotypic analyses of these cell lines, combined with manipulations that allowed for direct targeting of the *Bcr-Abl* fusion gene and/or targeting of upstream and/or downstream components of the associated signaling pathway, demonstrated that Bcr-Abl is able to drive cell proliferation. To confirm the ability of Bcr-Abl to mediate leukocyte proliferation in vivo and drive CML, several mouse models were developed, including xenograft models (human cell lines harboring the *Bcr-Abl* fusion gene were implanted into immunodeficient mice) and transgenic models (the hematopoietic progenitor cells of these mice were engineered to harbor the *Bcr-Abl* fusion gene).[30,32] In addition to confirming the importance of Bcr-Abl in driving CML, these mouse models were also subsequently used to test the efficacy of developmental drugs designed to target Bcr-Abl.

ASSESSING "DRUGGABILITY" OF A TARGET AND LEAD DRUG DISCOVERY STRATEGIES

Once a drug target has been identified and validated, then drugs may be developed to "hit" this target (by either inhibiting protein function and/or altering expression levels) and thereby negate its effect on mediating disease initiation and/or progression. A usual first step is to determine whether the target is in fact "druggable": that is, whether high affinity binding between the target and a pharmacological agent is possible, and therefore whether or not a suitable drug can be developed.[34-37] Currently, it is estimated that only 10% of the human genome is druggable.[38] Binding of a drug to its target can be affected by many factors, including protein structure, size, charge, and polarity. Additional genetic alterations in the target gene (genetic alterations other than those that are causing dysfunction and thereby disease, typically polymorphisms) can potentially affect all of these factors and thereby affect druggability and, if possible, their presence and frequency should be assessed in the patient population.[39] Different types of genetic alterations (genetic variants) may also exist and respond differently to drugs. For example, the genetic alteration that causes CML, Bcr-Abl, has three genetic variants: major (M-bcr), minor (m-bcr), and micro (mu-bcr).[40] It should be noted that the effect does not always decrease druggability, and that in some cases an increase in druggability of the target is observed. For example, gefitinib, an inhibitor of epidermal growth factor receptor (EGFR), has increased efficacy in patients who harbored EGFR-activating mutations; these mutations promote increased binding between gefitinib and EGFR: that is, increased druggability.[41] In a clinical trial of gefitinib, non–small cell lung cancer (NSCLC) patients who harbored the EGFR-activating mutations, typically patients from East Asia, had a median survival time of 3.1 years compared to 1.6 years in mutation-negative patients.

In silico (computer-based) analyses and/or screening assays are typically used for preliminary studies, and these can help reduce costs and expedite drug discovery (see Figure 4.2[42] for examples of *in silico* tests). *In silico* approaches to identifying drug candidates include similarity searching (using databases to identify drugs that have been shown to successfully target a family member of the gene of interest) and quantitative structure–activity relationships (QSARs, models that help predict the biological activity of a chemical structure).[35,43] Similarity searching is often preferred as it is more likely that additional information such as toxicity profile and efficacy data is available for these chemically and structurally similar drugs (often referred to as "me-too" drugs), and this information can be used to expedite drug development and reduce costs. Examples of "me-too" drugs include atenolol and timolol (structurally similar to propranolol), and ranitidine and nizatidine (structurally similar to cimetidine). A criticism of "me-too" drugs is that they often do not result in a significant improvement in patient outcomes compared to their parent drug.[44] In contrast to similarity searching, structure–activity models allow for the identification of novel drugs and have been instrumental in advancing the field of drug discovery. For example, QSAR has been used to identify ketolide derivatives (macrolide antibiotics) that have higher efficacy and lower toxicity.[45]

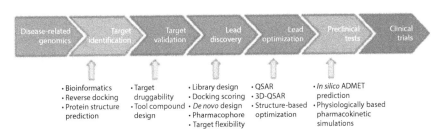

FIGURE 4.2 *In silico* studies can be used at several stages during the drug development process, including during target identification, target validation, druggability testing (lead discovery and optimization), and during preclinical tests. The types of *in silico* studies that may be performed are outlined in this diagram. (Adapted from Kore PP, et al., *Open J Med Chem*, 2, 139–148, 2012.)

Another advantage is that these models can also be used to assess the impact of polymorphisms on druggability. For example, QSAR has been used to develop functional screening analyses that evaluate the impact of polymorphisms on ABC transporter function and druggability.[46] Once lead drug candidates have been identified via *in silico* studies, druggability is then confirmed via protein binding analyses (NMR analysis[47] or Biacore;[48] www.biacore.com) and/or enzyme activity assays.[49] An alternative approach to identifying drug candidates for a particular target is the use of screening assays. Such studies involve the use of panels of existing drugs, including orphan drugs, and are used to establish the strength of binding between the drugs included in the panel and the target, or to assess the impact of the drug on enzyme activity.[50] Again, polymorphic target variant should be included in these analyses, particularly if the variant is known to have high frequency in the patient population and/or more potent effects on the disease process.

It is noteworthy that *in silico* and genomic analyses have led to the realization that several seemingly different diseases share common underlying genetic causes, and can in fact be treated with the same drugs. As a result, several drugs have been successfully "repurposed": that is, used for an indication that is different from that they were initially approved for. For example, genomic analyses identified significant similarities between catechol-*O*-methyltransferase (*COMT*), a molecule that mediates Parkinson disease; and enoyl-acyl carrier protein reductase, a bacterial protein in *Mycobacterium tuberculosis*. Genetic alterations in these molecules cause the associated disease in both instances. As the molecules are so similar, investigators decided to determine whether entacapone, a drug used to target COMT in patients with Parkinson disease, would be effective for the treatment of *M. tuberculosis*; entacapone is now used to successfully treat tuberculosis.[51] Another example is bexarotene, a drug that was originally developed to treat cutaneous T-cell lymphoma. Analysis of signaling pathways led to the discovery that bexarotene can modulate pathways that drive Alzheimer disease. Studies in mouse models of Alzheimer disease have shown bexarotene to be effective, although no human studies have been performed yet.[52] A major benefit of being able to utilize an existing drug to target a genetic alteration is that preclinical studies may not be required, and it is possible that the drug may go straight to clinical trials. This can save a significant amount of time and money in the development process, thereby making an effective drug available sooner to patients.

PHARMACOGENOMICS CONSIDERATIONS DURING PRECLINICAL STUDIES OF POTENTIAL LEAD DRUGS

Data from preclinical studies are usually needed to support an investigational new drug (IND) application that, if approved, will allow for a drug to be tested in patients. Initial studies typically use cell lines to assess drug efficacy and establish effective dose ranges. Animal studies are then used to establish drug efficacy in vivo, to optimize drug delivery and dosing, to confirm that the drug can hit its target in vivo, and to identify potential adverse drug reactions (ADRs). Figure 4.3 provides an overview of the types and sequence of preclinical studies that are used to generate data to support an IND.

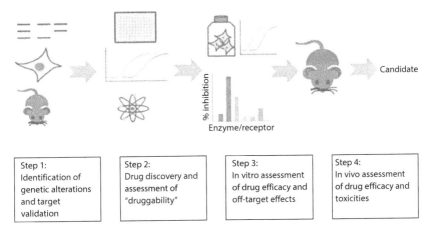

| Step 1: | Step 2: | Step 3: | Step 4: |
| Identification of genetic alterations and target validation | Drug discovery and assessment of "druggability" | In vitro assessment of drug efficacy and off-target effects | In vivo assessment of drug efficacy and toxicities |

FIGURE 4.3 Preclinical studies of potential lead drugs. The impact of pharmacogenomics on target identification (genomics analyses) and validation (cell line and animal studies), as well as drug discovery and assessment of druggability (*in silico*, high-throughput screening, and binding assays), were discussed in previous sections. Once these steps (steps 1 and 2, respectively) are completed, and a lead drug is identified, in vitro and in vivo analyses are performed to establish pharmacodynamic and pharmacokinetic characteristics for the drug. These studies test drug efficacy, that is, how well the drug can hit its intended target and inhibit disease initiation and/or progression. They also can help predict and/or identify drug-related toxicities. If these studies are successful (the drug has high efficacy and low toxicity), then data collected can be used to support an IND that if approved will allow for subsequent testing of the drug, in this diagram termed the candidate, in patients. (Adapted from Hughes JP, et al., *Br J Pharmacol*, 162, 1239–49, 2011.)

As mentioned above, cell line and animal models that were developed to validate a drug target as playing a causative role in disease initiation and/or progression are extremely useful for testing drug efficacy in preclinical studies; the presence of the target has already been confirmed in these models, and researchers will have already established the sequence and timing of molecular and pathophysiological changes that are associated with disease initiation and progression. Ideal cell lines and animal models that have polymorphic variants of the target that are prevalent in the patient population will have also been developed so that the impact of these on drug efficacy can be assessed. In addition to assessing drug efficacy, identification and minimization of the likelihood of ADR is a major focus during preclinical studies. ADRs not only cause a significant number of deaths each year, but also increase costs due to hospitalizations. In some instances, ADRs are severe and/or prevalent enough to result in removal of a marketed drug.[53] ADRs can result from "off-target" effects (the drug may target similar molecules that do not play a role in the disease process, but play a critical role in systemic and/or organ function), a broad tissue distribution of the target (the target molecule may be needed for the proper function of other body systems), and/or from polymorphisms in drug-metabolizing enzymes. The majority of ADRs are due to polymorphisms that cause dysfunction of phase I and II enzymes involved in drug metabolism, although polymorphisms in drug transporters often also play a role.[54–58] As discussed in the above sections, *in silico* studies that are used to identify lead drugs and assess "druggability" can be used to help predict ADRs, but these are not infallible and often false negatives slip through.[59] Cell line studies may also be used to help predict some ADRs; however, animal studies usually provide the strongest evidence for prediction of ADRs in patients as all absorption, distribution, metabolism, and excretion (ADME) effects can be assessed in a system that is similar to a patient.[60,61] In a typical animal model toxicity study, the impact of a high- and low-dose drug regimen on toxicity is compared to a vehicle control. Toxicity may be assessed by physical examination of the animals, serum biochemistry, hematological analysis, and urinalysis as well as histopathological analysis of certain tissues including the liver and kidney. If toxicity is observed, molecular studies may then be performed to better understand the underlying molecular mechanisms involved. If a particular enzyme responsible for the ADME of the drug is polymorphic, appropriate cell line and animal models should be tested to determine the impact of prevalent polymorphisms on drug toxicity.

The majority of ADRs that result from polymorphisms in ADME genes are identified after approval of the drug. This is usually due to lack of inclusion of pharmacogenomic studies in the drug development process, or because the pharmacogenomic studies that had been conducted were not sufficiently powered to detect these ADRs. Irinotecan, a DNA topoisomerase I inhibitor that is used to treat patients with lung or colorectal cancer, is a good example. Following FDA approval, a significant number of patients were found to experience severe leukopenia and/or diarrhea. These ADRs have not occurred at significant rates in clinical trials of irinotecan when it was first used in the general population. An initial pharmacogenomic analysis of specimens collected from patients who experienced these ADRs versus those who did not experience them revealed that 15% of ADR patients were homozygous for the *UGT1A1**28 allele and 33% were heterozygous.[62] UGT1A1, a phase II enzyme, glucoronidates the active metabolite of irinotecan (SN-38), and thereby mediates its excretion from the body. Subsequent studies confirmed that polymorphisms in *UGT1A1* are responsible for irinotecan ADRs,[63,64] and the FDA now strongly recommends pharmacogenomic testing for *UGT1A1**28 and corresponding dosage adjustments. It is very likely that the inclusion of pharmacogenomic testing during the development of irinotecan could predict these UGT1A1-related ADRs and hence would minimize or avoid them.

PHARMACOGENOMICS AND CLINICAL STUDIES

Once an IND gets approved, clinical studies can be performed to assess drug efficacy and toxicity in patients. Phase I studies are used to establish tolerated dose ranges, to assess pharmacokinetics, and to identify any major ADR. Phase II studies are primarily geared to test drug efficacy, and phase III studies test drug effectiveness although ADRs are monitored simultaneously. Ideally, pharmacogenomic samples and data should be collected throughout the clinical trial process to help maximize drug efficacy and minimize potential harm to patients by using the data to guide patient selection during subsequent stages of drug development and to establish guidelines for usage after the drug has been approved.[65,66] This strategy helps streamline the clinical trial process and reduce the number of patients needed for each phase; more patients are likely to respond to the drug and fewer are likely to drop from the study due to the reduced risk of developing ADRs. The collected pharmacogenomic data can also be used to identify biomarkers that predict whether patients could respond well to the drug.

Targeted therapies should be used only for patients who express the target of interest and/or have altered expression levels of the target. Because most diseases have multiple causes and because not all patients could harbor the target of interest, testing of the target is necessary. Pharmacogenomic testing can be used to determine whether the patient harbors the target and thereby predict whether the patient is likely to benefit from the drug. The codevelopment of pharmacogenomic tests is becoming increasingly common, particularly for anticancer drugs. This is undoubtedly related to the high cost, toxicity profiles, and the fact that many anticancer drugs are targeted therapies. An example is crizotinib, an ATP competitive kinase inhibitor that targets c-Met, ALK, ROS, and is used to treat NSCLC. A pharmacogenomic test (the Vysis ALK Break Apart FISH Probe Kit; Abbott Molecular; www.abbott-molecular.com), which detects genetic rearrangements involving the *ALK* gene, was codeveloped with crizotinib, and allowed for its rapid development and approval. The lead drug was developed in 2005, and, based on data from a phase III study that showed 88% clinical benefit, crizotinib received FDA approval in 2011.[67] As only 6% of NSCLC patients harbor *ALK* gene fusions,[68] it is essential that patients are screened for this target prior to prescription of crizotinib. The test was developed during preclinical studies (cell line and animal studies) based on the target kinase profiles in the molecularly defined target population; during these studies, *ALK* was identified as a predictive marker and a test developed that allowed for selection of patients in subsequent clinical studies.[68] Clearly, in this particular case, codevelopment of a pharmacogenomic test has proved to be hugely beneficial to both patients and the drug company.

Pharmacogenomic testing is also used to screen patients to be enrolled in phase I and II clinical trials for polymorphisms and other genetic alterations that have been shown to affect metabolism and/or efficacy of the drug in preclinical studies.[69] Patients should be excluded from clinical trials or have dose adjustments if they harbor polymorphisms that are more likely to place them at risk of developing treatment-related toxicity, or, in the case of prodrugs, harbor polymorphisms that affect drug efficacy. A number of tests are available to detect *CYP450* enzyme polymorphisms: for example, the Roche AmpliChip Cytochrome P450 genotyping test (www.roche.com).[39,70,71]

To date, over 80 FDA-approved drugs contain information regarding pharmacogenomic testing within their drug label (Table 4.1).[72] Ideally, pharmacogenomic tests should be codeveloped with the drug, beginning in preclinical studies. Such testing can allow clinical trials to be streamlined and benefit both patients and drug companies; fewer patients are needed for a clinical trial if the anticipated effect size is larger and patient drop rate is lower. The reality is that the majority of tests are developed and/or recommended post approval after severe ADRs have

TABLE 4.1
Pharmacogenomic Data and Testing Information for FDA-Approved Drugs

Clinical Specialty	Drug Used	Associated Genes
Allergy	Desloratadine and pseudoephedrine	*CYP2D6*
Analgesics	Celecoxib, codeine	*CYP2C9, CYP2D6*
	Tramadol and acetaminophen	*CYP2D6*
Antiarrhythmics	Quinidine	*CYP2D6*
Antifungals	Terbinafine, voriconazole	*CYP2D6, CYP2C19*
Anti-infectives	Chloroquine, rifampin, isoniazid, and pyrazinamide	*G6PD, NAT1; NAT2*
Antivirals	Abacavir, boceprevir, maraviroc, nelfinavir	*HLA-B*5701, IL28B, CCR5, CYP2C19, IL28B*
	Peginterferon alfa-2b, telaprevir	*IL28B*
Cardiovascular	Carvedilol, clopidogrel, isosorbide and hydralazine, metoprolol, prasugrel, pravastatin, propafenone, propranolol, ticagrelor	*CYP2D6, CYP2C19, NAT1, NAT2, CYP2D6 CYP2C19,* Genotype E2/E2 and Fredrickson Type III dysbetalipoproteinemia, *CYP2D6 CYP2D6, CYP2C19*
Dermatology and dental	Cevimeline, dapsone, fluorouracil, tretinoin	*CYP2D6, G6PD, DPD PML/ RARa*
Gastroenterology	Dexlansoprazole (1),[a] dexlansoprazole (2), esomeprazole, pantoprazole, rabeprazole, sodium phenylacetate and sodium benzoate, sodium phenylbutyrate	*CYP2C19, CYP1A2, CYP2C19, CYPC19, CYP2C19, UCD (NAGS; CPS; ASS; OTC; ASL; ARG), UCD (NAGS; CPS; ASS; OTC; ASL; ARG)*
Hematology	Lenalidomide, warfarin (1), warfarin (2)	5q Chromosome, *CYP2C9, VKORC1*
Metabolic and endocrinology	Atorvastatin	LDL receptor
Musculoskeletal	Carisoprodol, mivacurium	*CYP2C9*, Cholinesterase gene
Neurology	Carbamazepine, dextromethorphan and quinidine, galantamine, tetrabenazine	*HLA-B*1502, CYP2D6, CYP2D6, CYP2D6*
Oncology	Arsenic trioxide, brentuximab vedotin, busulfan, capecitabine, cetuximab (1), cetuximab (2), crizotinib, dasatinib, erlotinib, fulvestrant, gefitinib (1), gefitinib (2), imatinib (1), imatinib (2), imatinib (3), imatinib (4) irinotecan, lapatinib, mercaptopurine, nilotinib (1), nilotinib (2), panitumumab (1), panitumumab (2), rasburicase, tamoxifen, thioguanine, tositumomab, trastuzumab, vemurafenib	*PML/RARa,* CD30, Ph Chromosome, DPD *EGFR, KRAS, ALK,* Ph Chromosome, *EGFR* ER receptor, *CYP2D6, EGFR,* C-Kit, Ph Chromosome, *PDGFR, FIP ILI-PDGFRa, UGTIAI, Her2/neu, TPMT,* Ph Chromosome, *UGTIAI, EGFR, KRAS, G6PD,* ER receptor, *TPMT,* CD20 antigen, Her2/neu, *BRAF*

(Continued)

TABLE 4.1 *(Continued)*
Pharmacogenomic Data and Testing Information for FDA-Approved Drugs

Clinical Specialty	Drug Used	Associated Genes
Ophthalmology	Timolol	*CYP2D6*
Psychiatry	Aripiprazole, atomoxetine, chlordiazepoxide and amitriptyline, citalopram (1), citalopram (2), clomipramine, clozapine, desipramine, diazepam, doxepin, fluoxetine, fluoxetine and olanzapine, fluvoxamine (1), fluvoxamine (2), fluvoxamine (3), iloperidone, imipramine, modafinil (1), modafinil (2), nefazodone, nortriptyline, paroxetine, perphenazine, pimozide, protriptyline, risperidone, thioridazine, trimipramine, valproic acid, venlafaxine	*CYP2D6, CYP2D6, CPY2D6, CYP2C19, CYP2D6, CYP2D6, CYP2D6, CYP2D6, CYP2C19, CYP2D6, CYP2D6, CYP2D6, CYP2C9, CYP2C19, CYP2D6, CYP2D6, CYP2D6, CYP2C19, CYP2D6, CYP2D6, CYP2D6, CYP2D6, CYP2D6, CYP2D6, CYP2D6, CYP2D6, CYP2D6, CYP2D6, UCD (NAGS; CPS; ASS; OTC; ASL; ARG), CYP2D6*
Pulmonary	Tiotropium	*CYP2D6*
Reproductive	Drospirenone and ethinyl estradiol clomiphene, tolterodine	*CYP2C19*, Rh genotype, *CYP2D6*
Rheumatology	Azathioprine, flurbiprofen	*TPMT, CYP2C9*

Sources: Adapted from Gullapalli RR, et al., *J Pathol Inform*, 3, 40, 2012; data source: http://www. fda.gov/Drugs/ScienceResearch/ResearchAreas/Pharmacogenetics/ucm083378.htm.

[a] Numbers in the brackets indicate that the drug is affected by multiple genetic polymorphisms.

been identified. For example, it is now well known that the metabolism of tricyclic antidepressant drugs is dependent on *CYP2D6*, and to some extent *CYP2C19*, and polymorphisms in these enzymes can affect both efficacy and toxicity. As adverse effects occur quite frequently for this drug class, this has led to increased usage of pharmacogenomic testing for polymorphisms in these enzymes. An increasing number of pharmacogenomic tests are also being developed for existing drugs that are known to have substantial interpatient dose variability and high rates of adverse reactions.[71] A good example is warfarin. Approximately 20% of patients taking warfarin are hospitalized within the first six months of usage due to bleeding events. Warfarin is metabolized by *CYP2C9*, a highly polymorphic enzyme. The major two allele variants encoding this enzyme, *CYP2C9*2* and *CYP2C9*3*, are present in the patient population, and have been shown to account for approximately 12% of the interpatient variability for warfarin. Patients homozygous for the *CYP2C9*1* variant are extensive metabolizers, while patients who homozygous for *CYP2C9*3* are poor metabolizers. Patient with the following genotypes are intermediate metabolizers: *CYP2C9*1/*3* and *CYP2C9*1/*2*. While the results of three recent clinical studies are conflicting, pharmacogenomic testing has been approved and its usage is becoming increasingly common. Hopefully these successes will encourage other drug companies to develop pharmacogenomic testing services.

Pharmacogenomics can also be used during clinical studies to generate data that predict whether patients respond well to a drug, regardless of whether they have the target and/or genetic alterations that affect efficacy and toxicity. To allow for this, patient specimens must be collected during phase I, II, and III trials (and ideally postmarketing studies); once patient outcome is known (patient response to the drug in terms of effectiveness and/or toxicity and, in some instances, chemoresistance), correlations can be made between outcome and the presence/absence of pharmacogenomic biomarkers.[73,74] Similar strategies and tests that are used for target identification (target gene analysis and genome profiling) can be used to identify these biomarkers. Again, this information and the development of pharmacogenomic tests is beneficial to both patients and drug companies; it can be used to guide patient selection and thereby maximize effectiveness and minimize toxicity, making it more likely the drug is approved.[39,70] With regard to the latter, postmarketing studies are encouraged as they allow for analysis in a large and diverse population.

It is noteworthy that *de novo* mutations that occur in target genes, drug transporters, and enzymes responsible for drug metabolism can cause resistance to targeted therapies. This phenomenon occurs fairly frequently with cancer drugs due to high incidence of genetic instability in cancer cells. The collection of patient biospecimens throughout the clinical trial process can help identify these *de novo* mutations, and allow for correlations with resistance to be made. This knowledge can then be used to predict chemoresistance in patients and plan accordingly. For example, certain *de novo* mutations in Bcr-Abl that alter its kinase domain can result in resistance to imatinib.[75] This knowledge has led to the design of new tyrosine kinase inhibitors that are not impacted by these mutations and can be used to treat CML patients who have developed resistance to imatinib.[76,77]

The FDA and European Medicines Agency (EMA) are the regulatory agencies that oversee drug development and approval in the United States and in Europe, respectively. The benefits that result from the usage of pharmacogenomics during drug target identification and drug development studies support their mission to ensure drug safety and efficacy, and to support innovations in drug discovery and development; and as such, both agencies actively encourage the incorporation of pharmacogenomics into both drug development and monitoring. They have published several white papers that provide guidance to companies, and, in addition, the FDA offers drug companies access to review staff in their Office of Clinical Pharmacology who can help companies to devise a pharmacogenomic plan for inclusion in their drug development pipeline.* In some instances, the FDA may make approval of an NDA contingent on the inclusion of postmarketing

---

* EMA guidelines and concept papers: http://www.ema.europa.eu/ema/index.jsp?curl=pages/regulation/general/general_content_000411.jsp&mid=WC0b01ac058002958e; FDA guidelines and concept papers: http://www.fda.gov/Drugs/ScienceResearch/ResearchAreas/pharmacogenetics/default.htm

pharmacogenomic studies. As mentioned above, postmarketing studies are able to better detect rare ADRs due to a larger sample size and inclusion of more diverse populations. Undoubtedly, the guidance and support that these agencies offer have helped lead to the rapid increase in the number of drugs that have been discovered, developed, and/or repurposed based on pharmacogenomic studies.

CONCLUSIONS

The major advances in drug discovery and development have been made possible through the incorporation of pharmacogenomic data and testing. Pharmacogenomics can allow for the identification of novel drug targets, allow for drugs to be developed for specific subsets of patient populations that have high efficacy and low toxicity, decrease the time taken for a drug to reach the market, and improve cost–benefit ratios. As more drug companies incorporate pharmacogenomics into their drug discovery and development process, further improvements can be anticipated. The inclusion of these analyses during drug development will result in increased availability of pharmacogenomic data that clinicians can then use to help guide their clinical decision making.

STUDY QUESTIONS

1. Genomic profiling is more likely than target gene analysis to identify novel drug targets:
   a. True
   b. False
2. Which of the following types of study can be used to validate the involvement of a drug target in the disease process?
   I. *In silico*   II. Cell line   III. Animal model
   a. I only
   b. II only
   c. I and II
   d. II and III
   e. I, II, and III
3. Which of the following statements is FALSE?
   a. A polymorphism that is located in the drug binding site of a drug target will always decrease "druggability" of the target.
   b. It is estimated that only 10% of the human genome is druggable.
   c. Polymorphisms can affect binding of a drug to its target by altering target charge and polarity.
   d. Some drugs are designed to inhibit target function while others are designed to alter target expression levels.
4. Pharmacogenomic analyses are only useful during the preclinical study phase of drug development:
   a. True
   b. False

5. Pharmacogenomic testing can be used to:
    I. Predict drug efficacy   II. Predict drug toxicity
    III. Guide treatment decisions, including dosing
    a. I only
    b. II only
    c. I and II
    d. II and III
    e. I, II, and III

Answer Key
    1. a
    2. d
    3. a
    4. b
    5. e

REFERENCES

1. FDA. *Issues Advice to Make Earliest Stages of Clinical Drug Development More Efficient.* FDA; 2006.
2. Kola I, Landis J. Can the pharmaceutical industry reduce attrition rates? *Nat Rev Drug Discov* 2004;3:711–5.
3. Paul SM, Mytelka DS, Dunwiddie CT, Persinger CC, Munos BH, Lindborg SR, Schacht AL. How to improve R&D productivity: The pharmaceutical industry's grand challenge. *Nat Rev Drug Discov* 2010;9:203–14.
4. DiMasi JA, Feldman L, Seckler A, Wilson A. Trends in risks associated with new drug development: Success rates for investigational drugs. *Clin Pharmacol Ther* 2010;87:272–7.
5. Scannell JW, Blanckley A, Boldon H, Warrington B. Diagnosing the decline in pharmaceutical R&D efficiency. *Nat Rev Drug Discov* 2012;11:191–200.
6. McCarthy JJ, McLeod HL, Ginsburg GS. Genomic medicine: A decade of successes, challenges, and opportunities. *Sci Transl Med* 2013;5:189sr4.
7. Hudis CA. Trastuzumab—Mechanism of action and use in clinical practice. *N Engl J Med* 2007;357:39–51.
8. Moja L, Tagliabue L, Balduzzi S, Parmelli E, Pistotti V, Guarneri V, D'Amico R. Trastuzumab containing regimens for early breast cancer. *Cochrane Database Syst Rev* 2012;4:CD006243.
9. Pare G. Genome-wide association studies—Data generation, storage, interpretation, and bioinformatics. *J Cardiovasc Transl Res* 2010;3:183–8.
10. Baumgart DC, Sandborn WJ. Crohn's disease. *Lancet* 2012;380:1590–605.
11. Viatte S, Plant D, Raychaudhuri S. Genetics and epigenetics of rheumatoid arthritis. *Nat Rev Rheumatol* 2013;9:141–53.
12. Metzker ML. Sequencing technologies—The next generation. *Nat Rev Genet* 2010; 11:31–46.
13. Pavlopoulos GA, Oulas A, Iacucci E, Sifrim A, Moreau Y, Schneider R, Aerts J, Iliopoulos I. Unraveling genomic variation from next generation sequencing data. *BioData Min* 2013;6:13.
14. Kingsmore SF, Lindquist IE, Mudge J, Gessler DD, Beavis WD. Genome-wide association studies: Progress and potential for drug discovery and development. *Nat Rev Drug Discov* 2008;7:221–30.

15. Hageman GS, Anderson DH, Johnson LV, Hancox LS, Taiber AJ, Hardisty LI, Hageman JL, et al. A common haplotype in the complement regulatory gene factor H (HF1/CFH) predisposes individuals to age-related macular degeneration. *Proc Natl Acad Sci U S A* 2005;102:7227–32.

16. Yehoshua Z, de Amorim Garcia Filho CA, Nunes RP, Gregori G, Penha FM, Moshfeghi AA, Zhang K, Sadda S, Feuer W, Rosenfeld PJ. Systemic complement inhibition with eculizumab for geographic atrophy in age-related macular degeneration: The COMPLETE study. *Ophthalmology* 2014;121:693–701.

17. Ni Z, Hui P. Emerging pharmacologic therapies for wet age-related macular degeneration. *Ophthalmologica* 2009;223:401–10.

18. Ng SB, Buckingham KJ, Lee C, Bigham AW, Tabor HK, Dent KM, Huff CD, et al. Exome sequencing identifies the cause of a Mendelian disorder. *Nat Genet* 2010;42:30–5.

19. Simon R, Roychowdhury S. Implementing personalized cancer genomics in clinical trials. *Nat Rev Drug Discov* 2013;12:358–69.

20. L'Heureux J, Murray JC, Newbury E, Shinkunas L, Simon CM. Public perspectives on biospecimen procurement: What biorepositories should consider. *Biopreserv Biobank* 2013;11:137–43.

21. Kang B, Park J, Cho S, Lee M, Kim N, Min H, Lee S, Park O, Han B. Current status, challenges, policies, and bioethics of biobanks. *Genomics Inform* 2013;11:211–7.

22. Tong EK, Fung LC, Stewart SL, Paterniti DA, Dang JH, Chen MS, Jr. Impact of a biospecimen collection seminar on willingness to donate biospecimens among Chinese Americans: Results from a randomized, controlled community-based trial. *Cancer Epidemiol Biomarkers Prev* 2014;23:392–401.

23. Loffredo CA, Luta G, Wallington S, Makgoeng SB, Selsky C, Mandelblatt JS, Adams-Campbell LL. Knowledge and willingness to provide research biospecimens among foreign-born Latinos using safety-net clinics. *J Community Health* 2013; 38:652–9.

24. Dang JH, Rodriguez EM, Luque JS, Erwin DO, Meade CD, Chen MS, Jr. Engaging diverse populations about biospecimen donation for cancer research. *J Community Genet* 2014;5:313–27.

25. Koretzky GA. The legacy of the Philadelphia chromosome. *J Clin Invest* 2007; 117:2030–2.

26. Druker BJ, Guilhot F, O'Brien SG, Gathmann I, Kantarjian H, Gattermann N, Deininger MW, et al. Five-year follow-up of patients receiving imatinib for chronic myeloid leukemia. *N Engl J Med* 2006;355:2408–17.

27. Moore JH, Asselbergs FW, Williams SM. Bioinformatics challenges for genome-wide association studies. *Bioinformatics* 2014;26:445–55.

28. Chowdhury S, Pradhan RN, Sarkar RR. Structural and logical analysis of a comprehensive hedgehog signaling pathway to identify alternative drug targets for glioma, colon and pancreatic cancer. *PLoS One* 2013;8:e69132.

29. Schweizer L, Zhang L. Enhancing cancer drug discovery through novel cell signaling pathway panel strategy. *Cancer Growth Metastasis* 2013;6:53–9.

30. Daley GQ, Van Etten RA, Baltimore D. Induction of chronic myelogenous leukemia in mice by the P210bcr/abl gene of the Philadelphia chromosome. *Science* 1990; 247:824–30.

31. Lugo TG, Pendergast AM, Muller AJ, Witte ON. Tyrosine kinase activity and transformation potency of bcr-abl oncogene products. *Science* 1990;247:1079–82.

32. Honda H, Hirai H. Model mice for BCR/ABL-positive leukemias. *Blood Cells Mol Dis* 2001;27:265–78.

33. Salesse S, Verfaillie CM. BCR/ABL: From molecular mechanisms of leukemia induction to treatment of chronic myelogenous leukemia. *Oncogene* 2002;21:8547–59.

34. Jones G, Willett P. Docking small-molecule ligands into active sites. *Curr Opin Biotechnol* 1995;6:652–6.
35. Ekins S, Mestres J, Testa B. In silico pharmacology for drug discovery: Applications to targets and beyond. *Br J Pharmacol* 2007;152:21–37.
36. Keiser MJ, Roth BL, Armbruster BN, Ernsberger P, Irwin JJ, Shoichet BK. Relating protein pharmacology by ligand chemistry. *Nat Biotechnol* 2007;25:197–206.
37. Poroikov V, Filimonov D, Lagunin A, Gloriozova T, Zakharov A. PASS: Identification of probable targets and mechanisms of toxicity. *SAR QSAR Environ Res* 2007;18:101–10.
38. Gashaw I, Ellinghaus P, Sommer A, Asadullah K. What makes a good drug target? *Drug Discov Today* 2011;17(Suppl):S24–30.
39. Kitzmiller JP, Groen DK, Phelps MA, Sadee W. Pharmacogenomic testing: Relevance in medical practice: Why drugs work in some patients but not in others. *Cleve Clin J Med* 2011;78:243–57.
40. Melo JV. BCR-ABL gene variants. *Baillieres Clin Haematol* 1997;10:203–22.
41. Di Nicolantonio F, Arena S, Gallicchio M, Zecchin D, Martini M, Flonta SE, Stella GM, et al. Replacement of normal with mutant alleles in the genome of normal human cells unveils mutation-specific drug responses. *Proc Natl Acad Sci U S A* 2008;105:20864–9.
42. Kore PP, Mutha MM, Antre RV, Oswal RJ, Kshirsagar SS. Computer-aided drug design: An innovative tool for modeling. *Open J Med Chem* 2012;2:139–48.
43. Nisius B, Sha F, Gohlke H. Structure-based computational analysis of protein binding sites for function and druggability prediction. *J Biotechnol* 2012;159:123–34.
44. DiMasi JA, Faden LB. Competitiveness in follow-on drug R&D: A race or imitation? *Nat Rev Drug Discov* 2011;10:23–7.
45. Ruan ZX, Huangfu DS, Xu XJ, Sun PH, Chen WM. 3D-QSAR and molecular docking for the discovery of ketolide derivatives. *Expert Opin Drug Discov* 2013; 8:427–44.
46. Ishikawa T, Sakurai A, Kanamori Y, Nagakura M, Hirano H, Takarada Y, Yamada K, Fukushima K, Kitajima M. High-speed screening of human ATP-binding cassette transporter function and genetic polymorphisms: New strategies in pharmacogenomics. *Methods Enzymol* 2005;400:485–510.
47. Powers R. Advances in nuclear magnetic resonance for drug discovery. *Expert Opin Drug Discov* 2009;4:1077–98.
48. Jason-Moller L, Murphy M, Bruno J. Overview of Biacore systems and their applications. *Curr Protoc Protein Sci* 2006;Chapter 19:Unit 19.3.
49. Leon-Cachon RB, Ascacio-Martinez JA, Barrera-Saldana HA. Individual response to drug therapy: Bases and study approaches. *Rev Invest Clin* 2012;64:364–76.
50. Hughes JP, Rees S, Kalindjian SB, Philpott KL. Principles of early drug discovery. *Br J Pharmacol* 2011;162:1239–49.
51. Kinnings SL, Xie L, Fung KH, Jackson RM, Bourne PE. The *Mycobacterium tuberculosis* drugome and its polypharmacological implications. *PLoS Comput Biol* 2010;6:e1000976.
52. Cramer PE, Cirrito JR, Wesson DW, Lee CY, Karlo JC, Zinn AE, Casali BT, et al. ApoE-directed therapeutics rapidly clear beta-amyloid and reverse deficits in AD mouse models. *Science* 2012;335:1503–6.
53. Ernst FR, Grizzle AJ. Drug-related morbidity and mortality: Updating the cost-of-illness model. *J Am Pharm Assoc (Wash)* 2001;41:192–9.
54. Ono C, Kikkawa H, Suzuki A, Suzuki M, Yamamoto Y, Ichikawa K, Fukae M, Ieiri I. Clinical impact of genetic variants of drug transporters in different ethnic groups within and across regions. *Pharmacogenomics* 2013;14:1745–64.

55. Yiannakopoulou E. Pharmacogenomics of phase II metabolizing enzymes and drug transporters: Clinical implications. *Pharmacogenomics J* 2013;13:105–9.

56. Sim SC, Kacevska M, Ingelman-Sundberg M. Pharmacogenomics of drug-metabolizing enzymes: A recent update on clinical implications and endogenous effects. *Pharmacogenomics J* 2013;13:1–11.

57. Gardiner SJ, Begg EJ. Pharmacogenetics, drug-metabolizing enzymes, and clinical practice. *Pharmacol Rev* 2006;58:521–90.

58. King HC, Sinha AA. Gene expression profile analysis by DNA microarrays: Promise and pitfalls. *JAMA* 2001;286:2280–8.

59. Roncaglioni A, Toropov AA, Toropova AP, Benfenati E. In silico methods to predict drug toxicity. *Curr Opin Pharmacol* 2013;13:802–6.

60. Cheng F, Li W, Liu G, Tang Y. In silico ADMET prediction: Recent advances, current challenges and future trends. *Curr Top Med Chem* 2013;13:1273–89.

61. Knight A. Non-animal methodologies within biomedical research and toxicity testing. *ALTEX* 2008;25:213–31.

62. Ando Y, Saka H, Ando M, Sawa T, Muro K, Ueoka H, Yokoyama A, Saitoh S, Shimokata K, Hasegawa Y. Polymorphisms of UDP-glucuronosyltransferase gene and irinotecan toxicity: A pharmacogenetic analysis. *Cancer Res* 2000;60:6921–6.

63. Tadokoro J, Kakihata K, Shimazaki M, Shiozawa T, Masatani S, Yamaguchi F, Sakata Y, Ariyoshi Y, Fukuoka M. Post-marketing surveillance (PMS) of all patients treated with irinotecan in Japan: Clinical experience and ADR profile of 13,935 patients. *Jpn J Clin Oncol* 2011;41:1101–11.

64. Ando Y, Hasegawa Y. Clinical pharmacogenetics of irinotecan (CPT-11). *Drug Metab Rev* 2005;37:565–74.

65. Huang SM, Temple R. Is this the drug or dose for you? Impact and consideration of ethnic factors in global drug development, regulatory review, and clinical practice. *Clin Pharmacol Ther* 2008;84:287–94.

66. Antman E, Weiss S, Loscalzo J. Systems pharmacology, pharmacogenetics, and clinical trial design in network medicine. *Wiley Interdiscip Rev Syst Biol Med* 2012; 4:367–83.

67. Kwak EL, Bang YJ, Camidge DR, Shaw AT, Solomon B, Maki RG, Ou SH, et al. Anaplastic lymphoma kinase inhibition in non-small-cell lung cancer. *N Engl J Med* 2010;363:1693–703.

68. Soda M, Choi YL, Enomoto M, Takada S, Yamashita Y, Ishikawa S, Fujiwara S, et al. Identification of the transforming EML4-ALK fusion gene in non-small-cell lung cancer. *Nature* 2007;448:561–6.

69. Sheffield LJ, Phillimore HE. Clinical use of pharmacogenomic tests in 2009. *Clin Biochem Rev* 2009;30:55–65.

70. Scott SA. Clinical pharmacogenomics: Opportunities and challenges at point of care. *Clin Pharmacol Ther* 2013;93:33–5.

71. Samer CF, Lorenzini KI, Rollason V, Daali Y, Desmeules JA. Applications of CYP450 testing in the clinical setting. *Mol Diagn Ther* 2013;17:165–84.

72. Gullapalli RR, Desai KV, Santana-Santos L, Kant JA, Becich MJ. Next generation sequencing in clinical medicine: Challenges and lessons for pathology and biomedical informatics. *J Pathol Inform* 2012;3:40.

73. Poste G. Bring on the biomarkers. *Nature* 2011;469:156–7.

74. McDonald SA. Principles of research tissue banking and specimen evaluation from the pathologist's perspective. *Biopreserv Biobank* 2010;8:197–201.

75. Nestal de Moraes G, Souza PS, Costas FC, Vasconcelos FC, Reis FR, Maia RC. The interface between BCR-ABL-dependent and -independent resistance signaling pathways in chronic myeloid leukemia. *Leuk Res Treatment* 2012;2012:671702.

76. Bhatia R, Holtz M, Niu N, Gray R, Snyder DS, Sawyers CL, Arber DA, Slovak ML, Forman SJ. Persistence of malignant hematopoietic progenitors in chronic myelogenous leukemia patients in complete cytogenetic remission following imatinib mesylate treatment. *Blood* 2003;101:4701–7.
77. Dohse M, Scharenberg C, Shukla S, Robey RW, Volkmann T, Deeken JF, Brendel C, Ambudkar SV, Neubauer A, Bates SE. Comparison of ATP-binding cassette transporter interactions with the tyrosine kinase inhibitors imatinib, nilotinib, and dasatinib. *Drug Metab Dispos* 2010;38:1371–80.

5 Pharmacogenomics and Laboratory Medicine

Jianbo J. Song

CONTENTS

KEY CONCEPTS

- Genetics is estimated to account for 20–95% of differences in drug efficacy and adverse side effects.
- Pharmacogenomics studies and applies information of genetic variants into clinical practice of drug administration and disease treatment.

- Genetic variants associated with pharmacogenomics include SNPs, mutations, insertions, small deletions or duplications, large deletions or duplications, and chromosome rearrangements.
- Pharmacogenomic testing for clinical applications should be performed by CLIA-certified laboratories and handled by qualified personnel certified by an appropriate accrediting agency.
- Cytochrome P450 (CYP) superfamily is a class of the important drug-metabolizing enzymes. Polymorphisms of *CYP* genes are associated with variable drug response.
- Thiopurine methyltransferase (TPMT) is an important enzyme responsible for the metabolism of thiopurine drugs, such as 6-mercaptopurine (6-MP) used for the treatment of acute lymphoblastic leukemia (ALL) of childhood.
- Polymorphisms of other drug-metabolizing enzymes, including N-acetyltransferase (NAT) and uridine diphosphate glucuronosyltransferase 1A (UGT1A) are also associated with different drug responses among individual patients.
- Drug transporters, such as ATP-binding cassette subfamily B member 1 (ABCB1), breast cancer resistance protein (BCRP), and organic anion-transporting polypeptide (OATP), play important roles in transporting drugs across the cell membrane, and therefore polymorphisms of these transporters could be associated with variation in drug efficacy and adverse reactions.
- Pharmacodynamics studies the biochemical and physiological effects of the drug. There are more than 25 examples of drug targets with sequence variants associated with drug efficacy. One such example is the beta-2 adrenergic receptor (β2 adrenoceptor, or ADRB2).
- Pharmacogenomics has identified some novel cancer therapy targets, such as BCR/ABL in chronic myeloid leukemia (CML), HER2 in breast cancer, and ALK and EGFR in lung cancer.
- In addition to treating disease, pharmacogenomics can also provide information on disease prevention by genetic testing for inherited susceptibility to cancer. For example, testing of *BRCA1* and *BRCA2* is used to assess risk for developing breast cancer.
- Precision medicine through pharmacogenomics will revolutionize disease treatment and health improvement in the near future.

INTRODUCTION

It has been long recognized that there are often significant differences in drug efficacy and side effect profiles among patients. It is estimated that most drugs are effective for only about 30–60% of individual patients (Spear et al. 2001), and about 7% of patients receiving the same drugs suffer a serious adverse drug reaction (ADR) (Lazarou et al. 1998). Interindividual variations in drug efficacy and safety can be affected by either genetic or environmental factors, such as age, organ function, drug interactions, and comorbidities. Genetics is estimated to account for 20–95% of differences in drug disposition and effects (Kalow et al. 1998). Pharmacogenetics studies and applies into clinical practice of the genetic variants affecting drug metabolism, transport, molecular targets/pathway, and genetic susceptibility to diseases. The US Food and Drug Administration (FDA) defines pharmacogenomics (PGx) as "the study of variations of DNA and RNA characteristics as related to drug response" and pharmacogenetics (PGt) as "a subset of pharmacogenomics (PGx)" and "the study of variations in DNA sequence as related to drug response."* However, pharmacogenomics and pharmacogenetics are used interchangeably at most times. The ultimate goal of pharmacogenetics/pharmacogenomics is to improve drug safety and efficacy by applying the right drug to the right patient with the right dose at the right time, which is also known as personalized medicine (Hamburg and Collins 2010). Realization of this goal of precision medicine requires collaborative efforts from physicians, pharmacists, and staff working in laboratory medicine.

Laboratory medicine is the practice of a medical laboratory or clinical laboratory to perform tests on clinical specimens to obtain information that helps the diagnosis, treatment, and prevention of the disease. There are many fields of laboratory medicine, including chemistry, cytology, hematology, histology, pathology, and genetics. Dependent on the practical purpose, genetic testing can be diagnostic, prenatal, presymptomatic, predispositional, and pharmacogenetic. Genetic testing should be ordered by adequately trained healthcare providers who can give appropriate pretest and posttest counseling and also performed in a qualified laboratory. Appropriate laboratory should have Clinical Laboratory Improvement Amendments of 1988 (CLIA) certification and/or a necessary state license if required by the state where the healthcare providers perform the testing. CLIAs are federal regulatory standards that apply to all clinical laboratories testing performed on humans in the United States. An objective of the CLIA is to ensure quality laboratory testing, including accuracy, reliability, and timeliness of test results across all US facilities or sites that test human specimens for medical assessment or diagnosis, and treatment or prevention of the disease. Genetic testing laboratory personnel should be certified by the appropriate accrediting agency. The laboratory director should be an MD or PhD who is credentialed by the American Board of Medical Genetics (ABMG, now American Board of Medical Genetics and Genomics, ABMGG). Both the states New York and California also have specific credential requirements for a director of the genetic testing laboratory, respectively. Laboratory staff, such as technologists, are usually credentialed by the American Society of Clinical Pathology (ASCP). Genetic testing laboratories may also have genetic counselors certified by the American Board of Genetic Counseling, Inc. (ABGC) to provide pretest and posttest counseling.

Genetic variations affecting interindividual differences in drug response and safety could be sequence alterations in genes encoding drug-metabolizing enzymes, drug transporters, or drug targets. Drug-metabolizing enzymes catalyze many different types of chemical processes, including oxidation, hydroxylation, and hydrolysis, as well as conjugation reactions, such as acetylation, glucuronidation, or sulfation. Examples of drug-metabolizing enzymes include the cytochrome P450 (CYP) superfamily, N-acetyltransferases (NAT), UDP-glucuronosyltransferases (UGT),

---

* http://www.fda.gov/downloads/drugs/guidancecomplianceregulatoryinformation/guidances/ucm073162.pdf

thiopurine methyltransferase (TPMT), and more. There are mainly two types of drug transporters: the efflux transporters, such as the ATP-binding cassette (ABC) superfamily that transfers drug out of the cell; and the influx (or uptake) transporter, such as organic anion-transporting peptides (OATP) that transports drugs into the cell. Both drug-metabolizing enzymes and drug transporters are important for drug disposition (absorption, distribution, metabolism, and excretion). Genetic differences in drug targets, such as receptors, have profound influences on drug efficacy. In oncology, genetic variations include germline mutations (inherited) and somatic mutations (acquired). Germline mutations either affect genes responsible for drug deposition and pharmacokinetics or render the carrier more susceptible to developing certain types of cancer. One prominent example for cancer susceptibility mutations is familial breast cancer caused by mutations in the high-penetrance *BRCA1* and *BRCA2* genes. On the other hand, somatic mutations only in cancer cells often lead to alteration in drug response.

The aim of this chapter is to discuss the basic concepts of genetic variations and detection methods in pharmacogenetics/pharmacogenomics, provide examples of genetic variations associated with marked alterations in pharmacokinetics and pharmacodynamics, and review the novel targets of cancer treatment and genetic testing of cancer susceptibility genes. However, this chapter is not meant to be exhaustive but rather use the clinically relevant examples to illustrate how pharmacogenetics/pharmacogenomics and molecular (genetic) diagnostics improve drug efficacy and reduce toxicity. Refer to the relevant chapters for specific application of pharmacogenomics in certain diseases such as cardiovascular or neurologic diseases.

TYPES OF GENETIC VARIATION AND METHODS OF DETECTION

In this section, we discuss the basic types of genetic variations and briefly introduce the methods of detection. Refer to Chapter 3 for details in biotechnology and clinical testing. Genetic variations are differences in genetic sequence that occur at the levels of DNA, gene, or chromosome. The major genetic variations include polymorphism, mutation, small deletion and duplication, large deletion and duplication, insertion, and rearrangement at the chromosomal level, such as translocation and inversion (Table 5.1).

A mutation usually refers to the change of one or more base pairs at the DNA level. Based on the type of changes, a mutation can be classified as a missense mutation, nonsense mutation, insertion, deletion, duplication, or frameshift mutation. A missense mutation (also called nonsynonymous mutation) changes one DNA base pair, and results in the substitution of one amino acid for another in the protein encoded by a gene. A nonsense mutation refers to a change in one DNA base pair that causes the substitution of an amino acid–encoding codon to a stop codon, usually resulting in a truncated protein that may not function properly or may result in degradation of the mRNA by a mechanism called nonsense-mediated decay (NMD), which is a cellular process that specifically recognizes premature termination codon (PTC) carrying mRNA for degradation (Baker and Parker 2004). An insertion alters the DNA base number by adding a different nucleotide of DNA, often resulting in a nonfunctional protein or frameshift. A small deletion removes one or a few base pairs from a gene, whereas a small duplication adds one or a few base pairs by

TABLE 5.1

Types of Genetic Variations and Methods of Detection

Type	Feature	Method of Detection	Example
Polymorphism			
SNP	Single nucleotide	Sequencing, PCR	
CNV	Repeats of a few DNA base pairs	Sequencing, PCR	VNTR in TPMT CYP2D6
Mutations	Single or a few base pairs	Sequencing, PCR, RFLP	
Large deletions and duplications		Array CGH, MLPA	BCRA1 deletion
Gene amplification	One or more copies of a gene	FISH, RT-PCR	Her2
Chromosomal level rearrangements	Translocation	Chromosome study, FISH	BCR–ABL in CML
	Inversion	Chromosome study, FISH	ALK1

Note: CNV, copy-number variation.

copying one or more times. A frameshift mutation refers to the addition or loss of DNA base pairs that alter the reading frame (a group of three bases) of a gene, resulting in a nonfunctional protein or premature stop codon. Insertions, deletions, and duplications may all cause frameshift mutations. Because frameshifts and premature stop codons are typically deleterious in nature, these alterations are interpreted as disease-causing mutations per ACMGG recommendations for Standards for Interpretation and Reporting of Sequence Variations (Richards et al. 2008).

Polymorphisms are differences in individual DNA that are not mutations. Single nucleotide polymorphisms (SNPs) are the most common form of polymorphisms, defined as the occurrence at an allele frequency of at least 1/1000 bases in the general population (Sachidanandam et al. 2001). There are an estimated 1.4 million SNPs in the human genome, some of which contribute to the variability in drug response and adverse effects, including pharmacokinetic and pharmacodynamic processes. Mutations and SNPs can be detected by molecular biology methods, such as direct sequencing or PCR, followed by restriction enzyme digestion (also known as restriction fragment length polymorphism, or RFLP), and gel electrophoresis.

Genetic differences also occur in a manner of the small deletion or duplication, large deletion or duplication, or rearrangement. Larger deletions and duplications often include one or several genes. This type of genetic variation is also known as copy-number changes at the chromosome level. Other genetic rearrangements at the chromosomal level include translocation or inversion. A chromosome translocation is caused by rearrangement of parts between nonhomologous chromosomes. A gene fusion may be created when the translocation joins two otherwise-separated genes, whose occurrence is common in cancer. The most famous example is the *BCR-ABL* fusion gene that results from a translocation between chromosomes 9 and 22, which is also known as Philadelphia chromosome (Nowell and Hungerford 1960). Translocations can be balanced (in an even exchange of chromosome materials, with no extra or missing genes, and ideally full functionality) or unbalanced (where the exchange of chromosome materials is unequal, resulting in extra or missing genes). Translocation can be detected by chromosome study (also known as standard cytogenetics or a karyotype) of the affected cells or by fluorescence in situ hybridization (FISH) if probes are available.

Genes may also be amplified or overexpressed. One example is *HER2* (human epidermal growth factor receptor 2) amplification in breast cancer (Owens et al. 2004; Slamon et al. 1987; Yaziji et al. 2004). Common methods of detecting gene amplification include reverse transcription-polymerase chain reaction (RT-PCR), FISH, and SNP assay using next-generation sequencing (NGS) technologies.

A mutation, small deletion, or duplication can be detected by direct sequencing. Two most widely used sequencing technologies in genetic testing laboratory are Sanger sequencing and NGS. Sanger sequencing was developed by Frederick Sanger and colleagues in 1977, and it has become the classic DNA sequencing method (Sanger et al. 1977a, 1977b). This sequencing method is based on the selective incorporation of chain-terminating dideoxynucleotide (ddATP, ddGTP, ddCTP, or ddTTP) during in vitro DNA replication by DNA polymerase. Compared to NGS and other more recently developed sequencing technologies, Sanger sequencing is time consuming and expensive, but it is highly accurate. In contrast, the NGS technologies allow massive parallel sequencing of many DNA strands simultaneously; therefore they are faster and cheaper than Sanger sequencing. But the length of reading sequence is shorter, usually less than 300-bp compared to more than 500-bp by Sanger sequencing. However, NGS needs significantly less DNA and is more accurate and reliable.

Larger deletion and duplication can be detected by FISH or traditional cytogenetics (chromosome study or karyotyping). FISH can only detect deletion or duplication specifically targeted by the probe used, and the deletion or duplication is required to be larger than the probe size, usually lager than 100 kb. FISH can also be used to detect gene fusion, gene amplification, chromosome translocation, and inversion. Standard cytogenetic study (chromosome analysis) usually cannot detect deletions <5 Mb. Chromosomal microarray analysis (CMA) or comparative genomic hybridization (CGH) microarray testing is a technology that compares patient DNA with reference DNA from normal individuals to detect copy-number variation (CNV) at higher resolution than G-band chromosome analysis. Dependent on the purpose of genetic testing, these technologies are often used in combination to provide the best testing results.

PHARMACOKINETICS

Interindividual differences in genetic makeup can affect variability in drug responses at both pharmacokinetic and pharmacodynamic levels. Pharmacokinetics is the study of the absorption, distribution, metabolism, and excretion as well as transport of the drug. Pharmacodynamics studies the biochemical and physiological effects of the drug. Drug metabolism can be divided into three phases. Phase I metabolism often refers to oxidation, hydroxylation, and hydrolysis of the drug in the liver, and bioactivation of the prodrug. Sometimes, phase I metabolism could generate an intermediate metabolite in the inactivation and degradation of the drug. Phase II metabolism often produces a more easily excreted water-soluble compound through conjugation reactions, such as acetylation, glucuronidation, or sulfation. Last, in phase III, the conjugated drugs may be further processed, before being recognized by efflux transporters and pumped out of the cell. Genetic differences affect many steps in drug metabolism pathways. One of the most extensively studied pharmacokinetic examples is the CYP superfamily (Guengerich 2008; Zanger et al. 2008).

The CYP superfamily is a group of isoenzymes, primarily expressed in the hepatocytes and enterocytes, but also found in any other cells throughout the body. Inside the cell, the P450 enzymes are located in the endoplasmic reticulum and mitochondria. The mitochondria is the energy-producing center of the cell, and the CYP enzymes located in the mitochondria are generally the terminal oxidase enzymes in electron transfer chains. In contrast, the CYP enzymes located in the endoplasmic reticulum are involved in metabolizing a broad variety of drugs by catalyzing oxidative or reductive reactions.

The CYP enzymes are encoded by the superfamily of *CYP* genes (the gene name is *italicized* as required). There are approximately 60 *CYP* genes in the human genome that are classified into 18 families and 44 subfamilies based on sequence homology.* Among these families, the *CYP1* to *3* families are responsible for the major phase I drug metabolism, and the *CYP4* to *51* are associated with endobiotic metabolism. In this section, we will discuss the genetic polymorphisms of some major players in pharmacokinetics, including *CYP2D6, CYP2C9, CYP2C19, CYP3A4*, and *CYP3A5*. The number immediately following the *CYP* indicates the gene family, the capital letter indicates the subfamily, and the last number following the capital letter indicates the individual gene. The number after * indicates the allele (see the following section for examples).

The phenotypic distribution of genetic polymorphisms in a population for the *CYP* genes as well as other monogenes involved in pharmacokinetics may exhibit gene dosage effects. There are generally three types of genetic phenotype distributions associated with drug response: bimodal, multimodal, or broad without an apparent antimode. The bimodal phenotypic distribution exhibits continuous probability distribution with two different modes appearing as distinct peaks. An example of bimodal phenotypic distribution is NAT2 manifesting as fast acetylator or slow acetylator. The multimodal distribution has two or more modes. An example of multimodal phenotypic distribution is *CYP2D6* four distinct metabolizers (Ingelman-Sundberg et al. 2007; Zanger et al. 2004). The broad distribution has no apparent mode and an example is *CYP3A5*.

CYP2D6

The *CYP2D6* enzyme is responsible for the oxidative metabolism of up to 25% of commonly prescribed drugs, including the antidepressants, antipsychotics, opioids, antiarrhythmic agents, and tamoxifen. Many drugs metabolized by this enzyme

* www.http://drnelson.uthsc.edu/human.P450.table.html

TABLE 5.2

CYP2D6 **Metabolizer Group**

Phenotyping Group	Ethnic Group Frequency	Main Alleles
Poor metabolizer	Whites (5–10%)	Null alleles *3, *4, *6, and gene deletion *5
	East Asian (rare)	*4 (splice defect) 1–2%
	Africans (highly variable)	*4 (splice defect) 1–2%
Intermediate metabolizers	Whites (10–15%)	
	East Asian (up to 50%)	High prevalence of *10
	African (up to 30%)	High prevalence of *17
Ultrarapid metabolizers	Whites (1–10%)	Gene duplications or multiduplications
Extensive metabolizers	Whites (60–85%)	

have a narrow therapeutic window (Eichelbaum et al. 2006; Evans and Relling 1999; nomenclature; Zanger et al. 2004). The *CYP2D6* gene is located at chromosome band 22q13 and is highly polymorphic with more than 70 alleles and 130 genetic variations (nomenclature). The phenotypic distribution for *CYP2D6* is multimodal with four metabolizer groups by the predicted number of functional alleles: poor metabolizer (PM), intermediate metabolizer (IM), extensive metabolizer (EM), and ultrarapid metabolizer (UM) (see Table 5.2). The frequency of various alleles is significantly distinct among different ethnic groups (Ingelman-Sundberg et al. 2007; Zanger et al. 2004). The PMs have no CYP2D6 enzymatic activity due to carrying two null alleles. PM is more common in the white population with a frequency of 5–10%. It is rare in Asians and highly variable in those of African ancestry (Bradford 2002; McGraw and Waller 2012).

The IMs have reduced CYP2D6 enzymatic activity because they carry a combination of either a null allele or two deficient alleles. The common deficient alleles are *CYP2D6*9, *10, *17,* and *41* (Bradford 2002). In contrast to PMs, IMs are more frequent in Asians (up to 50%) than whites (10–15%) because of the high prevalence of the defective allele *CYP2D6*10*. The frequency of IMs in Africans (up to 30%) is also higher than that in whites and the allele *17* is more frequent in this ethnic group.

The UMs have increased CYP2D6 enzymatic activity due to gene duplications or multiduplications and are found in 1–10% of whites (Ingelman-Sundberg et al. 1999). Gene duplications are described in 20% of Saudi Arabians and 29% of Ethiopians. The EMs have normal enzymatic activity and represent 60–85% of the white populations.

CYP2C9

CYP2C9 is another gene of the CYP superfamily exhibiting a genetic polymorphism with more than 35 allelic variants. The two most common allelic variants are *CYP2C9*2* (Rettie et al. 1994) and *CYP2C9*3* (Sullivan-Klose et al. 1996). Both lead to reduced activity of *CYP2C9*. Patients with the *CYP2C9*2* and *CYP2C9*3* alleles required a low dose of warfarin to reduce the risk of bleeding because *CYP2C9* is the major inactivation pathway to metabolize this drug (Beyth et al. 2000; Higashi et al. 2002; Sanderson et al. 2005). Similar to *CYP2D6*, *CYP2C9* also exhibits allele frequency difference across ethnic groups. These two alleles are carried by about 35% of white subjects (Kirchheiner and Brockmoller 2005; Lee et al. 2002) but is relatively rare in East Asian and African populations (Xie et al. 2002).

CYP2C19

CYP2C19 catalyzes the metabolism of many commonly used drugs, including (*S*)-mephenytoin (anticonvulsant), omeprazole (antiulcerative), and diazepam (anxiolytic). *CYP2C19* plays an important role in the proton-pump inhibitor therapy for peptic ulcer and gastroesophageal reflux diseases. The *CYP2C19* enzyme is highly polymorphic. To date, more than 35 polymorphisms of *CYP2C19* have been reported (Ingelman-Sundberg 2011). PM of *CYP2C19* is attributable to be homozygous for *CYP2C19* *2 and/or *3 variant alleles, which are null alleles (Desta et al. 2002). About 15–25% of the Chinese, Japanese, and Korean populations are PMs of (*S*)-mephenytoin, whereas the PM frequency in white people is less than 5% (Martis et al. 2013; Scott et al. 2011). The effect of omeprazole on the intragastric pH value largely depends on the *CYP2C19* genotypes of the patients.

CYP3A4 AND *CYP3A5*

CYP3A4 is the most abundant P450 enzyme in human liver and is responsible for the metabolism of more than 50% of clinical drugs (de Wildt et al. 1999; Rendic and Di Carlo 1997). More than 20 *CYP3A4* variants have been identified, many of which have altered enzyme activities ranging from modest to significant loss in catalytic efficiency. There is also a large difference across ethnic groups in the frequency of *CYP3A4* variants (Ball et al. 1999; Walker et al. 1998). High frequencies of *CYP3A4*2* and *7* were found in white people and high frequencies of *CYP3A4*16* and *18* in Asian populations. The clinical significance of the *CYP3A4* variant alleles for many drugs metabolized by *CYP3A4* remains uncertain or is minimal to moderate based on the current data. These coding region variants are unlikely to account for the >10-fold differences in CYP3A4 activities observed in vivo, because the alleles cause only small changes in the enzyme activity and many of the alleles exist at a low frequency. One factor that may contribute to the complexity of the CYP3A4 puzzle is CYP3A5. Virtually all CYP3A4 substrates, with a few exceptions, are also metabolized by CYP3A5 (Wrighton et al. 1990). CYP3A5 metabolizes these drugs at slower rate in most cases. Some drugs can be metabolized by CYP3A5 as fast or faster than the CYP3A4 enzyme. Therefore, the metabolic rates of CYP3A4 drugs measured in vivo are likely to reflect combined activities of CYP3A4 and CYP3A5. About 35% of whites and 50% of blacks express functional CYP3A5. This dual pathway potentially obscures the clinical effects of *CYP3A4* variants in human studies. The *CYP3A5*3* allele (6986A>G) is the most frequently occurring allele of *CYP3A5* that results in a splicing defect and absence of enzyme activity (Roy et al. 2005). Individuals with at least one allele of 6986A>G designated as *CYP3A5*1* are classified as CYP3A5 expressors.

Polymorphisms of the *CYP* family members have been taken into consideration for adjusting drug doses. Practice guidelines need to be established to implement the application of genetic laboratory test results into actionable prescribing decisions for specific drugs. To address this need, a shared project, the Clinical Pharmacogenetics Implementation Consortium (CPIC), was established by the Pharmacogenomics Knowledgebase* and the National Institutes of Health–sponsored Pharmacogenomics Research Network (PGRN) in 2009 (Gonzalez-Covarrubias et al. 2009). Peer-reviewed gene–drug guidelines developed by this consortium are published and updated periodically[†] based on new developments in the field. For example, CPIC published a guideline for *CYP2D6* and *CYP2C19* genotype and dosing of the tricyclic antidepressants in 2013 (Hicks et al. 2013).

OTHER DRUG-METABOLIZING ENZYMES: TPMT, NAT, AND UGT1A1

Thiopurine Methyltransferase

Thiopurine methyltransferase (TPMT) is another example of an important genetic polymorphism responsible for drug metabolism. TPMT catalyzes the *S*-methylation of thiopurine drugs, such as 6-mercaptopurine (6-MP) and azathioprine (AZA), that are cytotoxic immunosuppressive agents used to treat acute lymphoblastic leukemia of childhood, inflammatory bowel disease, and organ transplant recipients (Weinshilboum and Sladek 1980). The thiopurine drugs have a narrow therapeutic window, and therefore the difference between the dose of the drug required to achieve desired therapeutic effect and that causing toxicity is relatively small. The most serious thiopurine-induced toxicity is life-threatening myelosuppression. The human *TPMT* cDNA and gene were cloned and characterized in the 1990s (Honchel et al. 1993). The most common *TPMT* variant allele in white populations is *TPMT*3A* (about 5%), an allele that is predominantly responsible for the trimodel frequency distribution of the levels of RBC TPMT activity (Krynetski et al. 1996; McLeod et al. 2000; Tai et al. 1996). *TPMT*3A* has two nonsynonymous SNPs, one in exon 7 and another in exon 10 of this 10-exon gene. The allozyme encoded by *TPMT*3A* is rapidly degraded by an ubiquitin–proteasome-mediated process. The level of TPMT in the RBC reflects the relative level of activity in other human tissues such as the liver and kidney. There are striking differences in the frequency of variant alleles for *TPMT*. *TPMT*3A* is rarely, if ever, found in East Asian populations but *TPMT*3C*, with only the exon 10 SNP, is the most common variant allele in East Asian populations (about 2%) (McLeod et al. 2000). Individuals with homozygous *TPMT*3A* are at greatly increased risk for life-threatening myelosuppression when treated with standard doses of the thiopurine drugs. Therefore, 1/10 to 1/15 of routine doses are prescribed to further avoid myelosuppression. *TPMT* is the first example selected by the US FDA for a public hearing on the inclusion of pharmacogenetic information in drug labeling. Clinical testing for *TPMT* genetic polymorphism is widely available (www.prometheuslabs.com).

---

* PharmGKB; www.pharmgkb.org
[†] http://www.pharmgkb.org

N-Acetyltransferase

N-Acetyltransferase (NAT) catalyzes acetylation of a diverse variety of aromatic amine drugs and carcinogens (Evans and White 1964). Interindividual difference in drug acetylation is one of the earliest examples of pharmacogenetic variation. Acetylation of many drugs, such as procainamide and isoniazid, exhibits bimodal distribution among individuals, with two distinct phenotypes as rapid and slow acetylators. Slow acetylators are at increased susceptibility to isoniazid- and hydralazine-associated toxicity and to certain cancers due to exposure to industrial chemicals such as α- and β-naphthylamine and benzidine. Individuals with a poor acetylator phenotype have an increased risk of developing lung, bladder, and gastric cancers if exposed to carcinogenic arylamines for a long period of time. The phenotypes of acetylation are attributed to differences in the enzymatic activities of NAT.

There are two arylamine acetyltransferase isozymes in humans: type I (NAT1) and type II (NAT2). Although these two isozymes share greater than 93% of their 290 amino acids, they have overlapping but different substrates. NAT2 catalyzes acetylation of hydralazine, isoniazid, and procainamide, while NAT1 catalyzes p-aminosalicylate. The NAT2 isozyme functions to both activate and deactivate arylamine and hydrazine drugs and carcinogens. Polymorphisms in NAT2 are also associated with higher incidences of cancer and drug toxicity. They also show distinct expression patterns, with NAT2 expressed mainly in the liver and gut, and NAT1 in many adult tissues as well as in early embryos.

NAT1 and NAT2 are encoded by two genes located on chromosome 8p22, a region often deleted in cancers. The two genes are separated by 870 bp and there is pseudogene (NATP) located in between. Polymorphisms in both genes are responsible for the N-acetylation polymorphisms. Most alleles of NAT1 and NAT2 are haplotypes of several point mutations with one signature mutation causing reduced enzymatic activity. These alleles usually cause unstable protein or affect the activity of the protein. A full description of NAT1 and NAT2 alleles can be found at http://nat.mbg.duth.gr/ although it was initially curated and hosted at the website of University of Louisville.

There are 15 NAT2 alleles identified in humans, and NAT2*5A, NAT2*6A, and NAT2*7A are associated with the slow acetylator phenotypes (Fretland et al. 2001; Zang et al. 2007). There are great differences of frequency of NAT2 alleles across ethnic groups. The slow acetylation form is present in up to 90% of some Arab populations, 40–60% of whites, and only to 25% of East Asians.

There are 26 NAT1 alleles (Hein et al. 2000). Interestingly, amino acid change at position 64 from arginine to tryptophan substitution (W64D) is found both in NAT1*17 and NAT2*19. This missense mutation results in an unfolded protein accumulating intracellularly, which is degraded through the ubiquitination pathway. Another interesting allele of NAT1 is the NAT1*10 allele with no amino acid change in the coding region. The NAT1*10 allele has deletions and insertions at the 3'-untranslated region (3'-UTR). It has been reported that the NAT1*10 allele is associated with increased activity in colon cancer (Bell et al. 1995; Zenser et al. 1996). However, the biological effect of this allele is still uncertain because it seems to have no correlation between the copy number of NAT1*10 allele and the level of NAT1 activity.

UDP-Glucuronosyltransferase 1A1

Uridine diphosphate (UDP) glucuronosyltransferase (UGT) 1A belongs to the super-family of uridine diphosphate glucuronosyltransferase enzymes that are responsible for the glucuronidation of a wide variety of affected substrates. The transfer of glucuronic acid (glucuronidation) by UGT1A renders the substrates, including small lipophilic molecules such as steroids, bilirubin, hormones, and drugs, into water-soluble, excretable metabolites. The UGT1A enzymes usually have considerable overlapping substrate specificities, but the *UGT1A1* enzyme is the sole enzyme responsible for bilirubin metabolism. *UGT1A1* catalyzes the glucuronidation of many commonly used drugs or metabolites, such as the active metabolic (7-ethyl-10-hydroxycamptothecin [SN-38]) of the anticancer drug irinotecan, and endogenous substrates, such as bilirubin. Irinotecan is a topoisomerase-I inhibitor widely used for the treatment of metastatic and recurrent colorectal cancer. More than 100 *UGT1A* polymorphisms have been identified (Lankisch et al. 2009; Strassburg 2008). The *UGT1A1* polymorphisms cause three forms of inherited, unconjugated hyperbilirubinemia in humans. The most serious one is Crigler–Najjar type I disease and results from a complete absence of bilirubin glucuronidation due to the presence of homozygous or compound heterozygous for inactive *UGT1A1* alleles. A mild condition is known as Crigler–Najjar type II with residual *UGT1A1* enzymatic activity (Farheen et al. 2006; Mackenzie et al. 2005). The least severe of the inherited unconjugated hyperbilirubinemia is Gilbert syndrome with about 30% of normal *UGT1A1* enzymatic activity. Genetic variations within the *UGT1A* are also associated with interindividual differences in the development of certain drug toxicities.[*]

The *UGT1A* family is located at chromosome 2q37 and this locus enables the transcription of nine unique enzymes through exon sharing. There are 13 unique alternate exon 1 at the 5'-end of the gene, followed by four common exons, exons 2–4 and exon 5a, at the 3'-end. Four of the alternative first exons are considered pseudogenes. Each of the remaining nine first exons may be spliced to the four common exons (exons 2–4 and exon 5a). Therefore, the nine proteins resulted from this exon sharing process have different N-termini but identical C-termini. Further, each of the nine first exons encodes the substrate binding site, and the transcription is regulated by its own promoter (Perera et al. 2008).

To date, there are over 100 different *UGT1A1* variants described. These variants confer increased, reduced, inactive, or normal enzymatic activity. The individual variants are described as alleles and named with a * symbol, followed by a number by the UGT nomenclature committee and a list of these variants is to be found.[†]

UGT1A1 alleles exhibit marked differences in the frequency across different ethnic groups.[‡] Two of the most well-studied alleles, *UGT1A1*6* and *28*, are discussed in details. The *UGT1A1*6* (Gly71Arg or G71R; *rs4148323*) allele is more common in the East Asian populations than the *28* allele. This allele is associated with Gilbert syndrome among East Asians, Crigler–Najjar syndrome type II (CN-II), and transient familial neonatal hyperbilirubinemia. The frequency of *UGT1A1*6*

---

[*] A summary of the genetic variants of *UGT1A1* can be found on the PharmGKB website at http://www.pharmgkb.org/gene/PA420.

[†] http://www.pharmacogenomics.pha.ulaval.ca/cms/ugt_alleles

[‡] http://www.ncbi.nlm.nih.gov/SNP/

polymorphism is high among Japanese and Chinese (16–23%) but is very low in the white population (<1%) (Ando et al. 2000). The high frequency of the *UGT1A1*6* allele may contribute to the high incidence of neonatal hyperbilirubinemia in the East Asian populations, consistent with a major role of *UGT1A1* in the glucuronidation of bilirubin. In addition, the *UGT1A1*6* allele is associated with the development of irinotecan-induced toxicities. Patients either heterozygous or homozygous for the *UGT1A1*6* allele are at a higher risk for developing neutropenia, diarrhea, dose modification, and increased exposure to the cytotoxic SN-38 metabolite as compared to those with the *UGT1A1*1/*1* genotype.

The *UGT1A1*28* allele (*rs8175347*) has a homozygous 2-bp insertion (TA) in the TATA box promoter region of the *UGT1A1* gene. The presence of an extra TA repeat in the TATA box significantly decreases *UGT1A1* transcription, resulting in reduced enzymatic activity and glucuronidation (Innocenti et al. 2006). The *UGT1A1*28* allele is found in many cases of Gilbert syndrome, but it is also associated with an increased risk for neutropenia in patients receiving irinotecan. Compared to carriers of the *UGT1A1*1/*1* or *1/*28* genotype, patients with the *28/*28* genotype are at an increased risk for developing neutropenia when treated with irinotecan. This may be contradictory since some studies found no link between the *28* allele and neutropenia. Patients homozygous for the *UGT1A1*28* allele are also at an increased risk for developing diarrhea after receiving irinotecan, but this is also contradictory because results from individual studies have been mixed. A meta-analysis with 1760 patients across 20 studies found that the correlation between *UGT1A1*28* allele and developing diarrhea after irinotecan treatment depends on dosage. Compared to the *1/*1* genotype, patients with either the *28/*28* or *1/*28* genotype are at a significantly greater risk for experiencing severe diarrhea after taking medium (150–250 mg/m²) or high doses (≥250 mg/m²) of irinotecan, with no associations found in patients taking a low (<150 mg/m²) dose of irinotecan.

Similar to the *UGT1A1*6* allele, the frequency of the *UGT1A1*28* allele also varies across ethnic groups. Approximately 10% of the US population is homozygous for *UGT1A1*28*. The highest frequency is found in the African population (~43%), followed by that in the white population (~39%), and in the Asian population (~16%). Variants with 5 or 8 TA repeats occur primarily in individuals of African descent at much lower frequencies (~3–7%) (Beutler et al. 1998).

In summary, pharmacogenetic knowledge of the *UGT1A1* polymorphism status could help reduce the risk of severe toxicity and improve the chance of maintaining therapy by guiding the selection of the appropriate starting dosages. In fact, the US FDA has recommended that patients with the *28/*28* genotype receive a lower starting dose of irinotecan on the label of the drug since 2004. But the Dutch Pharmacogenetics Working Group recommends reducing irinotecan dose by 30% only when high dose (250 mg/m²) is used for patients with the *28/*28* genotype.

DRUG TRANSPORTERS

Drug transporters play important roles in modulating the disposition and excretion of a large number of drugs. In many cases, drug transporters are also critical determinants of extent of drug entry into their target organs. In addition to drug–drug interactions and environmental factors, genetic polymorphisms of drug transporters have profound impacts on drug disposition, drug efficacy, and drug safety. Drug transporters are categorized into two classes: efflux transporters, such as ABC super-family that pumps substrates out of the cell; and uptake or influx transporters, such as OATP, which facilitate the entry of drugs into the cell. Here we discuss OATP and two members of ABC transporters—ABCB1 and BCRP—to understand how such drug transporters affect drug disposition, efficacy, and toxicity.

ABCB1

ATP-binding cassette subfamily B member 1 (ABCB1), also known as P-glycoprotein (permeability glycoprotein, abbreviated as P-gp) or multidrug resistance protein 1 (MDR1), is an important protein of the cell membrane that transports many foreign substances out of the cell. ABCB1 is extensively distributed and expressed in various tissues, including the enterocytes, hepatocytes, and testis barrier. ABCB1 is an ATP-dependent efflux pump with a wide variety of substrate specificities, including many important drugs such as colchicine and quinidine, the chemotherapeutic agents (such as etoposide, doxorubicin, and vinblastine), immunosuppressive agents, and HIV-type 1 antiretroviral agents (such as protease inhibitors).

ABCB1 has many functions, one function of which is to regulate the distribution and bioavailability of drugs. Increased intestinal expression of ABCB1 may decrease the absorption of its substrate drugs, which results in reduced bioavailability and drug plasma concentrations. In contrast, decreased intestinal ABCB1 expression may result in supratherapeutic drug plasma concentrations and drug toxicity. Unfortunately, some cancer cells also express large amounts of ABCB1, which makes these cancers multidrug resistant due to active efflux transport of the antineoplastics out of the cell. ABCB1 can also remove other toxic metabolites and xenobiotics from the cell into urine, bile, and the intestinal lumen for excretion from the body.

The ABCB1 protein is encoded by the *ABCB1* gene located at chromosome 7q21.12. *ABCB1* is highly polymorphic, and some alleles exhibit ethnicity-dependent distribution. There are two frequently discussed *ABCB1* SNPs: G2677T/A (*rs2033582*, in exon 21) and C3435T (*rs1045642*, in exon 26) (Komar 2007). Both SNPs have been examined for the association of the G2677T variant (also known as Ser893Thr) with increased risk for developing lung cancer. In addition, the C3435T variant may modulate this influence. Carriers of the 2677T/T plus heterozygous 3435C/T have the highest risk for developing lung cancer (up to 20-fold higher risk). The G2677T/T allele frequency is greatly different across the ethnic groups, with 13% in the Asian population, 2% in the white population, and 0% in the African population (Bournissen et al. 2009; Leschziner et al. 2007).

The SNP C3435T of *ABCB1* is a synonymous polymorphism without alteration in the amino acid. The frequency distribution varies across ethnic groups with the highest homozygous seen in the white population (~26%), followed by the Asian population (~17%), and the lowest in the black African population (~2.3%). There are conflicting results on functional impact of the C3435T variant in the disposition of the ABCB1 substrates. In one study, C3435T was shown as a "silent" polymorphism in determining the substrate specificity (e.g., verapamil). On the other hand, it has been shown that the *ABCB1* C3435 allele, but not T/T genotype, significantly modulated the concentrations of (*R*)-lansoprazole (a drug primarily used to treat ulcer and acid reflux disease) in *CYP2C19* EMs after renal transplantation (Li et al. 2014). However, the clinical relevance of this observation may be minor because these pharmacogenetic changes were not associated with the occurrence of gastroesophageal complications. Therefore, the functional impact of the C3435T variant on the pharmacokinetic and pharmacodynamic properties of ABCB1 substrates remains to be defined.

BREAST CANCER RESISTANCE PROTEIN

The human breast cancer resistance protein (BCRP) is the second member of the G subfamily of the ATP-binding cassette (ABC) efflux transporter superfamily and is therefore also known as ABCG2 (Ross et al. 1999). BCRP was originally discovered in human breast cancer cell line MCF-7/AdrVp, which exhibited multidrug resistance but had no expression of known multidrug efflux transporters such as ABCB. In this breast cancer cell line, BCRP was overexpressed and resulted in multidrug resistance to the chemotherapeutic agents, including mitoxantrone, topotecan, and methotrexate, by transporting these compounds out of the cell. In addition to the chemotherapeutic drugs, there are a wide variety of structurally and chemically diverse compounds, including non-chemotherapy drugs transported by BCRP in an ATP-dependent fashion. BCRP is highly expressed in various tissues, including the small intestine, liver, brain endothelium, and placenta (Endres et al. 2006). In these normal tissues, BCRP plays an important role in the absorption, elimination, and tissue distribution of drugs. BCRP has been documented to be highly enriched in a large number of stem cells of bone marrow and other organs, which is known as side-population phenotype. BCRP may enhance cell survival in hypoxic condition (hypoxia) by binding and interacting with heme to reduce the accumulation of toxic heme metabolites. It has been recognized that cancer cells frequently show side-population phenotype with expression of BCRP and other ABC family efflux transporters to survive hypoxia and to avoid exposure to chemotherapeutic agents. These subpopulations of so-called cancer stem cells are responsible for tumor self-renewal for many cancer types. Therefore, these ABC family efflux transporters may contribute to drug resistance and eventually cancer cure failure.

The BCRP efflux transporter is encoded by the *BCRP* gene located at chromosome 4q22.1. BCRP transporter is highly polymorphic, with over 80 SNPs identified in the *BCRP* gene. Some of these SNPs in the coding region of *BCRP* gene may have significantly physiological and pharmacological relevance. The most important *BCRP* variant is the C421A polymorphism (Imai et al. 2002), which causes a missense mutation in the BCRP protein (Gln141Lys, or Q141K), resulting in low protein expression and impaired transport activity, possibly due to increased lysosomal and proteasomal degradations of this variant than wild-type BCRP (Giacomini et al. 2013). This variant is associated with interindividual difference in pharmacokinetics and response or toxicity of many chemotherapeutic drugs. The frequency of C421A is markedly different across ethnic groups, with 30–60% in Asians and 5–10% in the white and black African Americans. Patients heterozygous for the *BCRP* C421A allele showed threefold higher plasma levels of diflomotecan, an anticancer drug, with an intravenous administration of the drug. In addition to chemotherapeutic drugs, BCRP also plays an important role in the disposition of rosuvastatin, a member of the statin class drugs used to treat high cholesterol and related conditions. The *BCRP* C421A polymorphism influences the pharmacokinetics and therapeutic effect of rosuvastatin in Chinese and white populations.

ORGANIC ANION–TRANSPORTING POLYPEPTIDES

Organic anion–transporting polypeptide (OATP) is a family of membrane proteins mediating the transport of mainly organic anions across the cell membrane. There are currently 11 members of this family, encoded by genes that are classified as the Solute Carrier Organic Anion (*SLCO*) gene subfamily. OATPs are expressed in various organs, such as the liver, intestine, and kidneys. OATPs have a wide variety of substrates, including endogenous substrates such as bile acids, bilirubin, and numerous hormones including thyroid and steroid hormones, and a diverse range of drug compounds such as antibiotics, lipid-lowering drugs, antidiabetic drugs, and various anticancer drugs, including pazopanib, vandetanib, nilotinib, canertinib, and erlotinib (Khurana et al. 2014).

The major members of the OATP family include OATP1B1, OATP1B3, and OATP2B1. They show overlapping but some different expression patterns. OATP1B1, OATP1B3, and OATP2B1 are expressed in the liver on the basolateral membrane of hepatocytes to facilitate uptake of their substrates from the portal circulation. OATP1B1 and OATP1B3 are expressed exclusively in the liver but OATP2B1 expression is more ubiquitous. Genetic variations in these OATPs contribute to the interindividual differences in drug response.

OAT1B1 is a liver-specific transporter protein encoded by the *SLCO1B1* gene (solute carrier organic anion transporter family, member 1B1) located at chromosome 12p12. This transporter is formerly known as organic anion transporter 2 (OATP2), OATPC, liver-specific transporter 1 (*LST1*), and *SLC21A6*. OATP1B1 is responsible for mediating active transport of many endogenous substrates such as bile acids, xenobiotics, and a broad range of pharmaceutical compounds including antidiabetic drugs and chemotherapeutic agents. The *SLCO1B1* gene is highly polymorphic, with 190 common variants identified. By designation, the *SLCO1B1*1a* represents the wild-type allele. There are two well-characterized common nonsynonymous alleles: *rs2306283* (*SLCO1B1*: 492A>G on NM_006446.4, previously referred to as 388A>G; encoding OATP1B1: N130D) and *rs4149056* (*SLCO1B1* 625T>C on NM_006446.4; commonly referred to as T521C, encoding OATP1B1: V174A). Due to partial linkage disequilibrium, there are four important haplotypes with these two variants: *SLCO1B1*1A* containing neither variant, *SLCO1B1*1B* (*rs2306283*), *SLCO1B1*5* (*rs4149056*), and *SLCO1B1*15* with both variants. In addition to variants in the coding region, there is also a promoter variant, termed g.-11187G>A, which is found in conjunction with *SLCO1B1*1B* (N130D) and *SLCO1B1*5* (V174A). The haplotype containing this promoter variant g.-11187G>A in combination with these two common nonsynonymous variants is designated as *SLCO1B1*17*.

SLCO1B1*5 has reduced uptake of OATP1B substrates, such as estrone sulfate and estradiol 17-β-ᴅ-glucuronide. *SLCO1B1*5* also has high plasma levels of the cholesterol-lowering drug pravastatin and the antidiabetic drug repaglinide. The frequency of this allele is also different across ethnic groups, present in 14% of European Americans and 9% of African Americans (Santos et al. 2011; Wilke et al. 2012).

PHARMACODYNAMICS

Pharmacodynamics studies the biochemical and physiological effects of the drug. The purpose of pharmacodynamics is to match the treatment to the pathophysiology of the disease. In addition, genetic variants can also be used to predict drug response. Patterns of gene expression can reveal disease subtypes. Pharmacogenetic information can also discover novel drug treatment targets.

In addition to drug disposition, interindividual differences in drug efficacy can be profoundly affected by genetic variations in drug targets such as receptors. There are more than 25 examples of drug targets with sequence variants that affect drug efficacy. For example, sequencing variants encoding angiotensin-converting enzyme (ACE) affect the renoprotective actions of ACE inhibitors (Narita et al. 2003).

The beta-2 adrenergic receptor (β_2 adrenoceptor, or ADRB2) is used as an example here to illustrate how genetic variants in the drug target affect efficacy and safety. ADRB2 is a member of the G protein–coupled receptor superfamily with seven transmembrane domains. ADRB2 reacts with adrenaline (epinephrine) as a hormone or neurotransmitter affecting muscles or organs, and is crucial for the functioning of the respiratory and circulatory systems as well as the immune and metabolic processes. ADRB2 also affects obesity and glucose intolerance due to its critical role in thermogenesis and energy balance. ADRB2 is encoded by the *ADRB2* gene located at chromosome 5q31-32. The *ADRB2* gene is highly polymorphic, with over 80 variants identified. Genetic polymorphism of *ADRB2* can affect the signal transduction by the G protein–coupled receptors and is associated with nocturnal asthma, obesity, and type 2 diabetes.

Three variant proteins of *ADRB2*, Arg16Gly, Gln27Glu, and Thr164Ile, show functional effects of altered expression, downregulation, or coupling of the receptor in response to β2-adrenoceptor agonists. The codons 16 and 27 are localized in the extracellular part of the receptor, and both Arg16Gly and Gln27Glu variants affect the downregulation and internalization of the receptors in vitro (Fenech and Hall 2002). The Gly16 allele was reported to be associated with enhanced agonist-related sensitivity, while the Glu27 allele was associated with its decreased downregulation. The Ile164 allele has been shown to be associated with impaired receptor binding and decreased basal and epinephrine-stimulated adenylyl cyclase activity. The Ile164 isoform has been revealed to be 3–4 times less responsive to agonist-induced stimulation. In addition to variants in the coding region of the *ADRB2* genes, Cys19 polymorphisms in the promoter region of the gene have been revealed to be related to the gene transcription. Carriers of Cys19 show a 72% higher expression of *ADRB2* together with increased activity of adenylyl cyclase.

ADRB2 polymorphisms are in linkage disequilibrium. According to Wang et al. (2001), the most common haplotypes are Arg19/Gly16/Glu27, Cys19/Gly16/Gln27, and Cys19/Arg16/Gln27. In a study by Cagliani et al. (2009), dominant clades were identified: Gly16/Gln27 Ha, Arg16Gln27 Hb, Arg16Gln27 Hc1, and Gly16/Glu27 Hc2. Drysdale et al. (2000) described 12 haplotypes and 5 pairs of haplotypes. These haplotypes have been shown to be responsible for genetic variability in 90% of the population studied. While any effect was observed for single polymorphisms in this study, only haplotype analysis brought some conclusion. The *ADRB2* gene is highly polymorphic and the allele prevalence differs among different ethnic groups. It has been studied in multiple populations and more than 80 polymorphisms have been identified. Four known SNPs are nonsynonymous polymorphisms: Val34Met, Arg16Gly, Gln27Glu, and Thr164Ile. Two of these SNPs (Arg16Gly and Gln27Glu) are common with minor allele frequencies (MAFs) of 40–50%. Thr164Ile occurs with MAFs of 1–3%. The rarely occurring Val34Met has MAFs of less than 1%. The genetic analysis also showed that the 3′-UTR of the gene contains a poly-C repeat of variable length that is interrupted by two polymorphisms. It was suggested that these polymorphisms could be responsible for the variable response to β-adrenergic therapy. However, a clinical study concerning therapy with a long-acting beta-agonist (LABA) plus inhaled corticosteroid did not show any association with this genotype.

PHARMACOGENOMICS OF TARGETED CANCER THERAPIES

In the past decade, advances in cancer biology, genetics, pharmacology, and biotechnology made novel cancer-targeted therapy possible. Targeted anticancer drugs have been designed to act on selected molecular targets/pathways to provide stratified treatment with the benefit of better antitumor efficacy and lower host toxicity based on a patient's unique germline (inherited) or cancer (somatic) genomic profile. These cancer-targeted treatment agents can be antibodies, small chemical molecules, natural or engineered peptides, proteins, or synthetic nucleic acids such as antisense oligonucleotides or ribozymes. Pharmacogenomics, particularly genomic-based diagnostics, plays a critical role in cancer-targeted treatment. Currently, the US FDA has recommended pharmacogenomic consideration or package-insert labeling for more than 120 drugs involving more than 50 genes (FDA Biomarker). These drugs are used for treatments of cancer, cardiovascular diseases, infectious and psychiatric diseases. In particular, anticancer pharmacogenomics is the most active area with 24 biomarkers available in the drug labels for 30 FDA-approved anticancer agents (FDA Biomarker). The US FDA defines a genomic biomarker as "a measurable DNA and/or RNA characteristic that is an indicator of normal biologic processes, pathogenic processes, and/or response to therapeutic or other interventions" (FDA Definition). Such genomic biomarkers can be gene variants, copy-number changes, chromosomal abnormalities, functional deficiencies, expression changes, and more. Drug labeling for genomic biomarkers can include description of drug exposure and clinical response variability, risk for adverse events, genotype-specific dosing, mechanisms of drug action, and polymorphic drug target and disposition genes.

This chapter will use *BCR-ABL* in CML, *HER2* amplification in breast cancer, *EGFR* mutation, and *ALK* rearrangement in lung cancer as examples to discuss genomic-based diagnostics. For more information regarding biomarkers, refer to Chapter 4.

BCR-ABL

The Philadelphia (Ph) chromosome is an abnormally short chromosome 22, resulting from a reciprocal translocation between chromosomes 9 and 22, which is specifically designated as t(9;22)(q34;q11) (Nowell and Hungerford 1960). This translocation occurs in a single bone marrow cell, which subsequently produces many cells (a process termed as clonal expansion by cytogenetics) and gives rise to leukemia. In fact, the seminal discovery of the Ph chromosome provides the first example of consistent chromosome abnormality found in cancer. The Ph chromosome is a hallmark of chronic myeloid leukemia (CML). About 95% of patients with CML have this abnormality and the remainder have either a cryptic translocation that is invisible by traditional chromosome study, or a variant translocation involving one or more other chromosomes in addition to chromosomes 9 and 22 (Talpaz et al. 2006). The presence of Ph chromosome is also found in 25–30% of adult acute lymphoblastic leukemia (ALL) and 2–10% of pediatric ALL cases, as well as occasionally in acute myelogenous leukemia (AML).

At the molecular level, a fusion gene *BCR-ABL* is generated by juxtapositioning the *ABL* gene on chromosome 9q34 to a part of the *BCR* ("breakpoint cluster region") on chromosome 22q11 (Heisterkamp and Groffen 1991). *ABL* gene encodes for a membrane-associated tyrosine kinase, and the *BCR-ABL* fusion transcript is also translated into a tyrosine kinase. Three clinically important variants of *BCR-ABL*—the p190, p210, and p230 isoforms—are produced, dependent on the precise location of fusion (Advani and Pendergast 2002). The p190 is typically associated with ALL, while p210 is usually associated with CML but can also be associated with ALL (Pakakasama et al. 2008). The p230 is generally associated with chronic neutrophilic leukemia. Additionally, the p190 isoform can also be expressed as a splice variant of p210 (Lichty et al. 1998).

Usually in normal cells, the tyrosine kinase activity of the *ABL* gene product is tightly regulated (controlled). However, the tyrosine kinase of the *BCR-ABL* gene is deregulated (uncontrolled) with continuous activity resulting in unregulated cell division and consequently malignant state. Understanding this process led to the development of the first genetically targeted drug imatinib mesylate (Gleevec), which specifically binds to and inhibits the tyrosine kinase activity of the BCR-ABL fusion protein (Druker et al. 1996). This discovery provides a prominent example of molecular targeted therapies against specific oncogenic events. Besides ABL, imatinib also directly inhibits other tyrosine kinases, such as ARG (ABL2), KIT, and PDGFR. This drug has had a major impact on the treatment of CML and other blood neoplasma and solid tumors with etiologies based on activation of these tyrosine kinases. Analyses of CML patients resistant to *BCR-ABL* suppression by imatinib coupled with the crystallographic structure of ABL complexed with this inhibitor have shown how structural mutations in *ABL* can circumvent an otherwise potent anticancer drug. The successes and limitations of imatinib hold general lessons for the development of alternative molecular targeted therapies in oncology.

BCR-ABL fusion can be detected by FISH using LSI *BCR/ABL* Dual Color, Dual Fusion probe (Abbott Molecular; www.abbottmolecular.com) (Figure 5.1b) or chromosome study (standard cytogenetics, Figure 5.1a). Figure 5.1c shows the crystal

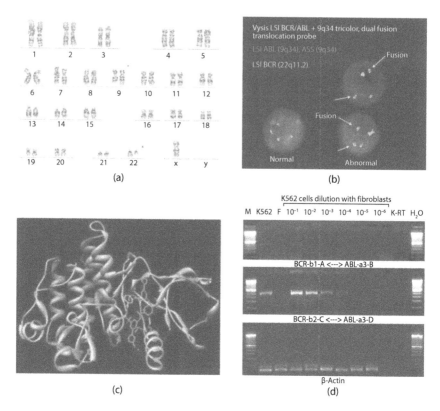

FIGURE 5.1 (See color insert.) Genetic testing results for *BCR-ABL* fusion. (a) Chromosome study at band level of 450. Arrows, derivative chromosomes 9 and 22 (Philadelphia chromosome). (b) FISH using Vysis LSI *BCR/ABL* Dual Color Dual Fusion probes. *ABL* (spectrum red), *BCR* (spectrum green), ASS (blue), yellow arrows indicate fusion. Each of the abnormal cells has two fusion signals, one for derivative chromosome 9 and the other for derivative chromosome 22. (c) Crystal structure of the kinase domain of ABL (blue) in complex with second-generation tyrosine kinase inhibitor nilotinib (red). (d) RT-PCR for *BCR-ABL*. K562, positive cell line for *BCR-ABL* fusion. (Parts [a] and [b] courtesy of Dr. Xinjin Xu, ARUP Laboratories and Department of Pathology, University of Utah School of Medicine.)

structure of ABL kinase domain (blue) in complex with the second-generation tyrosine kinase inhibitor nilotinib (red). *BCR/ABL* can also be detected by molecular method such as RT-PCR (Figure 5.1d) (Song et al. 2011) or Real-Time qRT-PCR to quantify the transcript. All methods can be used for initial diagnosis. FISH and qRT-PCR are used for monitoring treatment. RT-PCR-based methods may miss the rare form of fusions due to the specific primer design.

HER2 Testing in Breast Cancer

Breast cancer is categorized into two major tissue types: ductal carcinoma that develops from the cells of the lining of milk ducts and lobular carcinoma that develops from the cells of breast lobules. In addition, there are more than 18 other subtypes of breast cancer. Breast cancer is the most common invasive cancer in women worldwide, representing about 22.9% of invasive cancer in women.

About 18–20% of breast cancers are positive for *HER2* (human epidermal growth factor receptor 2) amplification (Owens et al. 2004; Slamon et al. 1987; Yaziji et al. 2004). The *HER2* gene encodes a cell surface receptor that can promote cell growth and division upon stimulation of growth signals. Amplification of the *HER2* gene results in abnormally high levels of *HER2* and increases the growth and spread of breast cancer. Compared to *HER2*-negative breast cancer, *HER2*-positive breast cancer tends to be much more aggressive and fast growing if untreated. Fortunately, there are several treatments available for *HER2*-positive breast cancer. Trastuzumab (Herceptin) is a monoclonal antibody that binds to *HER2* and prevents cell growth by blocking *HER2* receptors (Baselga et al. 1998; Pegram et al. 1998). In addition, trastuzumab can also stimulate the immune system to destroy cancer cells, and it is now a standard treatment along with adjuvant chemotherapy in patients with metastatic *HER2*-positive breast cancer after surgery, since this dramatically reduces the risk of recurrence. Another drug, lapatinib, a small molecule dual tyrosine kinase inhibitor for HER2 and EGFR pathways, is often given for advanced *HER2*-positive breast cancer if trastuzumab fails. Kadcyla, which is a trastuzumab connected to a drug called DM1 that interferes with cancer cell growth, is another drug that can be given to *HER2*-positive breast cancer patients previously treated with trastuzumab and taxanes. Pertuzumab, a monoclonal antibody that inhibits the dimerization of HER2 with other HER receptors, is also used for late-stage *HER2*-positive breast cancer in combination with trastuzumab and docetaxel.

Approximately 15% of all newly diagnosed breast cancers are *HER2*-positive, which means that the tumors either have extra copies of the *HER2* gene inside the cell and/or high levels of the HER2 protein on cell surfaced. *HER2*-positive breast cancer patients are most likely to benefit from HER2-targeted treatment, which substantially improves survival. On the other hand, *HER2*-negative patients are unlikely to benefit from HER2-targeted treatment, and these patients should be identified to avoid side effects as well as costs associated with these expensive drugs. It is therefore very important for the testing laboratories to provide high-quality testing to ensure accurate reporting of *HER2* status. In 2007, the American Society of Clinical Oncology (ASCO) and the College of American Pathologists (CAP) jointly issued a guideline to improve the accuracy and reporting of *HER2* testing in patients with invasive breast cancer, which was subsequently updated in 2013 (Wolff et al. 2007, 2013). The guideline recommends that *HER2* status must be tested for all newly diagnosed invasive breast cancers, including primary site and/or metastatic site, to determine whether *HER2*-targeted treatment is an option. Currently, two FDA-approved *HER2* testing methods are used in the United States, including immunohistochemistry (IHC) and in situ hybridization (ISH). The IHC HER2 test measures the amount of HER2 protein on the cancer cell surface; the results of IHC test can be: 0 (negative), 1+ (also negative), 2+ (equivocal), or 3+ (positive-HER2 overexpression). The ISH analyzes the number of copies of the *HER2* gene inside each tumor cell. The original ISH method is based on fluorescence (also known as FISH). The updated ASCO–CAP guideline also adds recommendations for a newer diagnostic method known as bright-field ISH that evaluates the amplification of the *HER2* gene using a regular bright microscope, rather than a fluorescent microscope. The FISH test result for the *HER2* gene can be:

- Positive (dual-probe *HER2/CEN17* ratio ≥2.0, with an average *HER2* copy number ≥4.0 signals/cell; dual-probe *HER2/CEN17* ratio ≥2.0, with an average *HER2* copy number <4.0 signals/cell; dual-probe *HER2/CEN17* ratio <2.0, with an average *HER2* copy numbers ≥6.0 signals/cell)
- Equivocal (single-probe ISH average *HER2* copy number ≥4.0 and <6.0 signals/cell or dual-probe *HER2/CEP17* ratio <2.0 with an average *HER2* copy number ≥4.0 and <6.0 signals/cell)
- Negative (single-probe ISH average *HER2* copy number <4.0 signals/cell or dual-probe *HER2/CEP17* ratio <2.0 with an average *HER2* copy number <4.0 signals/cell)

For an equivocal *HER2* test result, the decision to recommend HER2-targeted treatment should be delayed and mandatory retesting should be performed on an alternative specimen or on the same specimen using an alternative testing method. One example of *HER2* test by FISH is shown in Figure 5.2b. The patient was diagnosed with invasive breast cancer at the age of 52 years old, and the FISH test for *HER2* was performed using the FDA-approved PathVysion *HER2* DNA probe kit

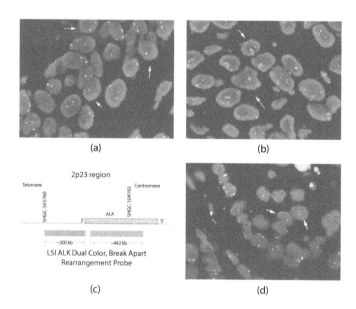

FIGURE 5.2 **(See color insert.)** FISH results for *HER2* (a and b) and *ALK* (c and d). (a) Control with normal cells (arrows) with two copies of centromere signals (green) and two copies of *HER2* signals (red) for chromosome 17; (b) a patient with abnormal cells (arrows), two copies of centromere signals (green), and amplified *HER2* signals (red). (c) LSI *ALK* Dual Color Break Apart probe (Abbott Molecular; www.abbottmolecular.com). (d) FISH results for *ALK* rearrangement. Arrow: abnormal cells with one fusion signal, separate one red and one green signal indicating *ALK* rearrangement.

(Abbott Molecular) with *HER2* labeled by fluorescence orange and centromere probe for chromosome 17 (*HER2* is located at 17q11.2q12) labeled by fluorescence green. Figure 5.2a shows a normal specimen with an average of two copies of *HER2* and two copies of the centromere signals for chromosome 17 in each cell (marked by arrows). Figure 5.2b shows amplification of the *HER2* gene in this breast cancer patient with an average ratio of *HER2/CEP17* 5.5/cell (arrows). Note that normal cells (arrowhead) with two copies of *HER2* gene and two copies of centromere signals for chromosome 17 were also observed in this specimen, reflecting the heterogeneity of cancer specimen. The *HER2* test for this patient was reported as positive for amplification, and HER2-targeted therapy was applied accordingly.

LUNG CANCER THERAPY TARGETS

Lung cancer is the leading cause of cancer-related death worldwide, with a mean five-year survival rate of <15%, mostly because the majority of patients have an advanced stage disease (stage IIIB or IV) at diagnosis. The majority of lung cancer patients have the non–small cell lung cancer (NSCLC) subtype. Lung cancer is histologically and biologically heterogeneous. The most common histology is adenocarcinoma (about 45–55%), followed by squamous histology (20–30%), and large cell histology (10–15%). Biologically, identification of *ALK* rearrangements and *EGFR* mutations in NSCLC not only further subdivides patients with advanced NSCLC but also leads to mutually exclusive targeted therapies. Patients carrying a known mutation in the *EGFR* gene that encodes the EGFR can be treated with FDA-approved TKI targeting at the EGFR as the first-line therapy option. The ALK inhibitors, such as crizotinib, are used to treat patients with an *ALK* rearrangement without regard to the line of therapy.

EGFR MUTATIONS IN LUNG CANCER

The *EGFR* gene is located at chromosome 7p12. EGFR, a transmembrane glycoprotein, is a member of the tyrosine kinase superfamily and can bind to a number of known ligands, including EGF, TGFA/TGF-α, amphiregulin, epigen/EPGN, BTC/betacellulin, epiregulin/EREG, and HBEGF/heparin-binding EGF. In a normal cell, binding of a ligand to EGFR induces receptor dimerization and tyrosine autophosphorylation, activation of the receptor, recruitment of adapter proteins such as GRB2, and subsequently triggering complex downstream signaling cascades that lead to cell proliferation. The EGFR signaling pathway activates at least four major downstream signaling cascades, including the RAS–RAF–MEK–ERK, PI3 kinase–AKT, PLC-gamma–PKC, and STATs modules. The presence of an *EGFR* gene mutation would result in the constant activation of the associated signaling pathways and consequently, cell proliferation and other cancer processes.

Mutations in the *EGFR* gene are associated with lung cancer. In fact, EGFR is the first identified target in NSCLC. In 2004, *EGFR*-activating mutations were identified in an adenocarcinoma subtype of lung cancer and were rapidly associated with response to EGFR-TKI. Clinical pathological features associated with these mutations include East Asian ethnicity, adenocarcinoma histology, female sex, and a history of not smoking. It is estimated that about 10–15% of all NSCLC patients have an *EGFR* mutation. However, the prevalence of *EGFR* mutations varies significantly across ethnic groups, from 10% in the white population to more than 40% in Asian populations (Lynch et al. 2004; Paez et al. 2004; Pao et al. 2004). In Asian NSCLC cancer patients who were nonsmokers or only light smokers, this percentage can be as high as 60%. For NSCLC with adenocarcinoma histology, the frequency for *EGFR* mutation is about 10–20%, whereas the rate for squamous histology is about 1–15%. It is therefore recommended to test *EGFR* mutations for all nonsquamous lung cancers, regardless of clinical characteristics.

The majority of *EGFR* mutations are located in the tyrosine kinase domain. The two most frequent *EGFR*-activating mutations, p.L858R point mutation in exon 21 and small deletions in exon 19 (Del19), are responsible for about 90% of cases (Ladanyi and Pao 2008). The less common or rare mutations (a varying frequency of 1–3%) include p.L861Q mutation, a missense mutation at codon 719 (p.G719X, about 3%) resulting in the substitution of the glycine at amino acid position 719 by a cysteine, alanine, or serine, and in-frame insertion mutations in exon 20. For patients with a known frequent mutation (Del19 and p.L858R), treatment with an EGFR-TKI, such as erlotinib, gefitinib, or afatinib, is a standard first-line therapy, and multiple phase III trials showed that EGFR-TKI treatment has improved objective response rate (ORR), progression-free survival (PFS), and health-related quality of life (HRQOL). However, the sensitivity to EGFR-TKI and PFS are globally lower for patients with less frequent or rare *EGFR* mutations.

The use of advanced molecular profiling of EGFR mutation status for patients with NSCLC to direct targeted therapy with the EGFR-TKIs significantly improves the treatment of this disease. However, patients undergoing EGFR-TKI treatments will eventually relapse by acquired resistance (progression after initial benefit). Acquired resistance may arise from different mechanisms, including pharmacological, biological, and evolutionary selection on molecularly diverse tumors. Polymorphisms in the previously discussed drug transporters such as ABCB1 (1236T>C, 2677G>T/A, and 3435C>T) and BCRP (ABCG2, 421C>A) may be relevant for the pharmacokinetics of the EGFR-TKIs. *EGFR* mutation heterogeneity could also contribute to the relapse. The emergence of *EGFR* exon 20 p.T790M mutation clones seems to be the most frequent mechanism for acquired resistance (Pao et al. 2005). It seems that there is the presence of a minor subclone (about 1%) of p.T790M mutation before treatment of EGFR-TKI is selected. Although patients with the *EGFR* exon 20 p.T790M mutation are resistant to EGFR-TKI treatment, the presence of this alteration predicts a favorable prognosis and indolent disease course, compared to the absence of it after TKI failure. Mutation in the genes of the downstream signaling cascade of the EGFR pathway is another mechanism for acquired resistance of EGFR-TKIs. One example is the *KRAS* gene, which has been implicated in the pathogenesis of several cancers. Mutations in the *KRAS* gene result in a constitutively activated KRAS protein that continually triggers these downstream signals. Although *EGFR* TKIs can block *EGFR* activation, they cannot block the activity of the mutated *KRAS* protein. Thus, patients with *KRAS* mutations tend to be resistant to erlotinib and gefitinib. *KRAS* mutations are more likely found in adenocarcinoma patients who are smokers, and white patients, rather than East Asians, and are prognostic for poor survival (Riely et al. 2009).

Once a patient experiences acquired resistance, there are several treatment options to maintain control of the disease, including local radiation to treat isolated areas of progression and continuation of the EGFR-TKI, and adding or switching to cytotoxic chemotherapy. In addition, there are successful novel approaches, including the development of second-generation and third-generation TKIs and the combination of TKIs with antibodies directly targeting the receptor. Different from the reversible first-generation TKIs, such as erlotinib and gefitinib, the second-generation EGFR-TKIs form covalent, irreversible binding to the target, and exhibit increased effectiveness through a prolonged inhibition of EGFR signaling. Preclinical studies showed that the second-generation irreversible TKIs effectively killed cells with acquired resistance to the first-generation TKIs. Currently, there are several irreversible TKIs, including afatinib, dacomitinib, and neratinib, in the clinical development for NSCLC. Another method to treat acquired resistance is the development of novel TKIs that inhibit both EGFR-activating mutations and T790M mutations. For example, the novel TKI rociletinib (CO-1686) has been shown to be active against both the EGFR T790M mutation and the activating mutations with only limited inhibition of the wild-type EGFR. Rociletinib is now in ongoing phase I/II trial for NSCLC patients with acquired resistance to the first-generation TKIs.

ALK REARRANGEMENT IN NSCLC

Anaplastic lymphoma kinase (ALK), also known as ALK tyrosine kinase receptor or cluster of differentiation 246 (CD246), is encoded by the *ALK* gene located at chromosome 2p23. The ALK tyrosine kinase receptor belongs to the insulin receptor superfamily of receptor tyrosine kinases (RTKs). The deduced amino acid sequences reveal that the ALK protein comprises an extracellular domain, a putative transmembrane region, and an intracellular kinase domain. It plays an important role in the development of the brain and exerts its effects on specific neurons in the nervous system. Alterations of the *ALK* gene, such as chromosomal rearrangements, mutations, and amplification, are oncogenic and have been found in various tumors, including anaplastic large cell lymphomas (ALCL) (Lamant et al. 1996; Shiota et al. 1995), neuroblastoma (Iwahara et al. 1997; Lamant et al. 2000), and NSCLC (Inamura et al. 2009; Martelli et al. 2009; Perner et al. 2008; Takeuchi et al. 2009).

The most common genetic alterations are chromosomal rearrangements, such as inversion and translocations that create multiple oncogenic fusion genes involving *ALK* and different partners, including *EML-ALK* (chromosome 2), *RANBP2-ALK* (chromosome 2), *ALK/TFG* (chromosome 3), *NPM1-ALK* (chromosome 5), *ALK/NPM1* (chromosome 5), *ALK/SQSTM1* (chromosome 5), *ALK/KIF5B* (chromosome 10), *ALK/CLTC* (chromosome 17), *ALK/TPM4* (chromosome 19), and *ALK/MSN* (chromosome X).

About 3–5% of NSCLC (the vast majority of which are adenocarcinoma) have an inversion of chromosome 2 that fuses the *ALK* gene with the *echinoderm microtubule-associated protein-like 4* (*EML4*) gene, and results in the EML4-ALK fusion protein (Choi et al. 2008; Inamura et al. 2008; Martelli et al. 2009; Shinmura et al. 2008; Soda et al. 2007). This fusion protein has functions of both ALK tyrosine kinase and the partner protein. The presence of the partner protein allows phosphorylation of ALK without stimulation, which results in constitutively activated ALK and signaling pathways that may abnormally increase cell proliferation and cancer formation. *EML4-ALK*-positive lung cancers are found in patients of all ages, although on average these patients may be somewhat younger. ALK lung cancers are more common in light cigarette smokers or nonsmokers, but a significant number of patients with this disease are current or former cigarette smokers. *EML4-ALK* rearrangement in NSCLC is exclusive and not found in EGFR- or KRAS-mutated tumors.

The gold standard test used to detect *ALK* rearrangements in tumor samples is FISH by the US FDA–approved Vysis *ALK* Break Apart Rearrangement Probe kit from Abbott Molecular. The FISH probe for the *ALK* gene region is labeled with dual colors—Spectrum Orange and Spectrum Green (Figure 5.2c). Normal cells have two signals with fusion color-yellow, whereas abnormal cells exhibit one fusion color (yellow) and separated color of green and orange due to inversion of the chromosome leading to break apart of the two-colored probe. A second method to detect *ALK* rearrangements is IHC using an antibody specifically bound to the ALK protein. Roche Ventana has got the approval of Chinese FDA and EMA to detect ALK protein by IHC using anti-ALK (D5F3) rabbit monoclonal primary antibody in formalin-fixed, paraffin-embedded neoplastic tissue. The result should be interpreted by a qualified pathologist in conjunction with histological examination, relevant clinical information, and appropriate controls. Molecular techniques, such as RT-PCR, can also be used to detect *ALK* gene fusion in lung cancers, but it is not recommended.

ALK inhibitors are available for the treatment of *EML4-ALK*-positive NSCLC. Crizotinib (Xalkori®), produced by Pfizer (www.pfizer.com), was approved by the US FDA for the treatment of late-stage lung cancer patients positive for an *ALK* rearrangement without regard to line of therapy. Compared to chemotherapy, patients receiving crizotinib treatment experienced a higher ORR, longer PFS, and better HRQOL (Kazandjian et al. 2014; Qian et al. 2014).

However, patients treated with ALK inhibitors may also experience acquired resistance due to various mechanisms, such as secondary mutations within the ALK tyrosine kinase domain, amplification of *ALK* fusion gene, amplification of *KIT*, mutations in *KRAS*, increased autophosphorylation of the EGFR, and activation of the EGFR signaling pathways. Another concern for crizotinib is a relatively low penetration of the blood–brain barrier (BBB).

The second-generation ALK inhibitors have potential therapeutic advantages compared with crizotinib. Recently, ceritinib (Zykadia™), produced by Novartis (www.novartis.com), was approved by the US FDA to treat ALK-positive metastatic NSCLC patients who experience acquired resistance or are intolerant to crizotinib. Ceritinib is a more potent and selective ALK inhibitor than crizotinib. Another second-generation ALK inhibitor, alectinib, has been recently approved by the Japanese Ministry of Health, Labour and Welfare (MHLW) for the treatment of ALK-positive NSCLC patients. Alectinib seems to show better penetration of the BBB based on studies for treatment of patients with tumors metastasized to the brain. This ALK inhibitor was granted Breakthrough Therapy Designation (BTD) by the US FDA for ALK-positive patients who progressed while on crizotinib. BTD is designed to expedite the development and review of medicines intended to treat serious diseases and to help ensure patients have access to them through FDA approval as soon as possible.

Additional ALK inhibitors currently undergoing clinical trials include AP26113 (Ariad), NMS-E628 (Nerviano, licensed by Ignyta and renamed RXDX-101), PF-0643922 (Pfizer), TSR-011 (Tesaro), CEP-37440 (Teva), and X-396 (Xcovery). Although these agents are still at the early development stage, the preliminary evidence indicates activity in patients with acquired resistance to crizotinib, increased ALK selectivity and greater potency, and the potential for intracranial disease responses.

In addition to EGFR, ALK, and KRAS, many studies revealed that NSCLC can also be stratified by recurrent "driver" mutations in multiple other oncogenes, including *AKT1*, *BRAF*, *HER2*, *MEK1*, *NRAS*, *PIK3CA*, *RET*, and *ROS1*. These "driver" mutations lead to constitutional activation of mutant signaling proteins that induce and sustain tumorigenesis. These mutations are rarely found concurrently in the same tumor. Mutations can be found in all NSCLC histologies, including adenocarcinoma, squamous cell carcinoma (SCC), and large cell carcinoma. The highest incidence of *EGFR*, *HER2*, *ALK*, *RET*, and *ROS1* mutations has been found in lifelong nonsmokers with adenocarcinoma. As discussed above, targeted small molecule inhibitors are currently available or being developed for specific subsets of patients with a specific molecular profile. Therefore, molecular profiling by NGS of these multiple genes with concurrent FISH for *ALK* and *ROS1* will stratify treatment for directing patients toward beneficial therapies and away from unfavorable ones.

GENETIC TESTING FOR INHERITED CANCER SUSCEPTIBILITY

Genetic susceptibility (also known as genetic predisposition) is defined as the increased likelihood or chance of developing a particular disease due to the presence of one or more gene mutations and/or a family history that indicates an increased risk of the disease. Cancer etiology is a result of interactions of multiple factors, including genetic, medical, environmental, and lifestyle factors. Although only a small portion of cancers are inherited, numerous cancer susceptibility syndromes and their causative genes have been identified (Lindor et al. 2008). Mutations and variants of the cancer susceptibility genes are associated with an increased risk of cancers, including breast cancer, ovarian cancer, colorectal cancer (CRC), thyroid cancer, hereditary diffuse gastric cancer (HDGC), and many other types of cancers (Garber and Offit 2005; Lindor et al. 2008) (see Table 5.3 for selected examples). Genetic testing provides the benefit of early detection and all aspects of cancer management, including prevention, screening, and treatment. An example for clinical intervention is prophylactic salpingo-oophorectomy for individuals carrying mutations in the *BRCA1* or *BRCA2* gene (Daly et al. 2010; Domchek et al. 2010; Kauff et al. 2002). The importance of inherited cancer risk has long been recognized, and the American Society of Clinical Oncology (ASCO) released its first statement on genetic testing in 1996 to give specific recommendations for cancer genetic testing (1996). This policy statement was updated in 2003, 2008, and 2010 (Robson et al. 2010).

Based on the penetrance, which refers to the proportion of individuals carrying a mutation or a variant who will develop cancer, there are three types of mutations

TABLE 5.3

Examples of Inherited Cancer Susceptibility Syndromes

Cancer	Hereditary Syndrome	Major Genes	Gene Function	Mode of Inheritance	Penetrance of Example Variants	Intervention
Breast and ovary	Hereditary breast and ovarian cancer	BRCA1, BRCA2	Tumor suppressor	Dominant	High-penetrance mutations	Mammography, MRI screening, risk-reducing surgery
	Li–Fraumeni syndrome	P53	Tumor suppressor	Dominant		
	Cowden syndrome (breast)	PTEN	Tumor suppressor	Dominant		
	HNPCC/Lynch syndrome (ovary)	MLH1, MSH2, MSH6, PMS2	DNA repair	Dominant	High-penetrance mutations	
	Ataxia telangiectasia (breast and ovary)	ATM	DNA repair	Recessive		
	Peutz–Jeghers syndrome (breast)	STK11	Tumor suppressor	Dominant		
		CHEK2	Tumor suppressor	Dominant	CHEK2 c.1100delC (intermediate-penetrance mutations)	None proven
Colorectal	HNPCC/Lynch syndrome (ovary)	MLH1, MSH2, MSH6, PMS2	DNA repair	Dominant	High-penetrance mutations	Endoscopy, prophylactic colectomy after diagnosis of malignancy

(Continued)

TABLE 5.3 (*Continued*)
Examples of Inherited Cancer Susceptibility Syndromes

Cancer	Hereditary Syndrome	Major Genes	Gene Function	Mode of Inheritance	Penetrance of Example Variants	Intervention
	Familial adenomatous polyposis (FAP)	*APC*	Tumor suppressor	Dominant	High-penetrance mutations	Endoscopy, prophylactic colectomy
					APC p.I1307K (intermediate-penetrance mutations)	None proven for APC p.I1307K
	Attenuated FAP	*APC*	Tumor suppressor	Dominant		
	MUTYH-associated polyposis	*MUTYH*	DNA repair	Recessive		
	Juvenile polyposis	*SMAD4/ DPC4*	Tumor suppressor	Dominant		
Thyroid		*RET*		Dominant	High-penetrance mutations	Prophylactic thyroidectomy
Prostate	Hereditary breast and ovarian cancer	*BRCA1, BRCA2*	Tumor suppressor	Dominant	High-penetrance mutations	
	Li-Fraumeni syndrome	*P53*	Tumor suppressor	Dominant		

and variants of cancer susceptibility genes and/or loci, including high-penetrance mutations, intermediate or moderate risk mutations, and low-penetrance variants (LPVs). High-penetrance mutations usually result in dramatic alteration of the function of the corresponding gene and are associated with a significantly increased risk of related cancer. High-penetrance mutations usually result in autosomal-dominant predispositions recognizable by pedigree analysis. There are also established medical managements for such high-penetrance mutations. For example, endoscopy and prophylactic colectomy are usually recommended for high-penetrance mutations in *APC* with risk of colorectal adenocarcinoma. Intermediate or moderate risk mutations refer to those that result in less dramatic increases in cancer risk: for example, the *APC* p.I1307K mutation in colon cancer risk and the *CHEK2* c.1100delC mutation in breast cancer. Unlike the high frequency of pharmacokinetic or pharmacodynamic polymorphisms, high-penetrance and moderate-penetrance mutations are rare or uncommon. Common in the context of cancer risk testing refers to a frequency of 1% or more, by convention. The third type of variants refers to LPVs that have been identified by large-scale GWASs. Compared to the high-penetrance and moderate-penetrance mutations that usually result in alteration of the function of the corresponding gene product, the LPVs are usually SNPs that are strongly associated with cancer risk in large case–control studies, without changes in the function of the relevant gene product. It is hypothesized that these LPVs are located in close proximity to unidentified causative variants. LPVs associated with cancer risk are common with allele frequencies as high as up to 50% in the population studied, but they only confer a modest increase in cancer risk, usually with per-allele odds ratio of <1.5. The examples of low-penetrance SNPs include *rs10505477* at 8q24 associated with risk for colon cancer and prostate cancer with lifetime relative risk at 1.27 and 1.43, respectively (Haerian et al. 2014), *rs13281615* at 8q24 with lifetime relative risk at 1.21 for breast cancer (Gong et al. 2013), and *rs1219648* at *FGFR* with lifetime relative risk at 1.23 for breast cancer (Andersen et al. 2013). Unlike the high-penetrance mutations, the impact of LPVs on clinical care is not proven, and the clinical validity for the genetic test on PLVs is uncertain. Therefore, they are not currently considered as part of standard oncology or preventive care (Robson et al. 2010).

Traditionally, cancer genetic testing is performed for the most likely genetic causes based on the evaluation of family history and/or personal clinical history such as disease histology and age at diagnosis. However, cancer susceptibility can be very complicated because of the fact that a mutation in one particular gene may increase risk for several types of cancers, or that the risk of one type of cancer may be elevated by mutations in one of a group of different genes. For example, multiple studies indicate that mutations in the *CHEK2* gene confer an increased risk of developing many types of cancer, including breast (Nevanlinna and Bartek 2006), prostate, colon, thyroid, and kidney (Nevanlinna and Bartek 2006). On the other hand, breast cancer risk can be increased by mutations in a number of different genes, including *BRCA1, BRCA2, ATM, CHEK2, CDH1, NF1, MUTYH,* and genes involved in the Fanconi anemia (FA)–BRCA pathway such as *BARD1, BRIP1, MRE11A, NBN, PALB2, RAD50,* and *RAD51C.* Given this complexity, serial testing is time consuming and expensive. Simultaneous testing of a panel of multiple cancer susceptibility genes (also known as multiplex testing) by NGS has the advantage of turnaround time and cost-effectiveness.

It is important to provide appropriate genetic counseling and obtain proper informed consent in the practice of responsible and effective genetic testing for patients with possible inherited susceptibility to cancer. The major indications for genetic counseling and testing include a strong family history of cancer, adequate interpretation of the testing result, and clinical management of the patient and/or family member. Genetic counseling should be provided by the qualified specialist who is credentialed by an appropriate organization, such as the American Board of Genetic Counseling, Inc. (www.abgc.net). Informed consent should include basic elements, such as information on the specific test, implication of a positive or negative test result, possibility that the test will not be informative, options for risk estimation without genetic testing, risk of passing a mutation or predisposition to children, technical accuracy of the test, risks of psychological distress, options and limitations of medical surveillance and screening following the testing, and confidentiality issues. In addition to possible benefits and risks of cancer's early detection and medical management, genetic counseling can also provide information on the possible participation of research studies and/or cooperative studies or registries.

In the following section, we will discuss examples of genetic test on high- and moderate-penetrance mutations of genes associated with the increased risk of breast cancer, CRC, and HDGC.

Breast cancer occurs in both men and women, although it is rare in males. Breast cancer is the most common cancer in women in developed countries with a lifetime risk of about 1 in 8 (~12.29%) women. The NCI estimates that there will be approximately 232,670 new cases of female breast cancer and 2360 new cases of male breast cancer diagnosed in the United States in 2014.[*]

Risk factors for breast cancer include genetic alterations and other factors, such as age, gender, reproductive and menstrual history, radiation, alcohol, hormone therapy, obesity, and benign breast diseases, including atypical ductal hyperplasia (ADH) and lobular carcinoma in situ (LCIS). Although the majority of breast cancer is sporadic, it is estimated that about 5–10% of breast cancers are due to a specific genetic cause. An additional 20–30% of breast cancers are "familial," which means more breast cancer found in a family than expected by chance. Compared to sporadic cases, hereditary breast cancers tend to occur earlier in life and are more likely to involve both breasts.

Germline mutations of the highly penetrant *BRCA1* and *BRCA2* genes cause hereditary breast–ovarian cancer (HBOC) syndrome, accounting for approximately 25–50% of hereditary breast cancer cases (Castera et al. 2014; Easton 1999; van der Groep et al. 2011; Walsh et al. 2010) and about 5–10% of all breast cancers (Campeau et al. 2008). *BRCA1* and *BRCA2* are tumor suppressor genes that have an essential role in both DNA repair and cell-cycle control systems. Mutations in these two highly penetrant genes would increase the risk for developing cancer of the breast, ovaries and fallopian tubes, pancreas, and prostate. Studies suggest that female carriers of *BRCA1* mutations would have a 57–87% risk to develop breast cancer, and a 39–40% risk to develop ovarian cancer by the age of 70 or higher (Antoniou et al. 2003;

---

[*] http://www.cancer.gov/cancertopics/types/breast

Chen and Parmigiani 2007; Ferla et al. 2007; Ford et al. 1998; Janavicius 2010; Tulinius et al. 2002). Similarly, male carriers of the *BRCA1* mutations would have a cumulative breast cancer risk of 1.2% by age 70 (Tai et al. 2007; Thompson et al. 2002). Similar studies suggest that female carriers of the *BRCA2* mutations would have a 45–84% risk of developing breast cancer, and an 11–18% risk of developing ovarian cancer (including primary peritoneal and fallopian tube) by the age of 70 (Antoniou et al. 2003; Chen and Parmigiani 2007; Folkins and Longacre 2013; Ford et al. 1998; Shannon and Chittenden 2012). Male carriers of the *BRCA2* mutations would have up to a 15% prostate cancer risk, and a cumulative breast cancer risk of 6.8% by the ages of 65 and 70, respectively (Kote-Jarai et al. 2011; Shannon and Chittenden 2012; Tai et al. 2007; Thompson et al. 2002). Furthermore, carriers of the *BRCA1/2* mutations are at an increased risk for melanoma and cancer of the pancreas, gallbladder, bile duct, and stomach. Cancer risks are further modified by family history, reproductive choices, lifestyle and environmental factors, and other genetic factors.

Similar to other genetic variations in pharmacogenomics, the frequency of *BRCA1/2* mutations also varies across ethnic groups. *BRCA1/2* mutations are more common in individuals of Ashkenazi Jewish (AJ) descent, with a carrier frequency of 1/40 or 2.6% (Ferla et al. 2007; Janavicius 2010) compared to a frequency of 0.2% or 1/500 in the non-AJ general population. Three founder mutations, c.68_69delAG (BIC: 185delAG), c.5266dupC (BIC: 5382insC) in *BRCA1* gene, and c.5946delT (BIC: 6174delT) in *BRCA2*, account for up to 99% of identified AJ mutations (Ferla et al. 2007; Janavicius 2010). *BRCA1* 185delAG has a frequency of 1% and attributes to 16–20% of breast cancer cases diagnosed before the age of 50; *BRCA1* 5382insC with a frequency of 0.13%; and *BRCA2* 6174delT with a frequency of 1.52% in the AJ population and attributes to 15% of breast cancer cases diagnosed before the age of 50 (Ferla et al. 2007; Janavicius 2010).

The major indications for *BRCA1/2* testing include a personal or family history of early-onset breast cancer (aged 45 years) or bilateral breast cancer, two primary breast cancers or clustering of breast and ovarian cancer, the presence of male breast cancer, ovarian cancer at any age, or at-risk populations such as AJ descent.

Mutations in *BRCA1* and *BRCA2* can explain about 25–50% of hereditary breast cancer (Castera et al. 2014; Easton 1999; van der Groep et al. 2011; Walsh et al. 2010). There are additional genes that have been shown to moderately increase the risk for breast cancer (Castera et al. 2014; Meindl et al. 2011; van der Groep et al. 2011; Walsh and King 2007; Walsh et al. 2010). These genes include *ATM, CHEK2, CDH1, NF1, MUTYH,* and genes involved in the FA-BRCA pathway, such as *BARD1, BRIP1, MRE11A, NBN, PALB2, RAD50,* and *RAD51C*.

Breast cancer susceptibility genetic testing can be performed for *BRCA1* and *BRCA2* by traditional Sanger DNA sequencing or for panels of multiple genes, including those listed above by NGS. Gross deletion or duplication is often disruptive to a gene function, and this can be detected by concurrent CMA. An example of breast cancer susceptibility genetic testing results is shown in Figure 5.3a and b. The female patient was diagnosed as LCIS at age 45, and previous testing showed that the cancer was triple negative for ER, PR, and HER/*neu*. Multiplex testing by NGS revealed that this patient carries a heterozygous germline mutation p.K1347* (c.4039G>T) in the *BRCA1* gene. The nucleotide guanine (G) at coding

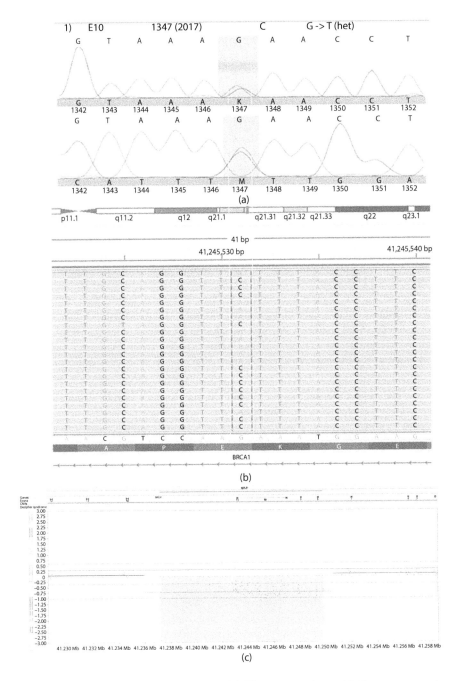

FIGURE 5.3 Sequencing results for *BRCA1*. The same patient with Sanger sequencing results c.4039G > T (a) and NGS sequencing results c.4039G > T (b, reverse strand). The gray arrows in (b) indicate the orientation of *BRCA1* gene on chromosome 17. (c) Chromosomal microarray analysis (CMA) results of *BRCA1* gene with a gross deletion, including coding exons (CDS5-9).

position 4039 is replaced by thymine (T), resulting in replacement of the amino acid lysine (K) at position 1347 by a premature stop codon (TAA). Figure 5.2a shows Sanger sequencing results and Figure 5.2b shows the same mutation by NGS. Please note that Sanger sequencing at the mutation position exhibits a mixed peak of G and T, whereas NGS can exhibit each read. The Sanger sequencing result is for the sense strand, whereas NGS displays the result for the antisense strand to better reflect that the *BRCA1* gene is oriented toward the centromere of chromosome 17. Figure 5.3c shows the CMA result for a different female patient diagnosed with LCIS at the age of 43. This patient carries a heterozygous germline gross deletion, including coding exons 5–9. Both patients are diagnosed with HBOC syndrome.

Similar to breast cancer, CRC can also be caused by a hereditary syndrome. CRC is the third most commonly diagnosed cancer in both men and women, and the second leading cause of cancer-related deaths when combined, affecting about 1 in 20 (5.1%) of men and women in their lifetime (NCI 2015). The NCI estimates that approximately 96,830 (colon) and 40,000 (rectum) new cases will be diagnosed, and that 50,310 CRC deaths will occur in the United States in 2014. Most colon and rectal cancers (over 95%) are adenocarcinoma. There are some other, more rare, types of tumors of the colon and rectum. Although the majority of CRC are sporadic, it is estimated that at least 30% of CRC are familial, a subset of which are caused by hereditary cancer syndromes (Hampel 2009), including Lynch syndrome (also known as hereditary nonpolyposis colorectal cancer, HNPCC), familial adenomatous polyposis (FAP), *MUTYH*-associated polyposis, *PTEN*-Related disorders, *CHEK2*-related cancers, HDGC, Li–Fraumeni syndrome (LFS), Peutz–Jegher syndrome, and juvenile polyposis syndrome.

Lynch syndrome, named after Dr. Henry Lynch (Lynch et al. 1966), is an autosomal-dominant inherited syndrome, accounting for 2–5% of all colon cancer and 2% of uterine cancer. Familial history is the second most common risk factor for CRC. Lynch syndrome is caused by mutations in mismatch repair (MMR) genes, including *EPCAM, MLH1, MSH2, MSH6,* and *PMS2*. Mutations in *MLH1* and *MSH2* account for more than 95% of Lynch syndrome. In addition to a significantly increased risk for colon cancer (60–80% lifetime risk), Lynch syndrome is associated with other cancers, including uterine/endometrial cancer (20–60% lifetime risk in women), stomach cancer (11–19% lifetime risk), and ovarian cancer (4–13% lifetime risk in women). Risks for cancers of the small intestine, hepatobiliary tract, upper urinary tract, and brain are also elevated (Bonadona et al. 2011; Capelle et al. 2010; Engel et al. 2012; Hegde and Roa 2009; Win et al. 2012). Dependent on the mutations in the *MMR* genes, the age incidence and spectrum of cancer in Lynch syndrome vary significantly. Individuals with Lynch syndrome have a higher risk of developing CRC, usually before the age of 50. The diagnosis of Lynch syndrome can be made by molecular genetic testing for germline mutations in the *MMR* genes or by the Amsterdam clinical criteria. The revised Amsterdam criteria (must meet all of them) include three or more relatives with a histologically verified Lynch syndrome (HNPCC)–related cancer, and one of whom is a first-degree relative of the two others; and two successively affected generations, and one or more Lynch syndrome (HNPCC)–related cancers diagnosed before the age of 50.

FAP, accounting for about 1% of all CRC cases (Lipton and Tomlinson 2006), is an autosomal-dominant colon cancer predisposition syndrome, characterized by hundreds to thousands of adenomatous polyps in the internal lining of the colon and rectum. Individuals affected with classic FAP may begin to develop multiple noncancerous colonic polyps (benign polyps) as early as their teenage years, and the number of polyps tends to increase with age (Petersen et al. 1991). However, colon cancer is inevitable without colectomy, and the average age of colon cancer diagnosis in untreated individuals is at 35–40 years of age (Pedace et al. 2008). FAP and attenuated familial adenomatous polyposis (AFAP) are caused by germline mutations in the adenomatous polyposis coli (*APC*) gene located at chromosome 5. Mutations in the *APC* gene can be detected in the peripheral blood of about 80–90% of families with FAP. The missense mutation p.I1307K of *APC* gene is a moderate risk mutation found in the subjects of AJ descent. The hypothesis is that this mutation does not directly cause colon cancer but creates a weak spot in the gene that renders more susceptible to additional genetic changes leading to colon cancer. Family members of the patient with the clinical syndrome of FAP should undergo clinical screening. Variants of FAP are Gardner syndrome, Turcot syndrome, and AFAP.

The *Cadherin-1* (*CDH1*) gene that encodes a calcium-dependent cell–cell adhesion glycoprotein belongs to the cadherin superfamily. Loss of function of the *CDH1* gene is associated with cancer progression by increasing proliferation, invasion, and/or metastasis. Mutations in the *CDH1* gene are associated with several types of cancer, including gastric, breast, colorectal, thyroid, and ovarian. *CDH1* germline mutations have been associated with HDGC and lobular breast cancer in women. The estimated cumulative risk of gastric cancer for *CDH1* mutation carriers by the age of 80 is 67% for men and 83% for women (Pharoah et al. 2001). HDGC patients typically present with diffuse-type gastric cancer with signet ring cells diffusely infiltrating the wall of the stomach and, at late stage, linitis plastica. An elevated risk of lobular breast cancer is also associated with HDGC, with an estimated lifetime breast cancer risk of 39–52% (Guilford et al. 2010).

Mutations in the *TP53* tumor suppressor gene cause LFS and Li–Fraumeni-like syndrome. An individual harboring a *TP53* mutation has a 50% risk of developing cancer by the age of 30 and a lifetime cancer risk of up to 90% (Hwang et al. 2003). The most common tumor types observed in LFS/LFL families include osteosarcoma, soft tissue sarcoma, breast cancer, brain tumor, and adrenocortical carcinoma. Several other types of cancer, including renal cell carcinoma (RCC) and pancreatic cancer, also occur frequently in individuals with LFS (Birch et al. 1994; Gonzalez et al. 2009; Olivier et al. 2003).

Molecular genetic testing can be performed for patients for germline mutations in *APC* gene or *MMR* genes if clinically indicated. Multiplex testing by NGS is also an option for testing patients at increased risk for CRC. For example, Ambry Genetics (www.ambrygen.com) utilizes NGS to offer a genetic testing panel for CRC (named ColoNext). Genes on this ColoNext panel include *APC, BMPR1A, CDH1, CHEK2, EPCAM, MLH1, MSH2, MSH6, MUTYH, PMS2, PTEN, SMAD4, STK11,* and *TP53*. Full gene sequencing and analysis of all coding domains plus at least 5 bases into the 5′- and 3′-ends of all the introns, 5′-UTR and 3′-UTR are performed for 13 of the 14 genes (excluding *EPCAM*). Gross deletion/duplication analysis is performed for all 14 genes. Specific-site analysis is available for individual gene mutations known to be in the family.

Currently, multiplex genetic testing is offered by Ambry Genetics and other commercial or academic laboratories for inherited cancer susceptibility for several types of cancers, including breast, colorectal, ovarian, pancreatic, and more. Multiplex genetic testing for inherited cancer susceptibility is particularly useful when there is significant genetic heterogeneity or there are difficulties to predict which gene may be mutated based on personal phenotype or family history. In some instances, NGS on the whole exome may also reveal mutations in genes that increase the risk of developing cancer. However, multiplex testing or even Sanger sequencing for a single gene or a few genes will also detect variants with limited supporting evidence or conflicting evidence to classify as mutation or benign variants. These variants are usually classified as variant with unknown significance (VUS). It is important for clinicians and patients to understand the significance of findings of mutations (including mutations and variants like pathogenic), VUS, and benign or negative. Pretest and posttest genetic counseling is recommended for all patients who undergo genetic testing.

SUMMARY

The first clinical observations of genetic variations in drug response were documented in the 1950s, involving the tuberculosis drug isoniazid that is metabolized by NAT2. These early studies gave rise to the field of pharmacogenetics and later pharmacogenomics. Pharmacogenomics is now a branch of pharmacology to study genetic variants affecting drug metabolism, transport, molecular targets/pathways, and genetic susceptibility to the diseases. Advances in genetic study and testing, particularly advances in DNA sequencing technologies, greatly increase the pace of pharmacogenomic study and application in clinical settings. The accomplishment of the whole human genome sequencing in 2003 provided the complete mapping and better understanding of all human genes and drove the development of DNA sequencing technologies. The cost to sequence the whole genome has dropped significantly due to the development and application of "next-generation" and "third-generation" sequencing technologies. Although the cost has stayed flat for the past couple of years, it is believed that sequencing cost per genome will drop below $1000 per genome in the near future. High throughput and lower cost in NGS technologies have profound influence on the realization of pharmacogenomics into the realm of clinical practice. Obviously, there are a dramatic number of challenges on interpreting and translating the genetic testing results into clinical practice. Regulations and guidelines are needed to implement pharmacogenomics into patient care. For example, a combined effort from the American Society of Clinical Pathology, College of American Pathologists (CAP), Association for Molecular Pathology, and American College of Medical Genetics and Genomics is currently working on guidelines for molecular testing for selection of CRC patients for targeted and conventional therapy. CAP and the American Society of Hematology published guidelines for molecular testing of acute leukemia. Whole exome sequencing is currently being performed in clinical genetic testing practice. The next step in pharmacogenomics is to analyze the whole exome and whole genome sequence to assess interpatient variability in drug response and side effects and discover novel targets for personalized therapy. In recognition of the importance of pharmacogenomics and revolution of personalized medicine, the Obama administration recently unveiled the Precision Medicine Initiative to improve health and treat disease.* Pharmacogenomics will shed light on producing more powerful drugs by targeting specific diseases with maximized therapeutic effects and minimized side effects. Appropriate drug doses will be determined by pharmacogenomics based on a patient's genetic makeup in addition to his or her body weight and age. In the near future, pharmacogenomics will make the dream of precision medicine a reality.

STUDY QUESTIONS

1. What percentage of differences in drug efficacy and adverse effect is due to genetics?
   a. 5%
   b. 10%
   c. 20–95%
   d. 100%
2. What are single nucleotide polymorphisms (SNPs)? How many SNPs are there in human genomes?
3. How many metabolizer groups are in the *CYP2D6* gene?
4. Give one example of a drug metabolized by thiopurine methyltransferase (TPMT)?
5. What does a drug transporter do?
6. What is a Philadelphia chromosome? What is the fusion gene on a Ph chromosome?
7. What is the gold standard test to detect *ALK* rearrangement?
8. List the two most common mutations in *EGFR* that are responsible for about 90% of NSCLC cases?
9. Germline mutations of which genes cause hereditary breast–ovarian cancer (HBOC) syndrome?

Answers
1. c
2. SNPs are the most common form of polymorphisms, defined as the occurrence at an allele frequency of at least 1/1000 bases in the general population. There are an estimated 1.4 million SNPs in the human genome.
3. Four. Poor metabolizer, intermediate metabolizer, ultrarapid metabolizer, and extensive metabolizer.
4. 6-Mercaptopurine is used to treat acute lymphoblastic leukemia of childhood.
5. To transport drug across the cell membrane.

---

* https://www.whitehouse.gov/the-press-office/2015/01/30/fact-sheet-president-obama-s-precision-medicine-initiative

6. The derivative chromosome 22 resulted from translocation between chromosomes 9 and 22 is known as Philadelphia chromosome. The fusion gene on Ph is *BCR/ABL*, which serves as drug target for the treatment of CML.
7. Fluorescence in situ hybridization (FISH).
8. p.L858R point mutation in exon 21 and small deletions in exon 19 (Del19).
9. *BRCA1* and *BRCA2*.

REFERENCES

Advani, A.S., and Pendergast, A.M. (2002). Bcr-Abl variants: Biological and clinical aspects. *Leukemia Research 26*, 713–720.
American Society of Clinical Oncology. (1996). Statement of the American Society of Clinical Oncology: Genetic testing for cancer susceptibility, Adopted on February 20, 1996. *Journal of Clinical Oncology 14*, 1730–1736.
American Society of Clinical Oncology. (2003). American Society of Clinical Oncology policy statement update: Genetic testing for cancer susceptibility. *Journal of Clinical Oncology 21*, 2397–2406.
Andersen, S.W., Trentham-Dietz, A., Figueroa, J.D., Titus, L.J., Cai, Q., Long, J., Hampton, J.M., Egan, K.M., and Newcomb, P.A. (2013). Breast cancer susceptibility associated with rs1219648 (fibroblast growth factor receptor 2) and postmenopausal hormone therapy use in a population-based United States study. *Menopause 20*, 354–358.
Ando, Y., Saka, H., Ando, M., Sawa, T., Muro, K., Ueoka, H., Yokoyama, A., Saitoh, S., Shimokata, K., and Hasegawa, Y. (2000). Polymorphisms of UDP-glucuronosyltransferase gene and irinotecan toxicity: A pharmacogenetic analysis. *Cancer Research 60*, 6921–6926.
Antoniou, A., Pharoah, P.D., Narod, S., Risch, H.A., Eyfjord, J.E., Hopper, J.L., Loman, N., et al. (2003). Average risks of breast and ovarian cancer associated with BRCA1 or BRCA2 mutations detected in case series unselected for family history: A combined analysis of 22 studies. *American Journal of Human Genetics 72*, 1117–1130.
Baker, K.E., and Parker, R. (2004). Nonsense-mediated mRNA decay: Terminating erroneous gene expression. *Current Opinion in Cell Biology 16*, 293–299.
Ball, S.E., Scatina, J., Kao, J., Ferron, G.M., Fruncillo, R., Mayer, P., Weinryb, I., et al. (1999). Population distribution and effects on drug metabolism of a genetic variant in the 5′ promoter region of CYP3A4. *Clinical Pharmacology and Therapeutics 66*, 288–294.
Baselga, J., Norton, L., Albanell, J., Kim, Y.M., and Mendelsohn, J. (1998). Recombinant humanized anti-HER2 antibody (Herceptin) enhances the antitumor activity of paclitaxel and doxorubicin against HER2/neu overexpressing human breast cancer xenografts. *Cancer Research 58*, 2825–2831.
Bell, D.A., Badawi, A.F., Lang, N.P., Ilett, K.F., Kadlubar, F.F., and Hirvonen, A. (1995). Polymorphism in the N-acetyltransferase 1 (NAT1) polyadenylation signal: Association of NAT1*10 allele with higher N-acetylation activity in bladder and colon tissue. *Cancer Research 55*, 5226–5229.
Beutler, E., Gelbart, T., and Demina, A. (1998). Racial variability in the UDP-glucuronosyltransferase 1 (UGT1A1) promoter: A balanced polymorphism for regulation of bilirubin metabolism? *Proceedings of the National Academy of Sciences of the United States of America 95*, 8170–8174.
Beyth, R.J., Quinn, L., and Landefeld, C.S. (2000). A multicomponent intervention to prevent major bleeding complications in older patients receiving warfarin. A randomized, controlled trial. *Annals of Internal Medicine 133*, 687–695.

Birch, J.M., Hartley, A.L., Tricker, K.J., Prosser, J., Condie, A., Kelsey, A.M., Harris, M., et al. (1994). Prevalence and diversity of constitutional mutations in the p53 gene among 21 Li-Fraumeni families. *Cancer Research 54*, 1298–1304.

Bonadona, V., Bonaiti, B., Olschwang, S., Grandjouan, S., Huiart, L., Longy, M., Guimbaud, R., et al. (2011). Cancer risks associated with germline mutations in MLH1, MSH2, and MSH6 genes in Lynch syndrome. *JAMA 305*, 2304–2310.

Bournissen, F.G., Moretti, M.E., Juurlink, D.N., Koren, G., Walker, M., and Finkelstein, Y. (2009). Polymorphism of the MDR1/ABCB1 C3435T drug-transporter and resistance to anticonvulsant drugs: A meta-analysis. *Epilepsia 50*, 898–903.

Bradford, L.D. (2002). CYP2D6 allele frequency in European Caucasians, Asians, Africans and their descendants. *Pharmacogenomics 3*, 229–243.

Cagliani, R., Fumagalli, M., Pozzoli, U., Riva, S., Comi, G.P., Torri, F., Macciardi, F., Bresolin, N., and Sironi, M. (2009). Diverse evolutionary histories for beta-adrenoreceptor genes in humans. *American Journal of Human Genetics 85*, 64–75.

Campeau, P.M., Foulkes, W.D., and Tischkowitz, M.D. (2008). Hereditary breast cancer: New genetic developments, new therapeutic avenues. *Human Genetics 124*, 31–42.

Capelle, L.G., Van Grieken, N.C., Lingsma, H.F., Steyerberg, E.W., Klokman, W.J., Bruno, M.J., Vasen, H.F., and Kuipers, E.J. (2010). Risk and epidemiological time trends of gastric cancer in Lynch syndrome carriers in the Netherlands. *Gastroenterology 138*, 487–492.

Castera, L., Krieger, S., Rousselin, A., Legros, A., Baumann, J.J., Bruet, O., Brault, B., et al. (2014). Next-generation sequencing for the diagnosis of hereditary breast and ovarian cancer using genomic capture targeting multiple candidate genes. *European Journal of Human Genetics 22*, 1305–1313.

Chen, S., and Parmigiani, G. (2007). Meta-analysis of BRCA1 and BRCA2 penetrance. *Journal of Clinical Oncology 25*, 1329–1333.

Choi, Y.L., Takeuchi, K., Soda, M., Inamura, K., Togashi, Y., Hatano, S., Enomoto, M., et al. (2008). Identification of novel isoforms of the EML4-ALK transforming gene in non-small cell lung cancer. *Cancer Research 68*, 4971–4976.

Daly, M.B., Axilbund, J.E., Buys, S., Crawford, B., Farrell, C.D., Friedman, S., Garber, J.E., et al. (2010). Genetic/familial high-risk assessment: Breast and ovarian. *Journal of the National Comprehensive Cancer Network 8*, 562–594.

de Wildt, S.N., Kearns, G.L., Leeder, J.S., and van den Anker, J.N. (1999). Cytochrome P450 3A: Ontogeny and drug disposition. *Clinical Pharmacokinetics 37*, 485–505.

Desta, Z., Zhao, X., Shin, J.G., and Flockhart, D.A. (2002). Clinical significance of the cytochrome P450 2C19 genetic polymorphism. *Clinical Pharmacokinetics 41*, 913–958.

Domchek, S.M., Friebel, T.M., Singer, C.F., Evans, D.G., Lynch, H.T., Isaacs, C., Garber, J.E., et al. (2010). Association of risk-reducing surgery in BRCA1 or BRCA2 mutation carriers with cancer risk and mortality. *JAMA 304*, 967–975.

Druker, B.J., Tamura, S., Buchdunger, E., Ohno, S., Segal, G.M., Fanning, S., Zimmermann, J., and Lydon, N.B. (1996). Effects of a selective inhibitor of the Abl tyrosine kinase on the growth of Bcr-Abl positive cells. *Nature Medicine 2*, 561–566.

Drysdale, C.M., McGraw, D.W., Stack, C.B., Stephens, J.C., Judson, R.S., Nandabalan, K., Arnold, K., Ruano, G., and Liggett, S.B. (2000). Complex promoter and coding region beta 2-adrenergic receptor haplotypes alter receptor expression and predict in vivo responsiveness. *Proceedings of the National Academy of Sciences of the United States of America 97*, 10483–10488.

Easton, D.F. (1999). How many more breast cancer predisposition genes are there? *Breast Cancer Research 1*, 14–17.

Eichelbaum, M., Ingelman-Sundberg, M., and Evans, W.E. (2006). Pharmacogenomics and individualized drug therapy. *Annual Review of Medicine 57*, 119–137.

Endres, C.J., Hsiao, P., Chung, F.S., and Unadkat, J.D. (2006). The role of transporters in drug interactions. *European Journal of Pharmaceutical Sciences 27*, 501–517.

Engel, C., Loeffler, M., Steinke, V., Rahner, N., Holinski-Feder, E., Dietmaier, W., Schackert, H.K., et al. (2012). Risks of less common cancers in proven mutation carriers with Lynch syndrome. *Journal of Clinical Oncology 30*, 4409–4415.

Evans, D.A., and White, T.A. (1964). Human acetylation polymorphism. *The Journal of Laboratory and Clinical Medicine 63*, 394–403.

Evans, W.E., and Relling, M.V. (1999). Pharmacogenomics: Translating functional genomics into rational therapeutics. *Science 286*, 487–491.

Farheen, S., Sengupta, S., Santra, A., Pal, S., Dhali, G.K., Chakravorty, M., Majumder, P.P., and Chowdhury, A. (2006). Gilbert's syndrome: High frequency of the (TA)7 TAA allele in India and its interaction with a novel CAT insertion in promoter of the gene for bilirubin UDP-glucuronosyltransferase 1 gene. *World Journal of Gastroenterology 12*, 2269–2275.

FDA Biomarker, D. US FDA-Approved pharmacogenomic biomarkers in drug labels. www.fdagov/Drugs/ScienceResearch/ResearchAreas/Pharmacogenetics/ucm083378htm

FDA Definition, G. E15 Definitions for genomic biomarkers, pharmacogenomics, pharmacogenetics, genomic data and sample coding catergories. http://www.fdagov/downloads/drugs/guidancecomplianceregulatoryinformation/guidances/ucm073162pdf

Fenech, A., and Hall, I.P. (2002). Pharmacogenetics of asthma. *British Journal of Clinical Pharmacology 53*, 3–15.

Ferla, R., Calo, V., Cascio, S., Rinaldi, G., Badalamenti, G., Carreca, I., Surmacz, E., Colucci, G., Bazan, V., and Russo, A. (2007). Founder mutations in BRCA1 and BRCA2 genes. *Annals of Oncology 18*(Suppl 6), vi93–98.

Folkins, A.K., and Longacre, T.A. (2013). Hereditary gynaecological malignancies: Advances in screening and treatment. *Histopathology 62*, 2–30.

Ford, D., Easton, D.F., Stratton, M., Narod, S., Goldgar, D., Devilee, P., Bishop, D.T., et al. (1998). Genetic heterogeneity and penetrance analysis of the BRCA1 and BRCA2 genes in breast cancer families. The Breast Cancer Linkage Consortium. *American Journal of Human Genetics 62*, 676–689.

Fretland, A.J., Leff, M.A., Doll, M.A., and Hein, D.W. (2001). Functional characterization of human N-acetyltransferase 2 (NAT2) single nucleotide polymorphisms. *Pharmacogenetics 11*, 207–215.

Garber, J.E., and Offit, K. (2005). Hereditary cancer predisposition syndromes. *Journal of Clinical Oncology 23*, 276–292.

Giacomini, K.M., Balimane, P.V., Cho, S.K., Eadon, M., Edeki, T., Hillgren, K.M., Huang, S.M., et al. (2013). International Transporter Consortium commentary on clinically important transporter polymorphisms. *Clinical Pharmacology and Therapeutics 94*, 23–26.

Gong, W.F., Zhong, J.H., Xiang, B.D., Ma, L., You, X.M., Zhang, Q.M., and Li, L.Q. (2013). Single nucleotide polymorphism 8q24 rs13281615 and risk of breast cancer: Meta-analysis of more than 100,000 cases. *PLoS One 8*, e60108.

Gonzalez, K.D., Noltner, K.A., Buzin, C.H., Gu, D., Wen-Fong, C.Y., Nguyen, V.Q., Han, J.H., et al. (2009). Beyond Li Fraumeni Syndrome: Clinical characteristics of families with p53 germline mutations. *Journal of Clinical Oncology 27*, 1250–1256.

Gonzalez-Covarrubias, V., Zhang, J., Kalabus, J.L., Relling, M.V., and Blanco, J.G. (2009). Pharmacogenetics of human carbonyl reductase 1 (CBR1) in livers from black and white donors. *Drug Metabolism and Disposition 37*, 400–407.

Guengerich, F.P. (2008). Cytochrome p450 and chemical toxicology. *Chemical Research in Toxicology 21*, 70–83.

Guilford, P., Humar, B., and Blair, V. (2010). Hereditary diffuse gastric cancer: Translation of CDH1 germline mutations into clinical practice. *Gastric Cancer 13*, 1–10.

Haerian, M.S., Haerian, B.S., Rooki, H., Molanaei, S., Kosari, F., Obohhat, M., Hosseinpour, P., et al. (2014). Association of 8q24.21 rs10505477-rs6983267 haplotype and age at diagnosis of colorectal cancer. *Asian Pacific Journal of Cancer Prevention 15*, 369–374.

Hamburg, M.A., and Collins, F.S. (2010). The path to personalized medicine. *The New England Journal of Medicine 363*, 301–304.

Hampel, H. (2009). Genetic testing for hereditary colorectal cancer. *Surgical Oncology Clinics of North America 18*, 687–703.

Hegde, M.R., and Roa, B.B. (2009). Genetic testing for hereditary nonpolyposis colorectal cancer (HNPCC). *Current Protocols in Human Genetics 61*, 10.12.1–10.12.28.

Hein, D.W., Grant, D.M., and Sim, E. (2000). Update on consensus arylamine N-acetyltransferase gene nomenclature. *Pharmacogenetics 10*, 291–292.

Heisterkamp, N., and Groffen, J. (1991). Molecular insights into the Philadelphia translocation. *Hematologic Pathology 5*, 1–10.

Hicks, J.K., Swen, J.J., Thorn, C.F., Sangkuhl, K., Kharasch, E.D., Ellingrod, V.L., Skaar, T.C., et al. (2013). Clinical Pharmacogenetics Implementation Consortium guideline for CYP2D6 and CYP2C19 genotypes and dosing of tricyclic antidepressants. *Clinical Pharmacology and Therapeutics 93*, 402–408.

Higashi, M.K., Veenstra, D.L., Kondo, L.M., Wittkowsky, A.K., Srinouanprachanh, S.L., Farin, F.M., and Rettie, A.E. (2002). Association between CYP2C9 genetic variants and anticoagulation-related outcomes during warfarin therapy. *JAMA 287*, 1690–1698.

Honchel, R., Aksoy, I.A., Szumlanski, C., Wood, T.C., Otterness, D.M., Wieben, E.D., and Weinshilboum, R.M. (1993). Human thiopurine methyltransferase: Molecular cloning and expression of T84 colon carcinoma cell cDNA. *Molecular Pharmacology 43*, 878–887.

Hwang, S.J., Lozano, G., Amos, C.I., and Strong, L.C. (2003). Germline p53 mutations in a cohort with childhood sarcoma: Sex differences in cancer risk. *American Journal of Human Genetics 72*, 975–983.

Imai, Y., Nakane, M., Kage, K., Tsukahara, S., Ishikawa, E., Tsuruo, T., Miki, Y., and Sugimoto, Y. (2002). C421A polymorphism in the human breast cancer resistance protein gene is associated with low expression of Q141K protein and low-level drug resistance. *Molecular Cancer Therapeutics 1*, 611–616.

Inamura, K., Takeuchi, K., Togashi, Y., Hatano, S., Ninomiya, H., Motoi, N., Mun, M.Y., et al. (2009). EML4-ALK lung cancers are characterized by rare other mutations, a TTF-1 cell lineage, an acinar histology, and young onset. *Modern Pathology 22*, 508–515.

Inamura, K., Takeuchi, K., Togashi, Y., Nomura, K., Ninomiya, H., Okui, M., Satoh, Y., et al. (2008). EML4-ALK fusion is linked to histological characteristics in a subset of lung cancers. *Journal of Thoracic Oncology 3*, 13–17.

Ingelman-Sundberg, M. (2011). The human cytochrome P450 (CYP) allele nomenclature database. CYP2C19 allele nomenclature. http://www.cypalleleskise/cyp2c19htm

Ingelman-Sundberg, M., Oscarson, M., and McLellan, R.A. (1999). Polymorphic human cytochrome P450 enzymes: An opportunity for individualized drug treatment. *Trends in Pharmacological Sciences 20*, 342–349.

Ingelman-Sundberg, M., Sim, S.C., Gomez, A., and Rodriguez-Antona, C. (2007). Influence of cytochrome P450 polymorphisms on drug therapies: Pharmacogenetic, pharmaco-epigenetic and clinical aspects. *Pharmacology & Therapeutics 116*, 496–526.

Innocenti, F., Vokes, E.E., and Ratain, M.J. (2006). Irinogenetics: What is the right star? *Journal of Clinical Oncology 24*, 2221–2224.

Iwahara, T., Fujimoto, J., Wen, D., Cupples, R., Bucay, N., Arakawa, T., Mori, S., Ratzkin, B., and Yamamoto, T. (1997). Molecular characterization of ALK, a receptor tyrosine kinase expressed specifically in the nervous system. *Oncogene 14*, 439–449.

Janavicius, R. (2010). Founder BRCA1/2 mutations in the Europe: Implications for hereditary breast-ovarian cancer prevention and control. *The EPMA Journal 1*, 397–412.

Kalow, W., Tang, B.K., and Endrenyi, L. (1998). Hypothesis: Comparisons of inter- and intra-individual variations can substitute for twin studies in drug research. *Pharmacogenetics 8*, 283–289.

Kauff, N.D., Satagopan, J.M., Robson, M.E., Scheuer, L., Hensley, M., Hudis, C.A., Ellis, N.A., et al. (2002). Risk-reducing salpingo-oophorectomy in women with a BRCA1 or BRCA2 mutation. *The New England Journal of Medicine 346*, 1609–1615.

Kazandjian, D., Blumenthal, G.M., Chen, H.Y., He, K., Patel, M., Justice, R., Keegan, P., and Pazdur, R. (2014). FDA approval summary: Crizotinib for the treatment of metastatic non-small cell lung cancer with anaplastic lymphoma kinase rearrangements. *The Oncologist 19*, e5–11.

Khurana, V., Minocha, M., Pal, D., and Mitra, A.K. (2014). Role of OATP-1B1 and/or OATP-1B3 in hepatic disposition of tyrosine kinase inhibitors. *Drug Metabolism and Drug Interactions 29*, 179–190.

Kirchheiner, J., and Brockmoller, J. (2005). Clinical consequences of cytochrome P450 2C9 polymorphisms. *Clinical Pharmacology and Therapeutics 77*, 1–16.

Komar, A.A. (2007). Silent SNPs: Impact on gene function and phenotype. *Pharmacogenomics 8*, 1075–1080.

Kote-Jarai, Z., Leongamornlert, D., Saunders, E., Tymrakiewicz, M., Castro, E., Mahmud, N., Guy, M., et al. (2011). BRCA2 is a moderate penetrance gene contributing to young-onset prostate cancer: Implications for genetic testing in prostate cancer patients. *British Journal of Cancer 105*, 1230–1234.

Krynetski, E.Y., Tai, H.L., Yates, C.R., Fessing, M.Y., Loennechen, T., Schuetz, J.D., Relling, M.V., and Evans, W.E. (1996). Genetic polymorphism of thiopurine S-methyltransferase: Clinical importance and molecular mechanisms. *Pharmacogenetics 6*, 279–290.

Ladanyi, M., and Pao, W. (2008). Lung adenocarcinoma: Guiding EGFR-targeted therapy and beyond. *Modern Pathology 21*(Suppl 2), S16–22.

Lamant, L., Meggetto, F., al Saati, T., Brugieres, L., de Paillerets, B.B., Dastugue, N., Bernheim, A., et al. (1996). High incidence of the t(2;5)(p23;q35) translocation in anaplastic large cell lymphoma and its lack of detection in Hodgkin's disease. Comparison of cytogenetic analysis, reverse transcriptase-polymerase chain reaction, and P-80 immunostaining. *Blood 87*, 284–291.

Lamant, L., Pulford, K., Bischof, D., Morris, S.W., Mason, D.Y., Delsol, G., and Mariame, B. (2000). Expression of the ALK tyrosine kinase gene in neuroblastoma. *The American Journal of Pathology 156*, 1711–1721.

Lankisch, T.O., Behrens, G., Ehmer, U., Mobius, U., Rockstroh, J., Wehmeier, M., Kalthoff, S., et al. (2009). Gilbert's syndrome and hyperbilirubinemia in protease inhibitor therapy—An extended haplotype of genetic variants increases risk in indinavir treatment. *Journal of Hepatology 50*, 1010–1018.

Lazarou, J., Pomeranz, B.H., and Corey, P.N. (1998). Incidence of adverse drug reactions in hospitalized patients: A meta-analysis of prospective studies. *JAMA 279*, 1200–1205.

Lee, C.R., Goldstein, J.A., and Pieper, J.A. (2002). Cytochrome P450 2C9 polymorphisms: A comprehensive review of the in-vitro and human data. *Pharmacogenetics 12*, 251–263.

Leschziner, G.D., Andrew, T., Leach, J.P., Chadwick, D., Coffey, A.J., Balding, D.J., Bentley, D.R., Pirmohamed, M., and Johnson, M.R. (2007). Common ABCB1 polymorphisms are not associated with multidrug resistance in epilepsy using a gene-wide tagging approach. *Pharmacogenetics and Genomics 17*, 217–220.

Li, C.Y., Zhang, J., Chu, J.H., Xu, M.J., Ju, W.Z., Liu, F., and Jian-Dong, Z. (2014). A correlative study of polymorphisms of CYP2C19 and MDR1 C3435T with the pharmacokinetic profiles of lansoprazole and its main metabolites following single oral administration in healthy adult Chinese subjects. *European Journal of Drug Metabolism and Pharmacokinetics 39*, 121–128.

Lichty, B.D., Keating, A., Callum, J., Yee, K., Croxford, R., Corpus, G., Nwachukwu, B., Kim, P., Guo, J., and Kamel-Reid, S. (1998). Expression of p210 and p190 BCR-ABL due to alternative splicing in chronic myelogenous leukaemia. *British Journal of Haematology 103*, 711–715.

Lindor, N.M., McMaster, M.L., Lindor, C.J., Greene, M.H., National Cancer Institute, Division of Cancer Prevention, Community Oncology, and Prevention Trials Research Group. (2008). Concise handbook of familial cancer susceptibility syndromes—Second edition. *Journal of the National Cancer Institute Monographs 38*, 1–93.

Lipton, L., and Tomlinson, I. (2006). The genetics of FAP and FAP-like syndromes. *Familial Cancer 5*, 221–226.

Lynch, H.T., Shaw, M.W., Magnuson, C.W., Larsen, A.L., and Krush, A.J. (1966). Hereditary factors in cancer. Study of two large midwestern kindreds. *Archives of Internal Medicine 117*, 206–212.

Lynch, T.J., Bell, D.W., Sordella, R., Gurubhagavatula, S., Okimoto, R.A., Brannigan, B.W., Harris, P.L., et al. (2004). Activating mutations in the epidermal growth factor receptor underlying responsiveness of non-small-cell lung cancer to gefitinib. *The New England Journal of Medicine 350*, 2129–2139.

Mackenzie, P.I., Bock, K.W., Burchell, B., Guillemette, C., Ikushiro, S., Iyanagi, T., Miners, J.O., Owens, I.S., and Nebert, D.W. (2005). Nomenclature update for the mammalian UDP glycosyltransferase (UGT) gene superfamily. *Pharmacogenetics and Genomics 15*, 677–685.

Martelli, M.P., Sozzi, G., Hernandez, L., Pettirossi, V., Navarro, A., Conte, D., Gasparini, P., et al. (2009). EML4-ALK rearrangement in non-small cell lung cancer and non-tumor lung tissues. *The American Journal of Pathology 174*, 661–670.

Martis, S., Peter, I., Hulot, J.S., Kornreich, R., Desnick, R.J., and Scott, S.A. (2013). Multi-ethnic distribution of clinically relevant CYP2C genotypes and haplotypes. *The Pharmacogenomics Journal 13*, 369–377.

McGraw, J., and Waller, D. (2012). Cytochrome P450 variations in different ethnic populations. *Expert Opinion on Drug Metabolism & Toxicology 8*, 371–382.

McLeod, H.L., Krynetski, E.Y., Relling, M.V., and Evans, W.E. (2000). Genetic polymorphism of thiopurine methyltransferase and its clinical relevance for childhood acute lymphoblastic leukemia. *Leukemia 14*, 567–572.

Meindl, A., Ditsch, N., Kast, K., Rhiem, K., and Schmutzler, R.K. (2011). Hereditary breast and ovarian cancer: New genes, new treatments, new concepts. *Deutsches Arzteblatt International 108*, 323–330.

Narita, I., Goto, S., Saito, N., Song, J., Omori, K., Kondo, D., Sakatsume, M., and Gejyo, F. (2003). Angiotensinogen gene variation and renoprotective efficacy of renin–angiotensin system blockade in IgA nephropathy. *Kidney International 64*, 1050–1058.

NCI (National Cancer Institute). (2015). Cancer Stat Fact Sheets. http://seercancergov/

Nevanlinna, H., and Bartek, J. (2006). The CHEK2 gene and inherited breast cancer susceptibility. *Oncogene 25*, 5912–5919.

Nowell, P.C., and Hungerford, D.A. (1960). A minute chromosome in human chronic granulocytic leukemia. *Science 132*, 1497–1501.

Olivier, M., Goldgar, D.E., Sodha, N., Ohgaki, H., Kleihues, P., Hainaut, P., and Eeles, R.A. (2003). Li-Fraumeni and related syndromes: Correlation between tumor type, family structure, and TP53 genotype. *Cancer Research 63*, 6643–6650.

Owens, M.A., Horten, B.C., and Da Silva, M.M. (2004). HER2 amplification ratios by fluorescence in situ hybridization and correlation with immunohistochemistry in a cohort of 6556 breast cancer tissues. *Clinical Breast Cancer 5*, 63–69.

Paez, J.G., Janne, P.A., Lee, J.C., Tracy, S., Greulich, H., Gabriel, S., Herman, P., et al. (2004). EGFR mutations in lung cancer: Correlation with clinical response to gefitinib therapy. *Science 304*, 1497–1500.

Pakakasama, S., Kajanachumpol, S., Kanjanapongkul, S., Sirachainan, N., Meekaewkunchorn, A., Ningsanond, V., and Hongeng, S. (2008). Simple multiplex RT-PCR for identifying common fusion transcripts in childhood acute leukemia. *International Journal of Laboratory Hematology 30*, 286–291.

Pao, W., Miller, V., Zakowski, M., Doherty, J., Politi, K., Sarkaria, I., Singh, B., et al. (2004). EGF receptor gene mutations are common in lung cancers from "never smokers" and are associated with sensitivity of tumors to gefitinib and erlotinib. *Proceedings of the National Academy of Sciences of the United States of America 101*, 13306–13311.

Pao, W., Miller, V.A., Politi, K.A., Riely, G.J., Somwar, R., Zakowski, M.F., Kris, M.G., and Varmus, H. (2005). Acquired resistance of lung adenocarcinomas to gefitinib or erlotinib is associated with a second mutation in the EGFR kinase domain. *PLoS Medicine 2*, e73.

Pedace, L., Majore, S., Megiorni, F., Binni, F., De Bernardo, C., Antigoni, I., Preziosi, N., Mazzilli, M.C., and Grammatico, P. (2008). Identification of a novel duplication in the APC gene using multiple ligation probe amplification in a patient with familial adenomatous polyposis. *Cancer Genetics and Cytogenetics 182*, 130–135.

Pegram, M.D., Lipton, A., Hayes, D.F., Weber, B.L., Baselga, J.M., Tripathy, D., Baly, D., et al. (1998). Phase II study of receptor-enhanced chemosensitivity using recombinant humanized anti-p185HER2/neu monoclonal antibody plus cisplatin in patients with HER2/neu-overexpressing metastatic breast cancer refractory to chemotherapy treatment. *Journal of Clinical Oncology 16*, 2659–2671.

Perera, M.A., Innocenti, F., and Ratain, M.J. (2008). Pharmacogenetic testing for uridine diphosphate glucuronosyltransferase 1A1 polymorphisms: Are we there yet? *Pharmacotherapy 28*, 755–768.

Perner, S., Wagner, P.L., Demichelis, F., Mehra, R., Lafargue, C.J., Moss, B.J., Arbogast, S., et al. (2008). EML4-ALK fusion lung cancer: A rare acquired event. *Neoplasia 10*, 298–302.

Petersen, G.M., Slack, J., and Nakamura, Y. (1991). Screening guidelines and premorbid diagnosis of familial adenomatous polyposis using linkage. *Gastroenterology 100*, 1658–1664.

Pharoah, P.D., Guilford, P., Caldas, C., and International Gastric Cancer Linkage Consortium. (2001). Incidence of gastric cancer and breast cancer in CDH1 (E-cadherin) mutation carriers from hereditary diffuse gastric cancer families. *Gastroenterology 121*, 1348–1353.

Qian, H., Gao, F., Wang, H., and Ma, F. (2014). The efficacy and safety of crizotinib in the treatment of anaplastic lymphoma kinase-positive non-small cell lung cancer: A meta-analysis of clinical trials. *BMC Cancer 14*, 683.

Rendic, S., and Di Carlo, F.J. (1997). Human cytochrome P450 enzymes: A status report summarizing their reactions, substrates, inducers, and inhibitors. *Drug Metabolism Reviews 29*, 413–580.

Rettie, A.E., Wienkers, L.C., Gonzalez, F.J., Trager, W.F., and Korzekwa, K.R. (1994). Impaired (S)-warfarin metabolism catalysed by the R144C allelic variant of CYP2C9. *Pharmacogenetics 4*, 39–42.

Richards, C.S., Bale, S., Bellissimo, D.B., Das, S., Grody, W.W., Hegde, M.R., Lyon, E., Ward, B.E., and Molecular Subcommittee of the ACMG Laboratory Quality Assurance Committee. (2008). ACMG recommendations for standards for interpretation and reporting of sequence variations: Revisions 2007. *Genetics in Medicine 10*, 294–300.

Riely, G.J., Marks, J., and Pao, W. (2009). KRAS mutations in non-small cell lung cancer. *Proceedings of the American Thoracic Society 6*, 201–205.

Robson, M.E., Storm, C.D., Weitzel, J., Wollins, D.S., Offit, K., and American Society of Clinical Oncology. (2010). American Society of Clinical Oncology policy statement update: Genetic and genomic testing for cancer susceptibility. *Journal of Clinical Oncology 28*, 893–901.

Ross, D.D., Yang, W., Abruzzo, L.V., Dalton, W.S., Schneider, E., Lage, H., Dietel, M., Greenberger, L., Cole, S.P., and Doyle, L.A. (1999). Atypical multidrug resistance: Breast cancer resistance protein messenger RNA expression in mitoxantrone-selected cell lines. *Journal of the National Cancer Institute 91*, 429–433.

Roy, J.N., Lajoie, J., Zijenah, L.S., Barama, A., Poirier, C., Ward, B.J., and Roger, M. (2005). CYP3A5 genetic polymorphisms in different ethnic populations. *Drug Metabolism and Disposition 33*, 884–887.

Sachidanandam, R., Weissman, D., Schmidt, S.C., Kakol, J.M., Stein, L.D., Marth, G., Sherry, S., et al. (2001). A map of human genome sequence variation containing 1.42 million single nucleotide polymorphisms. *Nature 409*, 928–933.

Sanderson, S., Emery, J., and Higgins, J. (2005). CYP2C9 gene variants, drug dose, and bleeding risk in warfarin-treated patients: A HuGEnet systematic review and meta-analysis. *Genetics in Medicine 7*, 97–104.

Sanger, F., Air, G.M., Barrell, B.G., Brown, N.L., Coulson, A.R., Fiddes, C.A., Hutchison, C.A., Slocombe, P.M., and Smith, M. (1977a). Nucleotide sequence of bacteriophage phi X174 DNA. *Nature 265*, 687–695.

Sanger, F., Nicklen, S., and Coulson, A.R. (1977b). DNA sequencing with chain-terminating inhibitors. *Proceedings of the National Academy of Sciences of the United States of America 74*, 5463–5467.

Santos, P.C., Soares, R.A., Nascimento, R.M., Machado-Coelho, G.L., Mill, J.G., Krieger, J.E., and Pereira, A.C. (2011). SLCO1B1 rs4149056 polymorphism associated with statin-induced myopathy is differently distributed according to ethnicity in the Brazilian general population: Amerindians as a high risk ethnic group. *BMC Medical Genetics 12*, 136.

Scott, S.A., Sangkuhl, K., Gardner, E.E., Stein, C.M., Hulot, J.S., Johnson, J.A., Roden, D.M., Klein, T.E., Shuldiner, A.R., and Clinical Pharmacogenetics Implementation Consortium. (2011). Clinical Pharmacogenetics Implementation Consortium guidelines for cytochrome P450-2C19 (CYP2C19) genotype and clopidogrel therapy. *Clinical Pharmacology and Therapeutics 90*, 328–332.

Shannon, K.M., and Chittenden, A. (2012). Genetic testing by cancer site: Breast. *Cancer Journal 18*, 310–319.

Shinmura, K., Kageyama, S., Tao, H., Bunai, T., Suzuki, M., Kamo, T., Takamochi, K., et al. (2008). EML4-ALK fusion transcripts, but no NPM-, TPM3-, CLTC-, ATIC-, or TFG-ALK fusion transcripts, in non-small cell lung carcinomas. *Lung Cancer 61*, 163–169.

Shiota, M., Nakamura, S., Ichinohasama, R., Abe, M., Akagi, T., Takeshita, M., Mori, N., et al. (1995). Anaplastic large cell lymphomas expressing the novel chimeric protein p80NPM/ALK: A distinct clinicopathologic entity. *Blood 86*, 1954–1960.

Slamon, D.J., Clark, G.M., Wong, S.G., Levin, W.J., Ullrich, A., and McGuire, W.L. (1987). Human breast cancer: Correlation of relapse and survival with amplification of the HER-2/neu oncogene. *Science 235*, 177–182.

Soda, M., Choi, Y.L., Enomoto, M., Takada, S., Yamashita, Y., Ishikawa, S., Fujiwara, S., et al. (2007). Identification of the transforming EML4-ALK fusion gene in non-small-cell lung cancer. *Nature 448*, 561–566.

Song, J., Mercer, D., Hu, X., Liu, H., and Li, M.M. (2011). Common leukemia- and lymphoma-associated genetic aberrations in healthy individuals. *The Journal of Molecular Diagnostics 13*, 213–219.

Spear, B.B., Heath-Chiozzi, M., and Huff, J. (2001). Clinical application of pharmacogenetics. *Trends in Molecular Medicine 7*, 201–204.

Strassburg, C.P. (2008). Pharmacogenetics of Gilbert's syndrome. *Pharmacogenomics 9*, 703–715.

Sullivan-Klose, T.H., Ghanayem, B.I., Bell, D.A., Zhang, Z.Y., Kaminsky, L.S., Shenfield, G.M., Miners, J.O., Birkett, D.J., and Goldstein, J.A. (1996). The role of the CYP2C9-Leu359 allelic variant in the tolbutamide polymorphism. *Pharmacogenetics 6*, 341–349.

Tai, H.L., Krynetski, E.Y., Yates, C.R., Loennechen, T., Fessing, M.Y., Krynetskaia, N.F., and Evans, W.E. (1996). Thiopurine S-methyltransferase deficiency: Two nucleotide transitions define the most prevalent mutant allele associated with loss of catalytic activity in Caucasians. *American Journal of Human Genetics 58*, 694–702.

Tai, Y.C., Domchek, S., Parmigiani, G., and Chen, S. (2007). Breast cancer risk among male BRCA1 and BRCA2 mutation carriers. *Journal of the National Cancer Institute 99*, 1811–1814.

Takeuchi, K., Choi, Y.L., Togashi, Y., Soda, M., Hatano, S., Inamura, K., Takada, S., et al. (2009). KIF5B-ALK, a novel fusion oncokinase identified by an immunohistochemistry-based diagnostic system for ALK-positive lung cancer. *Clinical Cancer Research 15*, 3143–3149.

Talpaz, M., Shah, N.P., Kantarjian, H., Donato, N., Nicoll, J., Paquette, R., Cortes, J., et al. (2006). Dasatinib in imatinib-resistant Philadelphia chromosome-positive leukemias. *The New England Journal of Medicine 354*, 2531–2541.

Thompson, D., Easton, D.F., and Breast Cancer Linkage Consortium. (2002). Cancer incidence in BRCA1 mutation carriers. *Journal of the National Cancer Institute 94*, 1358–1365.

Tulinius, H., Olafsdottir, G.H., Sigvaldason, H., Arason, A., Barkardottir, R.B., Egilsson, V., Ogmundsdottir, H.M., Tryggvadottir, L., Gudlaugsdottir, S., and Eyfjord, J.E. (2002). The effect of a single BRCA2 mutation on cancer in Iceland. *Journal of Medical Genetics 39*, 457–462.

van der Groep, P., van der Wall, E., and van Diest, P.J. (2011). Pathology of hereditary breast cancer. *Cellular Oncology 34*, 71–88.

Walker, A.H., Jaffe, J.M., Gunasegaram, S., Cummings, S.A., Huang, C.S., Chern, H.D., Olopade, O.I., Weber, B.L., and Rebbeck, T.R. (1998). Characterization of an allelic variant in the nifedipine-specific element of CYP3A4: Ethnic distribution and implications for prostate cancer risk. Mutations in brief no. 191. Online. *Human Mutation 12*, 289.

Walsh, T., and King, M.C. (2007). Ten genes for inherited breast cancer. *Cancer Cell 11*, 103–105.

Walsh, T., Lee, M.K., Casadei, S., Thornton, A.M., Stray, S.M., Pennil, C., Nord, A.S., Mandell, J.B., Swisher, E.M., and King, M.C. (2010). Detection of inherited mutations for breast and ovarian cancer using genomic capture and massively parallel sequencing. *Proceedings of the National Academy of Sciences of the United States of America 107*, 12629–12633.

Wang, J., Mougey, E.B., David, C.J., Humma, L.M., Johnson, J.A., Lima, J.J., and Sylvester, J.E. (2001). Determination of human beta(2)-adrenoceptor haplotypes by denaturation selective amplification and subtractive genotyping. *American Journal of Pharmacogenomics 1*, 315–322.

Weinshilboum, R.M., and Sladek, S.L. (1980). Mercaptopurine pharmacogenetics: Monogenic inheritance of erythrocyte thiopurine methyltransferase activity. *American Journal of Human Genetics 32*, 651–662.

Wilke, R.A., Ramsey, L.B., Johnson, S.G., Maxwell, W.D., McLeod, H.L., Voora, D., Krauss, R.M., et al. (2012). The clinical pharmacogenomics implementation consortium: CPIC guideline for SLCO1B1 and simvastatin-induced myopathy. *Clinical Pharmacology and Therapeutics 92*, 112–117.

Win, A.K., Young, J.P., Lindor, N.M., Tucker, K.M., Ahnen, D.J., Young, G.P., Buchanan, D.D., et al. (2012). Colorectal and other cancer risks for carriers and noncarriers from families with a DNA mismatch repair gene mutation: A prospective cohort study. *Journal of Clinical Oncology 30*, 958–964.

Wolff, A.C., Hammond, M.E., Hicks, D.G., Dowsett, M., McShane, L.M., Allison, K.H., Allred, D.C., et al. (2013). Recommendations for human epidermal growth factor receptor 2 testing in breast cancer: American Society of Clinical Oncology/College of American Pathologists clinical practice guideline update. *Journal of Clinical Oncology 31*, 3997–4013.

Wolff, A.C., Hammond, M.E., Schwartz, J.N., Hagerty, K.L., Allred, D.C., Cote, R.J., Dowsett, M., et al. (2007). American Society of Clinical Oncology/College of American Pathologists guideline recommendations for human epidermal growth factor receptor 2 testing in breast cancer. *Archives of Pathology & Laboratory Medicine 131*, 18–43.

Wrighton, S.A., Brian, W.R., Sari, M.A., Iwasaki, M., Guengerich, F.P., Raucy, J.L., Molowa, D.T., and Vandenbranden, M. (1990). Studies on the expression and metabolic capabilities of human liver cytochrome P450IIIA5 (HLp3). *Molecular Pharmacology 38*, 207–213.

Xie, H.G., Prasad, H.C., Kim, R.B., and Stein, C.M. (2002). CYP2C9 allelic variants: Ethnic distribution and functional significance. *Advanced Drug Delivery Reviews 54*, 1257–1270.

Yaziji, H., Goldstein, L.C., Barry, T.S., Werling, R., Hwang, H., Ellis, G.K., Gralow, J.R., Livingston, R.B., and Gown, A.M. (2004). HER-2 testing in breast cancer using parallel tissue-based methods. *JAMA 291*, 1972–1977.

Zang, Y., Doll, M.A., Zhao, S., States, J.C., and Hein, D.W. (2007). Functional characterization of single-nucleotide polymorphisms and haplotypes of human N-acetyltransferase 2. *Carcinogenesis 28*, 1665–1671.

Zanger, U.M., Raimundo, S., and Eichelbaum, M. (2004). Cytochrome P450 2D6: Overview and update on pharmacology, genetics, biochemistry. *Naunyn-Schmiedeberg's Archives of Pharmacology 369*, 23–37.

Zanger, U.M., Turpeinen, M., Klein, K., and Schwab, M. (2008). Functional pharmacogenetics/genomics of human cytochromes P450 involved in drug biotransformation. *Analytical and Bioanalytical Chemistry 392*, 1093–1108.

Zenser, T.V., Lakshmi, V.M., Rustan, T.D., Doll, M.A., Deitz, A.C., Davis, B.B., and Hein, D.W. (1996). Human N-acetylation of benzidine: Role of NAT1 and NAT2. *Cancer Research 56*, 3941–3947.

6 Clinical Applications of Pharmacogenomics in Cancer Therapy

Mario Listiawan, Max Feng, and Xiaodong Feng

CONTENTS

KEY CONCEPTS

- Better management of cancer patients is a big challenge clinicians have to face in patient care as many cancer chemotherapies are associated with severe drug toxicities. For these patients, these toxicities may affect their quality of life and also impair their adherence to subsequent therapy.
- Factors that could affect drug efficacy and toxicities include demographic, physiological, pathophysiological, and pharmacogenomic (PGx) factors.
- PGx factors that include genetic polymorphisms have been associated with interindividual differences in toxicity of many chemotherapy agents. With modern-day research, biomarkers for these genetic polymorphisms have been identified.
- There are three different types of biomarkers that have been associated with drug toxicities: drug-metabolizing enzymes, drug transporters, and drug targets. The majority of the biomarkers that have been identified are drug-metabolizing enzymes.
- Genetic polymorphisms of drug-metabolizing enzymes may have important effects on both drug efficacy and toxicities. Based on these genetic polymorphisms, various dosing recommendations have been published.
- Genetic polymorphisms of drug transporters can lead to decreased efficacy of therapeutic medication and unpredictable toxicities with drug therapy.
- Drug target-related biomarkers may lead to drug toxicities for drug therapies: for example, patients with *EGFR* intron 1 short allele polymorphisms can have a higher risk of skin toxicities.
- Various polymorphisms or biomarkers related to repair and detoxifying mechanisms can lead to reduced response, poor progression-free survival (PFS), and increased toxicities from chemotherapy. An example of this includes lower levels of glutathione *S*-transferase leading to increased toxicities from cyclophosphamide.
- Integration of PGx in cancer therapy can offer the ability to maximize therapy while decreasing chances of adverse drug reactions (ADRs) in patients.

INTRODUCTION

Although cancer and heart disease remain the top two causes of death (representing nearly 25% of all deaths in the United States alone), the total mortality of cancer has already fallen more than 20% in the past two decades. Evidence credits this steady decline mainly due to a reduction in the smoking population, improved cancer treatments with more specific targets, increased drug efficacies and minimized toxicities, and earlier diagnosis of cancer.[1] When a patient is recommended to undergo cancer chemotherapies, narrow therapeutic indices, low overall response rates (ORRs), rapid and severe systemic toxicity, and unpredictable clinical outcomes are all hallmarks of these cancer therapies. In addition, missed early diagnosis and staging, rapid progress and wild metastasis, considerable heterogeneity across the cancers, and frequent drug resistance are some of the reasons to indicate why many malignancies, especially advanced ones, are extremely difficult to treat.[2-4] For example, most of the standard chemotherapy, radiotherapy, surgical therapy, and even biological therapies (such as interleukin and interferon) have failed to significantly improve the overall survival (OS) for patients with advanced melanoma.[5] The new strategy is to target the specific mutations at the molecular level to slow down or block the process of tumor growth and metastasis. Since the genetic mutations in melanoma tumors are heterogeneous, it is important to stratify the patient population based on their genetic profiles and to further individualize the treatment.[5] Pharmacogenomics (PGx) plays an important role in targeted therapy through genetic subgrouping to define who would respond well to the specific treatment. Therefore, nowhere is PGx research needed more than in cancer treatment to guide clinicians to better predict the differences in drug response, resistance, efficacy, and toxicity among patients treated with chemotherapy or targeted therapy, and to further optimize the treatment regimens based on these differences.[2–4]

The application of PGx in oncology is in the discovery of biomarkers that guide selective therapy, predict drug toxicities, screen and detect high-risk patients, and target the mechanisms of drug resistance. Because adverse drug reactions (ADRs) to prescribed drugs are one of the leading causes of death in the United States, according to the National Council on Patient Information and Education, applying PGx in therapeutics has the potential to save lives and increase the quality of life for patients.[6] One of the most significant challenges in cancer therapy is to manage the severe or fatal toxicities associated with cancer treatments. Many of these severe toxicities, such as myelosuppression, chronic renal insufficiency, acute renal failure, elevated transaminases, acute left ventricular failure, heart failure, diarrhea, constipation, pneumonitis, thrombosis, pulmonary fibrosis, and seizures, can be dose limiting and even life-threatening. These toxicities can compromise both the patients' quality of life and the carefully designed curative therapy plan for the patients.[4] The most important example of a biomarker associated with cancer therapy is thiopurine methyltransferase (TPMT), which is a drug-metabolizing enzyme (DME) responsible for the inactivation of 6-mercaptoputine (6-MP) in the liver. Clinical evidence indicates that the steady-state levels of 6-MP in acute lymphocytic leukemia (ALL) patients vary up to 10-fold or higher among cancer patients treated with the same dose because of the highly variable and polymorphic TPMT activity levels. Patients with low or absent TPMT activity have an increased risk for developing severe, life-threatening myelotoxicitiy. In contrast, the error margin for dose calculation of the affected drugs in oncology practice is less than 3%. Thus, it is critical for clinicians to identify those patients with *TPMT* polymorphisms and then adjust their dose of 6-MP accordingly.[3,4] The ultimate goal of application of PGx in oncology is to focus therapy on specific biomarkers to identify interracial, interethnic, and interindividual genetic polymorphisms related to tumor molecular targets/signal transduction pathways, DMEs, transporters, and drug resistance, in order to reduce ADRs and improve therapeutic outcomes in cancer patients.[4] The factors that could affect drug efficacy and toxicities are summarized in Figure 6.1, including demographic, physiological, morphometric, pathophysiological, pharmacological, and PGx factors.[4] Clinicians should personalize cancer treatment by evaluating the impact of these essential factors on the pharmacokinetics and pharmacodynamics of the pharmacotherapy, and adjust the treatment based on the scientific evidence and clinical recommendations.[4]

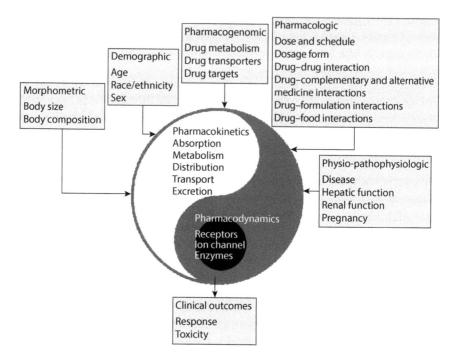

FIGURE 6.1   Factors that may affect drugs' efficacy and toxicities. (Adapted from Feng X, et al., *US Pharmacist*, 37(1), 2–7, 2012. With permission from U.S. Pharmacist/Jobson Medical Information LLC.)

PGx BIOMARKERS FOR TOXICITIES ASSOCIATED WITH CANCER THERAPY

Cutting-edge pharmacogenetic research plays an essential role in identifying these biomarkers that are critical for personalized medicine and enable clinicians to minimize the risk of "one-size-fits-all" or "trial and error" patient care approaches. Since the US FDA approved the first pharmacogenetic test (AmpliChip CYP450 Test, Roche Diagnostics; www.roche.com) in 2004 to identify a patient's *CYP2D6* and *CYP2C19* genotype by analyzing DNA extracted from a whole blood sample, there are currently over 30 cancer treatment agents with recommended genetic testing. Based on the clinical evidence, the pharmacogenetic testing is usually mandatory for those cancer drugs with boxed warning or contraindication for specific genetic polymorphisms. Some of the pharmacogenetic testing is usually recommended, which is either clearly recommended or identified at the section "Indications and Usages" of drug package insert. In addition, some PGx instructions are placed at other sections of the package insert.[7–8] The clinical recommendations for PGx associated with cancer treatments are also publicly available from clinical guidelines published by the Clinical Pharmacogenetics Implementation Consortium (CPIC), National Comprehensive Cancer Network (NCCN), and American Society of Clinical Oncology (ASCO). Table 6.1 lists some key biomarkers commonly associated with

TABLE 6.1

Most Common Cancer PGx Biomarkers

Tumor Type	Biomarker	Drug (Brand)
Breast	*ER*	Fulvestrant (Faslodex®), tamoxifen (Nolvadex®), letrozole (Femara®), anastrozole (Arimidex®)
	HER2	Lapatinib (Tykerb®), trastuzumab (Herceptin®), pertuzumab (Perjeta®), ado-trastuzumab (Kadcyla®)
	DPD	Capecitabine (Xeloda®), fluorouracil (5-FU)
CRC	*EGFR*	Cetuximab (Erbitux®), panitumumab (Vectibix®)
	K-RAS	Cetuximab (Erbitux®), panitumumab (Vectibix®)
	UGT1A1	Nilotinib (Tasigna®), irinotecan (Camptosar®)
	DPD	Capecitabine (Xeloda®), fluorouracil (5-FU)
GISTs	*c-Kit*	Imatinib (Gleevec®)
Leukemia and lymphoma	*G6PD*	Rasburicase (Elitek®)
	Philadelphia chromosome	Imatinib (Gleevec®), nilotinib (Tasigna®), dasatinib (Sprycel®), bosutinib (Bosulif®), ponatinib (Iclusig®)
	PML-RAR-α translocation	Arsenic trioxide (Trisenox®)
	TPMT	6-Mercaptopurine (6-MP)
	UGT1A1	Nilotinib (Tasigna®), belinostat (Beleodaq®)
Lung (NSCLC)	*EGFR*	Erlotinib (Tarceva®), gefitinib (Iressa®), afatinib (Gilotrif®)
	K-RAS	Erlotinib (Tarceva®), gefitinib (Iressa®), afatinib (Gilotrif®)
	EML4-ALK	Crizotinib (Xalkori®), ceritinib (Zykadia®)
Pancreatic	*EGFR*	Erlotinib (Tarceva®)
Head and neck	*EGFR*	Cetuximab (Erbitux®)
Melanoma	*BRAF*	Vemurafenib (Zelboraf®), dabrafenib (Tafinlar®)
	MEK	Trametinib (Mekinist®)
	PD-1	Nivolumab (Opdivo®), pembrolizumab (Keytruda®)

Sources: FDA Drug Package Inserts, Lexi-Comp Drug Monographs. (Adapted from Weng L, et al., *Pharmacogenomics*, 14(3), 15–24, 2013; Feng X, et al., *US Pharmacist*, 36(11), 5–12, 2011. With permission from U.S. Pharmacist/Jobson Medical Information LLC.)

Note: CRC, colorectal cancer; *c-Kit*, stem cell growth factor receptor; *DPD*, dihydropyrimidine dehydrogenase; *EGFR*, epidermal growth factor receptor; *EML4-ALK*, fusion genes are resulted from inversions between the echinoderm microtubule-associated protein-like 4 (EML4) and the anaplastic lymphoma receptor tyrosine kinase (ALK) genes; *ER*, estrogen receptor; *GIST*, gastrointestinal stromal tumor; *G6PD*, glucose-6-phosphate dehydrogenase; *HER2*, human epidermal growth factor receptor-2; *K-RAS*, Ki-ras2 Kirsten rat sarcoma viral oncogene homolog; *MEK*, mitogen-activated protein kinase; NSCLC, non–small cell lung cancer; *PD-1*, programmed cell death protein 1; *PML-RAR-α*, promyelocytic leukemia protein-retinoic acid receptor alpha; *TPMT*, thiopurine *S*-methyltransferase; *UGT1A1*, UDP glucuronosyltransferase 1 family, polypeptide A1.

severe cancer treatment toxicities.[8] The PGx instructions in the drug package insert of these drugs have clearly explained the potential clinical impact of these gene variants on toxicities associated with cancer treatments or potential drug interactions. There are over 120 FDA-approved drugs with recommended PGx information linked to over 50 genes in the drug package insert. About 30 of these PGx drugs are commonly indicated for cancer patients.[7]

Although over 2 million single nucleotide polymorphisms (SNP, resulting from the deletion, insertion, or exchange of a single nucleotide) have been discovered in the human genome, the majority of these SNPs have no clinical significance. Some of these SNPs (*TPMT, UGT1A1, CYP2D6*, and *G6PD*) contribute to clinically significant changes in the pharmacokinetics and/or pharmacodynamics of cancer drugs. Most of these SNPs are identified for the DMEs. Based on their catalytic activity levels, patients with SNPs can be classified as: (1) extensive metabolizers (EM); (2) poor metabolizers (PMs); (3) intermediate metabolizers (IM); and (4) ultrarapid metabolizers (UMs).[4] Standard doses of drugs with a steep dose–response curve or a narrow therapeutic window are more likely to produce ADRs in PMs who are taking that drug, or have decreased efficacy when taking a prodrug. In contrast, when taken by UMs, that standard dose may be inadequate to exert the desired therapeutic effect. PGx plays an important role in the pharmacotherapy of cancer due to the presence of narrow therapeutic windows and low ORRs for cancer drugs. Therefore, acute and severe systemic toxicity along with unpredictable efficacy are all hallmarks of cancer therapies. Approximately 10% of Caucasians have partial deficiency in TPMT enzyme activity, while 0.3% of the patients have a complete deficiency in the enzyme activity. However, these 0.3% TPMT PMs account for over 25% of those life-threatening toxicities associated with 6-MP treatment. Therefore, PM of TPMT is at significant risk for severe drug toxicity associated with the use of 6-MP. The FDA recommends a dosage reduction for patients with heterozygous or homozygous mutations in *TPMT*, but the FDA does not provide the exact scale of dose reduction. The NCCN Acute Lymphoblastic Leukemia Guidelines require a 10- to 15-fold reduction in 6-MP dosage for patients homozygous for *TPMT* loss-of-function variant alleles to alleviate hematopoietic toxicity. Heterozygosity at the *TPMT* gene locus occurs in 5–10% of the population and has been shown to have intermediate enzyme activity; therefore, a 10–15% reduction in 6-MP dose is recommended in these patients to minimize the risk of drug toxicity. Determination of patient *TPMT* genotype using genomic DNA is recommended to optimize 6-MP dosing, especially in patients who have experienced myelosuppression at standard doses.[3,4]

On the other hand, *CYP2D6* is an important DME to activate the prodrug tamoxifen (Table 6.2). Although *CYP2D6* genetic testing is not recommended by the US FDA and NCCN and ASCO guidelines for breast cancer patients treated with tamoxifen due to limited clinical evidence, it is recommended to avoid strong *CYP2D6* inhibitors, such as most selective serotonin reuptake inhibitor antidepressants, such as fluoxetine, paroxetine, and sertraline.[4,20] Moreover, clinical evidence indicates that the risk of morphine overdose is higher for patients who were UMs of CYP2D6 taking codeine, since CYP2D6 is essential to activate codeine, the prodrug of morphine. For patients with UMs of CYP2D6, it is recommended to select an alternative analgesic or to closely monitor for ADRs due to morphine overdose. It is recommended to either

TABLE 6.2
PGx Biomarkers for the Prevention of Toxicity in Cancer Therapy

PGx Biomarkers	Therapeutic Agents	PK/PD Impact	Frequency of Affected Phenotype	Clinical Significance
UGT1A1[9]	Irinotecan, nilotinib	UGT1A1*28: increased systemic exposure to SN-38.	Deficiency of the enzyme may occur in 35% of white and black Americans.	UGT1A1*28 /*28: increased risk for developing severe diarrhea, and neutropenia at doses >200 mg/m². (TA)$_7$/(TA)$_7$: increased risk of nilotinib-induced hyperbilirubinemia.
DPD[10,11]	Fluorouracil (5-FU), capecitabine	Heterozygosity of DYPD*2A allele increases the systemic exposure to 5-FU.	3% of the white population may have deficient enzyme activity.	Deficiency can lead to fatal neurological and hematological toxicities, grade 3 diarrhea, and hand–foot syndrome associated with FU plasma levels >3 µg/mL in males.
TPMT[3,12]	6-Mercaptopurine (6-MP)	TPMT inactivates 6-MP. Low or absent TPMT activity increases systemic exposure to 6-MP.	Approximately 10% of white patients are PMs of this enzyme, and 0.3 % have complete deficiency of the enzyme activity	Patients with low or absent TPMT activity are at increased risk for developing severe, life-threatening myelotoxicity. Dose adjustment is required.
CYP3A4/3A5[13–15]	Cyclophosphamide	Activates the prodrug to its active form.	CYP3A5*3: a variant allele occurs in 45% of black Americans and 9% of whites.	Deficiency of this enzyme causes interindividual variability in efficacy and enhanced toxicity.
CYP2B6[13–15]	Cyclophosphamide, ifosfamide	CYP2B6 is a major drug-metabolizing enzyme for cyclophosphamide and ifosfamide.	Exact frequency is unknown, but 48 different alleles have been identified.	Standard dosing for patients with dysfunctional CYP2B6 may be at increased risk for nephrotoxicity.

(Continued)

TABLE 6.2 (Continued)
PGx Biomarkers for the Prevention of Toxicity in Cancer Therapy

PGx Biomarkers	Therapeutic Agents	PK/PD Impact	Frequency of Affected Phenotype	Clinical Significance
CYP2D6[14–16]	Tamoxifen	Tamoxifen is converted to the potent active metabolite endoxifen.	PMs causing absent CYP2D6 activity are found in 7% of whites, less than 3% of black Americans and 1% of Asians.	Deficiency of CYP2D6 causes a reduction in plasma endoxifen levels and may change toxicity; however, routine genotyping is not recommended due to variability.
GST1[17]	Cyclophosphamide, busulfan	A family of enzyme detoxifying electrophilic group of some chemotherapeutic agents.	About 57% of white and 13% of black Americans have deficient enzyme.	Deficiency of this enzyme may result in enhanced drug toxicity.
ERCC1 and ERCC2[18]	Cisplatin, carboplatin, oxaliplatin	ERCC1 is an endonuclease repairing damaged DNA segments.	No data	Not conclusive and high gene expression associated with inferior outcomes for bladder, gastric, and colorectal cancers.
ABCB1 (also known as MDR1), ABCB2 (MRP2)[19]	Paclitaxel	P-glycoprotein is responsible for efflux of drugs from the cell.	Three common SNPs have been identified with ABCB1.	Overexpression of ABCB1 leads to drug resistance, and patients with wild type demonstrate reduced neuropathy.

Source: Adapted from Feng X, et al., *US Pharmacist*, 37(1), 2–7, 2012. With permission.

Note: DPD, dihydropyrimidine dehydrogenase; ERCC1, excision repair cross complementing rodent repair; GST, glutathione *S*-transferase; TPMT, thiopurine methyltransferase; UGT, UDP-glucuronyltransferase.

select an alternative analgesic for patients who are PMs of CYP2D6 or monitor for insufficient pain control.[21] Clinicians can use PGx biomarkers to predict the severity of toxicity based on the patient's genetic profile, enabling a more individualized therapy, which is thought to have higher selection for cancer cells than non-cancer cells. The PGx-based monitoring can also significantly improve the prognosis of cancer patients and potentially decrease the toxic effects of anticancer drugs on normal cells.[4,8,21,22]

By identifying the biomarkers associated with severe and potentially life-threatening drug toxicity for cancer patients, PGx may hold the potential to minimize drug toxicity while maximizing drug response and improving patients' outcomes. These biomarkers can be divided into three major categories:[20–23]

1. Drug-metabolizing enzyme: DME plays an important role in PGx by inactivating many chemotherapeutic drugs and activating some chemotherapeutic prodrugs. The genetic polymorphisms of these drug metabolism-related potential biomarkers have important effects on drug efficacy and sensitivity to toxicity. Table 6.2 lists some key biomarkers associated with potentially severe cancer treatment toxicities, many of which are cited in FDA-approved drug labels.[4,9–19] Table 6.3 outlines the clinical implications of genetic polymorphisms in chemotherapeutic agent–metabolizing enzymes, such as CYP450, glutathione S-transferase (GST), uridine diphosphate–glucuronosyltransferase (UGT), TPMT, and dihydropyrimidine dehydrogenase (DPD).[4,24–38]

2. Drug transporters: Genetic polymorphisms of drug transporters are key contributors to multidrug resistance (MDR) to cancer treatments, which may lead to decreased efficacy and unpredictable toxicity associated with the drug therapy. The key players of MDR are a group of membrane transporters known as ATP-binding cassette (ABC), and organic cation transporter (OCT), both of which play a critical role in drug efflux, especially for chemotherapeutic agents.[4,39]

3. Drug targets and associated signal transduction: Biomarkers for the selective therapy are usually associated with targeted therapeutic agents directed at tumor cells with particular protein characteristics that significantly differ from their normal cell counterparts. By identifying specific PGx biomarkers present in tumors, physicians can select and tailor a patient's treatment based on his or her genetic profile. Thus, targeted therapy guided by PGx biomarkers has the potential to be more selective for cancer cells than for normal cells, which can significantly improve the prognosis of cancer patients and potentially decrease the toxic effects of anticancer drugs on normal cells. Examples of these indicators include epidermal growth factor receptor (EGFR), K-RAS (v-Ki-ras2 Kirsten rat sarcoma viral oncogene homolog), human epidermal growth factor receptor-2 (HER2), and stem cell growth factor receptor (c-Kit).[7,22] Table 6.4 lists the PGx biomarkers associated with cancer treatments cited in FDA-approved drug labels. These common PGx biomarkers play an important role in cancer treatment by identifying responders from nonresponders to medications, avoiding ADRs, and optimizing drug dose.[8,40–45]

TABLE 6.3
Clinical Recommendations for Biomarkers Related to Drug-Metabolizing Enzymes

PGx Biomarkers/ Affected Cancer Drugs	Clinical Significant Polymorphisms	Genetic Testing	Clinical Recommendations
UGT1A1/	UGT1A1*28 (insertion of an additional seventh TA)	UGT1A1 Gene Polymorphism (TA Repeat) testing by PCR	A reduction in the starting dose of irinotecan for any patient homozygous for the UGT1A1*28 allele.[24]
Irinotecan	sequence in the promoter region of UGT		Routine use of genotyping when dosing irinotecan at high dose (>200 mg/m²) may be useful.[26]
Nilotinib	1A1 gene.[25]		The FDA recommends, but does not require, genetic testing for UGT1A1 variants prior to initiating or reinitiating treatment with nilotinib but recommends frequent liver function monitoring.[27]
DPD/ Fluorouracil (5-FU) Capecitabine	DPD*2A	Amplichip	Genetic screening recommended to be performed before starting 5-FU therapy.[28,29]
TPMT/ 6-Mercaptopurine (6-MP)	TPMT*2 TPMT*3A TPMT*3C[30]	TPMT Activity Assay	Substantial dose reductions are generally required for homozygous TPMT-deficient patients.[31]
CYP2D6/ Tamoxifen	CYP2D6*3 CYP2D6*4 CYP2D6*5 CYP2D6*6[32]	AmpliChip CYP450 Test[33]	No clinical recommendations regarding CYP2D6 polymorphisms and tamoxifen therapy.[34]
CYP2B6/ Cyclophosphamide Ifosfamide	CYP2B6*5 CYP2B6*6	Pyrosequencing assays for CYP2B6 polymorphisms[35]	CYP2B6 is highly inducible. The interindividual difference of CYP2B6 enzyme activity can be 20- to 250-fold, which can affect drug efficacy and toxicity.[36]
MTHFR Methotrexate	C677T	Methotrexate sensitivity 2 Mutations assay (ARIP lab)	C677T homozygous variant with 35% of normal enzyme activity occurs in up to 10% of whites. The rate of oral mucositis is 36% higher for chronic myelogenous leukemia patients with the C677T polymorphism than those without.[37,38]

(Continued)

TABLE 6.3 *(Continued)*
Clinical Recommendations for Biomarkers Related to Drug-Metabolizing Enzymes

PGx Biomarkers/ Affected Cancer Drugs	Clinical Significant Polymorphisms	Genetic Testing	Clinical Recommendations
G6PD/ rasburicase dabrafenib	G6PD deficiency	G6PD activity assay	FDA recommends not to administer rasburicase to patient with G6PD deficiency, which may trigger acute hemolysis. FDA recommends a close monitoring for patients with G6PD deficiency that is started with Dabrafenib.

Sources: References 24–38, FDA-approved drug package inserts, and Lexi-Comp drug monographs. Adapted from Feng X, et al., *US Pharmacist*, 37(1), 2–7, 2012. With permission.

Note: MTHFR, methylenetetrahydrofolate reductase.

PATIENT CASE 1

AC is a 46-year-old man who is undergoing treatment of FOLFIRI + bevacizumab for Stage 4 colon cancer. AC has undergone 1 cycle treatment of FOLFIRI + bevacizumab. Today is AC's cycle 2 day 1 treatment. He has reported complaints of severe bloody diarrhea (grade 3), and thus his physician has reduced his dose of irinotecan from 180 to 140 mg/m^2. On AC's cycle 3 treatment, he still reports complaints of severe bloody diarrhea (grade 3).

Question: What should have been considered for the patient at the beginning of therapy?

Answer: UGT1A1 genetic testing can determine the presence of the *UGT1A1*28 variant allele that is associated with an increased toxicity for irinotecan, which may lead to an increase in cholinergic symptoms, diarrhea, and hematological toxicities, such as anemia, thrombocytopenia, and neutropenia.

UGT1A1 AND IRINOTECAN

Irinotecan, a topoisomerase I inhibitor, is approved by the US FDA for the treatment of advanced colorectal cancer and other various solid tumors. Irinotecan is a prodrug that requires activation by carboxylesterase to its active metabolite, SN-38. Furthermore, SN-38 is converted to inactive metabolite SN-38G by glucuronidation via UGT1A1 isoform. Diarrhea and neutropenia are the result of dose-limiting toxicities of irinotecan, which are associated with increased levels of SN-38.[46–48] The *UGT1A1*28 polymorphism may occur in 35% of white and black Americans. *UGT1A1*28 allele is associated with reduced *UGT1A1* gene expression, which leads to reduced SN-38

TABLE 6.4

PGx Biomarkers for the Selection of Cancer Therapy

PGx Biomarker	Therapeutic Agents	Mutations to Be Detected	Potential Clinical Impact	ASCO or NCCN Guidelines
EGFR (HER1) in NSCLC	Erlotinib (Tarceva®), gefitinib (Iressa®)	Activating tumor EGFR mutations: mainly deletions in exon 19 and L858R Resistance tumor mutation: T790M	The presence of EGFR-activating mutations predicts response to gefitinib and erlotinib. The presence of EGFR T790M mutation predicts resistance to gefitinib.	Recommends testing for EGFR mutations before use of gefitinib or erlotinib.[40]
HER2/neu (ErbB2) in breast cancer and gastric tumor	Trastuzumab (Herceptin®), lapatinib (Tykerb®)	Use IHC or FISH to detect HER2 gene overexpression	Overexpression of HER2 (+3 by IHC or FISH) predicts response to trastuzumab and lapatinib.	Recommends testing HER2 expression for all breast cancer tumors and to use trastuzumab for patients with HER2 overexpression.[41] List lapatinib in combination with capecitabine as an option for trastuzumab-refractory breast cancer patients.[7]
K-RAS in metastatic colon cancer and SCCHN	Cetuximab (Erbitux®), panitumumab (Vectibix®)	Activating tumor K-RAS mutations: mainly exon 2 codon 12 and 13	Presence of K-RAS mutations predicts nonresponse to cetuximab and panitumumab. Absence of K-RAS mutations predicts response to cetuximab and panitumumab.	Recommends genotyping tumor tissue for K-RAS mutation in all patients with metastatic colorectal cancer.[42] Patients with known codon 12 and 13 K-RAS gene mutation are unlikely to respond to EGFR inhibitors and should not receive cetuximab.[42]

(Continued)

TABLE 6.4 (*Continued*)
PGx Biomarkers for the Selection of Cancer Therapy

PGx Biomarker	Therapeutic Agents	Mutations to Be Detected	Potential Clinical Impact	ASCO or NCCN Guidelines
BCR-ABL or Philadelphia chromosome in CML	Imatinib (Gleevec®), nilotinib (Tasigna®), dasatinib (Sprycel®) Bosutinib (Bosulif®), ponatinib (Iclusig®)	Detecting Philadelphia chromosome FISH BCR-ABL mutations	Presence of BCR-ABL or Philadelphia chromosome predicts response to imatinib and nilotinib. Presence of BCR-ABL mutation predicts resistance to imatinib; dasatinib overcome most BCR-ABL mutation (except T315I).	Recommends cytogenetics and mutation analysis for patients receiving imatinib, nilotinib, or dasatinib therapy and an 18-month follow-up evaluation with treatment recommendations based upon cytogenetic response.[43] Recommends dasatinib for the treatment of adults with chronic, accelerated, or myeloid or lymphoid blast phase chronic myeloid leukemia with resistance or intolerance to prior therapy, including imatinib.[43] Ponatinib is approved to treat BCR-ABL positive CML or ALL that is resistant or intolerant to prior TKI therapy due to T315I mutation.[43]
c-Kit in GIST	Imatinib (Gleevec®)	Oncogenic c-Kit mutation in exon 9 and 11 D816V mutation of c-Kit	Presence of a c-Kit mutation in exon 11 is associated with a more favorable prognosis and greater likelihood of response to imatinib therapy in patients with advanced GIST. Presence of D816V mutation of c-Kit predicts resistance to imatinib.	Mutational analysis of c-Kit is strongly recommended in the diagnostic work-up of GIST patients. In locally advanced, inoperable and metastatic GIST, imatinib 400 mg daily is the standard of care. In patients whose GIST harbors c-Kit exon 9 mutations, imatinib 800 mg daily is the recommended dose.[44]

(*Continued*)

TABLE 6.4 (Continued)
PGx Biomarkers for the Selection of Cancer Therapy

PGx Biomarker	Therapeutic Agents	Mutations to Be Detected	Potential Clinical Impact	ASCO or NCCN Guidelines
PML–RAR-α translocation in APL	Arsenic trioxide (Trisenox®)	t(15:17) translocation determined by FISH or PML–RAR-α gene expression	Presence of PML–RAR-α fusion gene predicts clinical outcome following arsenic trioxide treatment.	Arsenic trioxide induces PML–RAR-α degradation. Diagnostic testing of PML–RAR-α is required for treatment with arsenic trioxide. Used for remission induction and consolidation in patients with relapsed or refractory APL characterized by PML–RAR-α expression.[45]
BRAF MEK	Vemurafenib (Zelboraf®), dabrafenib (Tafinlar®), trametinib (Mekinist®)	Substitution of glutamic acid for valine at amino acid 600 (V600E) Substitution of lysine for valine at the amino acid 600 (V600K)	Presence of the substitution mutations in the BRAF gene predicts and indicates clinical outcome following the BRAF-inhibitor therapies such as vemurafenib, dabrafenib, and trametinib.	Recommends genetic testing for metastatic melanoma to detect the presence of BRAF mutation to consider whether or not BRAF inhibitors are valid therapy options.

Sources: FDA Drug Package Inserts, Lexi-Comp Drug Monographs, ASCO Clinical Guidelines, NCCN Clinical Guidelines. (Adapted from Feng X, et al., *US Pharmacist*, 36(11), 5–12, 2011. With permission.)

Note: APL, acute promyelocytic leukemia; CML, chronic myelogenous leukemia; EGFR, epidermal growth factor receptor; FISH, fluorescence in situ hybridization; GIST, gastrointestinal stromal tumor; HER2, human epidermal growth factor receptor-2; IHC, immunohistochemistry; K-RAS, Ki-ras2 Kirsten rat sarcoma viral oncogene homolog; NSCLC, non–small cell lung cancer; PML–RAR-α, promyelocytic leukemia protein–retinoic acid receptor alpha; SCCHN, squamous cell carcinoma of head and neck.

glucuronidation, prolonged SN-38 half-life, increased exposure to SN-38 (AUC), and late-onset diarrhea. *UGT1A1* expression is highly variable. Clinical evidence indicates that the steady-state concentrations of SN-38 may vary by up to 50-fold between *UGT1A1*28 homozygous and heterozygous groups. Patients have a five- to ninefold greater risk of grade 4 leukopenia with the standard dosage of irinotecan.[46–48]

The US FDA recommends dose reduction in patients homozygous for the *UGT1A1*28 allele, and some studies recommend 20% dose reduction for patients homozygous for *UGT1A1*28. Caution and possible dose reduction are advised for patients with heterozygous *UGT1A1*28 allele (Table 6.2).[25,26,46,47]

Nilotinib is indicated for the treatment of chronic myeloid leukemia (CML) (Ph +). Nilotinib is an inhibitor of the Bcr-Abl kinase that binds to and stabilizes the inactive conformation of the kinase domain of the Abl protein. It is also an inhibitor of *UGT1A1* in vitro, and individuals with the *UGT1A1*28 $(TA)_7/(TA)_7$ are at an increased risk of hyperbilirubinemia when taking nilotinib compared to those with the $(TA)_6/(TA)_6$ or $(TA)_6/(TA)_7$ genotypes. The largest increases in bilirubin were observed in the *UGT1A1*28 $(TA)_7/(TA)_7$ patients. $(TA)_7/(TA)_7$ polymorphism occurs at up to 42% in individuals with African or South Asian ancestry.[27,49]

The mechanism by which hyperbilirubinemia occurs with both nilotinib and *UGT1A1*$(TA)_7/(TA)_7$ is still unclear. Hypothetically, nilotinib inhibits *UGT1A1* activity, which includes the glucuronidation of bilirubin. The already sensitized $(TA)_7/(TA)_7$ genotype will further push patients into a hyperbilirubinemia state. The US FDA recommends, but does not require, genetic testing for *UGT1A1* variants prior to initiating or reinitiating treatment with nilotinib.[27,48,49]

PATIENT CASE 2

KC is a 13-year-old boy who was diagnosed with ALL. He has finished his induction and consolidation therapies and will need to move on to his maintenance therapy.

Question: Knowing that the maintenance portion of his therapy requires mercaptopurine (6-MP), what should be considered for KC?

Answer: TPMT genetic testing should be considered. If KC is a PM of TPMT, a reduction of 6-MP dose should be warranted in order to reduce side effects. 6-MP toxicity may include drug fever, hyperpigmentation, myelosuppression (anemia, thrombocytopenia, and neutropenia).

TPMT POLYMORPHISMS AND 6-MP

6-MP is extensively used in the maintenance chemotherapy for ALL in children. TPMT is responsible for the *S*-methylation of 6-MP to 6-methylmercaptopurine, which diverts the prodrug 6-MP from converting to 6-thioguanine for its incorporation into DNA and suppresses tumor DNA synthesis. 6-Thioguanine is the active metabolite of 6-MP that affects drug efficacy and toxicity.

Most large-scale PGx studies indicate that TPMT activity is highly variable and polymorphic. Approximately 90% of individuals have high activity (EMs), 10% have intermediate activity (IMs), and 0.3% have low activity or complete deficiency of

TABLE 6.5
Clinical Recommendation of 6-MP Based on TPMT Phenotype

Phenotype	Implications	Dosing Recommendations
Homozygous wild-type (normal TPMT activity, or EMs)	Lower concentrations of toxic metabolites of 6-MP	Start with 6-MP normal dose Allow 2 weeks to reach steady state before adjusting doses
Heterozygous (intermediate TPMT activity, or IMs)	Moderate concentrations of toxic metabolites of 6-MP	Start with reduced dose of 6-MP (30–70% of full dose) and adjust doses based on toxic events Allow 2–4 weeks to reach steady state before adjusting doses
Homozygous mutant or variant type (low or deficient TPMT activity, or PMs)	High concentrations of toxic metabolites of 6-MP	Start with 1/10 standard dose and adjust doses based on toxic events Allow 4–6 weeks to reach steady state before adjusting doses

Source: Adapted from Relling MV, et al., *Clin Pharmacol Ther*, 93(4), 324–5, 2013. With permission.

the enzyme activity (PMs).[22] At the steady state, the dosages of 6-MP vary up to 10-fold between PMs and EMs. IMs have up to threefold differences in 6-MP concentrations compared to EMs. Clinical evidence indicates that a 10- to 15-fold less dose than conventional in TPMT PM patients enables successful treatment without substantial toxicity.[3,12,30,31,50]

Although this polymorphism is relatively rare, PMs would have severe, life-threatening hepatic toxicity, gastrointestinal toxicity, and myelosuppression due to excessive levels of 6-thioguanine when exposed to standard doses of 6-MP. TPMT deficiency also increases risk of secondary cancers associated with 6-MP standard dose treatment. Based on this, the US FDA recommends genotype testing for TPMT for patients with leukopenia and frequent monitoring of liver function.

Measurement of TPMT activity or genotyping of *TPMT* before administration of 6-MP may be useful to minimize or avoid toxicity associated with use of 6-MP. The US FDA recommends *TPMT* testing to be considered when a patient has clinical or laboratory evidence of severe meylosuppression after the use of 6-MP.[51] Although it is shown that *TPMT* genotyping is clinically beneficial, currently there is no general correlation regarding *TPMT* genotyping and the cost-effectiveness of such a test.[52]

Current dosing recommendations with different TPMT phenotype and 6-MP can be found in the CPIC Guidelines, whose summary can be found in Table 6.5.[53]

TPMT POLYMORPHISMS AND CISPLATIN

Cisplatin is a platinum-containing compound that acts as an alkylating agent. Cisplatin is a widely used chemotherapeutic agent but has been associated with a high incidence of ototoxicity. Cisplatin ototoxicity may lead to dose reduction and also termination of therapy. Genetic variants of *TPMT* and *COMT* have been identified to be highly associated with cisplatin-induced ototoxicity in several cohort studies.

TPMT (rs12201199) and *COMT* (rs9332377) are associated with higher incidences of ototoxicity. Cisplatin ototoxicity is associated with the PGx because the decreased TPMT enzyme activity causes an increase in cisplatin cross-linking efficiency, thus increasing cisplatin toxicity.[54,55]

Although there is currently no recommendation from the US FDA, with the result of the cohort study, it may be possible to identify individuals at a higher risk of ototoxicity with the use of cisplatin.[54,55] Therefore, treatment alternatives, such as lower doses of cisplatin or treatment with other agents such as carboplatin, may be warranted for patients who could face this higher risk.

PATIENT CASE 3

AR is a 68-year-old African American man with Stage 4 colon cancer metastatic to the liver and lungs. He is treated with the CAPEOX regimen. AR has been tested for DPD deficiency through *DPYD**2A.

Question: What does this mean for AR's therapy?

Answer: As AR's therapy includes capecitabine, which is an oral prodrug for 5-FU, the reduced expression of DPD will result in severe or even life-threatening toxicity, such as diarrhea, neutropenia, and hand–foot syndrome. Careful monitoring and dose reduction of capecitabine may be warranted for AR.

DIHYDROPYRIMIDINE AND 5-FU

5-FU is an injectable antimetabolite pyrimidine analog drug used in the treatment of colorectal cancer, breast cancer, ovarian cancer, gastrointestinal cancers, and head and neck cancers. The enzyme responsible for the catabolism of 5-FU is DPD, which is the rate-limiting step involved in the nucleotide metabolism of pyrimidines uracil and thymine. DPD is the first of the three enzymes in the fluoropyrimidine metabolic pathway, and its activity varies widely due to genetic polymorphisms.

DPD deficiency is an autosomal recessive metabolic disorder in which DPD activity is absent or significantly diminished in cells. Patients with partial or complete DPD deficiency may have increased toxicity from 5-FU. Individuals with DPD deficiency may develop life-threatening toxicity following exposure to 5-FU or the widely prescribed oral prodrug fluoropyrimidine capecitabine. Such toxicities include severe diarrhea, mucositis, and pancytopenia. More than 40 SNPs have been identified in the DPD gene.

DPD deficiency may be detected with the assay of TheraGuide 5-FU. This assay analyzes DNA samples taken from peripheral blood cells. The cost of the test is approximately $1000 with a turnaround time of seven days. The results will classify patients as high, moderate, low risk, less or no risk.[56]

The most common variant, *DPYD**2A, has been reported in approximately 45% of people with partial or complete DPD deficiency.[16] Complete DPD deficiency is relatively uncommon, occurring in less than 5% of white populations. Black Americans, especially black women, have a markedly reduced mean DPD activity and a higher partial DPD deficiency when compared with their white counterparts.

TABLE 6.6

Clinical Recommendation of 5-FU Dosing Based on DPD Phenotype

Phenotype	Implications	Dosing Recommendations
Homozygous for wild-type allele (normal DPD activity)	Normal DPD activity, less or no risk of 5-FU toxicity	Use standard dose
Heterozygote (intermediate DPD activity)	Decreased DPD activity, increased risk for severe or fatal 5-FU toxicities	Start with at least 50% reduction of starting dose May titrate up the dose based on toxicities
Homozygous for variant or mutant (deficiency of DPD activity)	No DPD activity, high risk for severe or fatal 5-FU toxicities	Use alternative drugs

Source: Adapted from Caudle KE, et al., *Clin Pharmacol Ther*, 94(6), 640, 2013.

Approximately 30% of patients with reduced mean DPD activity could experience grade 3 or 4 toxicities from 5-FU. Screening for *DPYD**2A is commercially available but accounts for approximately 45% of all DPD deficiency cases (Table 6.2). In addition, patients with partial or complete DPD activity can be identified by blood lymphocytes, which can also determine whether patients would be at increased risk for developing 5-FU-related toxicity.[10,11,16,28,29]

Although the manufacturer recommends that 5-FU and capecitabine should not be prescribed for patients with DPD deficiency, preemptive genetic testing is not currently recommended by the US FDA because the frequency of the DPD-deficient variant alleles is relatively low, and patients not carrying these deficient variants may still have grade 3 and 4 toxicities from 5-FU. Clinical guidelines regarding the dose of 5-FU for carriers of the *DPD* variants have been published and summarized in Table 6.6.[57]

CYP PGx

PGx of cytochrome P450 (CYP) plays a critical role in regulating the activation and inactivation of many cancer therapeutic agents (Table 6.1). Several chemotherapeutic agents are prodrugs that need to be activated by CYPs. *CYP2B6* is one of the key enzymes involved in the activation of cyclophosphamide, ifosfamide, procarbazine, and thiotepa, and thus, *CYP2B6* polymorphism could affect the plasma levels of their active metabolites, drug efficacy, and toxicities.

Although 50 polymorphisms of *CYP2B6* have been identified, the most common functional variants are *CYP2B6**5 and *CYP2B6**6 with decreased gene expression and enzyme activity. Takada and colleagues reported higher rates of nephrotoxicity associated with *CYP2B6**5 in lupus patients treated with pulse cyclophosphamide. However, the role of *CYP2B6**5 is still not clearly defined.[14–16,35,50,58]

Constituting only 2–4% of total CYP enzymes in the human liver, *CYP2D6* metabolizes 25–30% of all clinically used drugs. Currently, there are four known phenotypes of the CYP2D6: extensive metabolizer (EM), intermediate metabolizer (IM), poor metabolizer (PM), and ultrarapid metabolizer (UM). Because CYP2D6 exhibits genetic and phenotypic polymorphisms, it plays an essential role in the metabolism of tamoxifen, a selective estrogen receptor modulator, widely used for estrogen receptor-positive premenopausal breast cancer. Tamoxifen is a substrate of CYP3A, CYP2C19, and CYP2D6, and an inhibitor of P-glycoprotein (P-gp). CYP2D6 is the predominant enzyme producing the active metabolite endotoxifen, whose binding affinity for estrogen receptors is 100 times stronger than tamoxifen.

Although studies show genetic polymorphisms of *CYP2D6* are associated with decreased plasma concentrations of endotoxifen, their clinical impact on breast cancer patients is still inconclusive.[33,34]

Several studies have determined whether different variant alleles of *CYP2D6* could be associated with both the side effects and clinical efficacy of the relevant chemotherapy (measured with disease-free survival). Goetz et al. reported that the *CY2PD6*4 variant allele was associated with a higher risk of relapse and a lower incidence of hot flashes.[59] The NCCTG 89-30-52 trial reported that the PM of CYP2D6 was associated with a higher risk of relapse of breast cancer.[60] In the study conducted by Schroth et al, patients with two *CYP2D6* null alleles or presumed CYP2D6-reduced activity showed reduced OS compared to other individuals with the wild-type alleles,[61] and patients lacking CYP2D6 activity (PM) had a twofold increased risk of breast cancer recurrence compared with those with two functional CYP2D6 alleles.[61]

Currently, most PGx studies suggest that the number of CYP2D6 allele variants could affect the plasma concentrations of tamoxifen metabolites. The decrease in metabolites in turn may negatively affect tamoxifen efficacy and its treatment outcome.[56] The relationship between CYP2D6 polymorphisms and the treatment outcome of tamoxifen points to a possible benefit from detecting CYP2D6 genotype prior to making a decision on an adjuvant endocrine therapy. At present, there is no enough data to justify the routine testing of CYP2D6 to incorporate it to the guideline of clinical decision making.[62,63]

PATIENT CASE 4

GC is a patient who was diagnosed with acute myeloid leukemia (AML), and she was to be started on the 7+3 regimen. GC's uric acid after 1 treatment cycle of the 7+3 regimen has increased from 7.5 to 11.5 (unit). Her white blood cell count has decreased from 14.9 to 4.8 (unit). Her K+ increased from 3.5 to 6.7 (unit).

Question: What genetic test should be considered for this patient at this time before therapy is started?

Answer: *G6PD* testing should be considered. Patients with AML and a high tumor burden, like GC, are most likely to experience tumor lysis syndrome, which can increase uric acid, potassium, and phosphorus levels in these patients. When patients have a high uric acid level, rasburicase can be used. Rasburicase is contraindicated with the patients with G6PD deficiency as it might cause hemolytic anemia and methemoglobinemia. Thus, a *G6PD* test should be considered prior to the start of the treatment for AML patients.

G6PD and Rasburicase and Dabrafenib

Rasburicase, a recombinant urate oxidase, has been approved by the US FDA for the initial management of plasma uric acid in pediatric and adult patients with leukemia, lymphoma, and solid tumors who are receiving anticancer therapy and are expected to have tumor lysis syndrome. Rasburicase is usually well tolerated; however, several ADRs are of particular concern. The drug is contraindicated in patients with G6PD deficiency. Due to the contraindication, the US FDA recommends that patients at higher risk for G6PD deficiency (e.g., patients of African or Mediterranean ancestry) should be screened prior to starting rasburicase therapy.[64] The enzyme activity is easily tested from a blood sample of the patient. Genetic testing can be performed to check for mutations within families.

G6PD deficiency is the most common disease-causing enzymopathy in humans. It is a metabolic enzyme involved in the pentose phosphate pathway, which is important in red blood cell metabolism, and is inherited as an X-linked recessive disorder that affects 400 million people worldwide. The gene encoding G6PD is highly polymorphic, with more than 300 variants reported. The excess peroxide due to the deficiency poses a risk for both hemolytic anemia and methemoglobinemia. Hemolysis occurred in <1% of the patients who received rasburicase with severe hemolytic reactions presenting within 2–4 days of the start of rasburicase.[65,66]

A new drug for metastatic melanoma was approved by the US FDA in 2014. Dabrafenib is a kinase inhibitor that blocks the activity of the mutated form of the V600E BRAF protein, which ultimately inhibits tumor growth. Dabrafenib is indicated as a single agent for the treatment of patients with unresectable or metastatic melanoma. As dabrafenib contains a sulfonamide moiety, patients with G6PD deficiency should be monitored closely, as they are at higher risk of hemolytic anemia. There is currently no recommendation to test patients with G6PD deficiency prior to starting dabrafenib.[67]

Drug Transporter–Related Biomarkers

Genetic polymorphisms of drug transporters are key contributors to MDR to cancer treatments, which may lead to decreased efficacy and unpredictable toxicity associated with the drug therapy. The key players of MDR are a group of membrane transporters known as ABC and organic cation transporter (OCT), both of which play a critical role in drug efflux, especially for chemotherapeutic agents.

There are seven families of ABC transporters, among which three ABC efflux pumps are particularly important for chemoresistance, including ABCB1 (encoding P-gp), ABCC1, and ABCG2. P-gp is the most studied drug transporter for MDR in cancer therapy due to it being a major obstacle for efficient and safe chemotherapy. ABC transporters are ubiquitously expressed in many normal tissues, and they are also overexpressed in numerous tumor cells. For example, epithelial cells of GI and biliary tract, and normal hematopoietic stem cells express ABCB1 and ABCG2. ABCB1 and ABCC1 expression is shown in GIST, while ABCB1, ABCG2, and OCT1 are found in mononuclear cells in CML patients. Interestingly, it has been proven that one of the revolutionary CML treatment agents, imatinib, is an inhibitor of ABCG2. Imatinib is also a substrate for hOCT1. However, it is still not clear whether *hOCT1* polymorphism is associated with resistance to imatinib in CML.[68]

Overexpression of P-gp in tumor cells will allow for the efflux of a large variety of structurally unrelated cancer drugs from the cell, such as corticosteroids and paclitaxel, resulting in decreased oral bioavailability of the drugs and increased resistance. Clinical evidence indicates that polymorphisms in ABC drug transporters may contribute to the varying individual response to standard cancer treatment regimens. Over 48 SNPs of the *ABCB1* gene have been identified. Hoffmeyer et al. demonstrated that the 3435T variant is associated with decreased expression of P-gp.[69] Illmer et al. reported the association of P-gp polymorphisms with complete remission, but not OS of AML.[70] Furthermore, Kao et al. reported that P-gp overexpression contributes to the resistance to paclitaxel in NSCLC.[71] In addition, Robey et al. used cytotoxicity assays to prove that cancer cells transfected with mutant *ABCG2* were more resistant to anthracyclines, such as doxorubicin, daunorubicin, and epirubicin, compared to wild-type *ABCG2*. ABCG2 is also known as breast-cancer resistance protein or mitoxantrone resistance-associated protein.[72]

PATIENT CASE 5

JG is a 46-year-old woman who had been diagnosed with early-stage breast cancer. Mammography confirmed the presence of a 2.5 cm lesion in the upper outer quadrant of her right breast. Malignant cells were also detected in two axillary lymph nodes, and JG was ultimately diagnosed with stage IIA (T1N1M0) infiltrating ductal adenocarcinoma.

Question: What genetic test should be considered for this patient at this time before therapy is started?

Answer: A tumor biopsy was positive for both the estrogen receptor (ER) and progesterone receptor (PR), and HER2-positivity was confirmed by both immunohistochemistry (IHC) and fluorescence in situ hybridization (FISH).

DRUG TARGET–RELATED BIOMARKERS

Over the past decade, PGx data have greatly affected all aspects of oncology research and practice. Emerging clinical evidence from PGx studies has enabled healthcare professionals to use PGx indicators to predict tumor cell susceptibility to certain therapeutic agents and the patient's prognosis after treatment, and to increase the drug response and decrease the dose-limiting toxicities associated with cancer treatment. Table 6.4 summarizes the key PGx biomarkers for the selection of therapy cited by the US FDA cancer drug labels for therapeutic indications with clinical evidence from PGx studies. These common PGx biomarkers play an important role in cancer treatment by identifying responders from nonresponders to medications, avoiding ADRs, and optimizing drug dose.

HER2 BLOCKING MONOCLONAL ANTIBODY AND BREAST CANCER

HER2/erb-B2, a membrane glycoprotein belonging to the epidermal growth factor (EGF) receptor family, is commonly overexpressed in approximately one-fourth of breast cancer patients and less frequently in gastric, ovarian, and lung cancers.

Overexpression of the HER2 oncogene is correlated with a poor prognosis, increased tumor growth and metastasis, and resistance to chemotherapeutic agents.[22] Assessment of HER2 status by IHC or FISH is the standard for the evaluation of newly diagnosed carcinomas of the breast. Trastuzumab, pertuzumab, ado-trastuzumab emtansine, and lapatinib are HER2 antagonists approved for the treatment of HER2-positive cancers. Several FDA-approved HER2 tests are available, including InSite HER2/neu (CB11), HER2 FISH PharmDx kit, HercepTest kit, Spot-Light HER2 CISH kit, Inform HER22 Dual ISH DNA Probe Cocktail, Pathway, and so on. Positive HER2 is evaluated by FISH or IHC with an average copy number of the HER2 gene > 6 or a ratio of HER2 gene/chromosome 17 > 2.2 by FISH, or scored 3+ by IHC.

For patients with positive HER2/neu, there are currently four HER2 targeted therapy regimens in the market: trastuzumab (Herceptin®), lapatinib (Tykerb®), pertuzumab (Perjeta®), and ado-trastuzumab emtansine (Kadcyla®).

Trastuzumab (Herceptin) is a recombinant DNA-derived humanized monoclonal antibody that selectively binds with high affinity to the extracellular domain of the human EGFR 2 protein (HER2). Assessment of HER2 status by IHC or FISH is the gold standard for the evaluation of newly diagnosed carcinomas of the breast. HER2 overexpression is an indicator for use of trastuzumab therapy. Trastuzumab has been approved for monotherapy or combined therapy in the treatment of HER2-positive metastatic breast cancer and for HER2-positive metastatic gastric or gastroesophagel junction adenocarcinoma as it can improve OS and PFS in HER2-positive advanced gastric cancer.[73,74]

Lapatinib (Tykerb) is an oral dual tyrosine kinase inhibitor (TKI) that targets EGFR and HER2. Lapatinib can inhibit HER2 activation via ligand-induced heterodimerization or truncated HER2 receptors. Clinical studies have shown that trastuzumab-refractory breast cancer patients respond well to lapatinib. The NCCN lists lapatinib in combination with capecitabine as an option for trastuzumab-refractory breast cancer patients.[75,76] Blackwell and colleagues demonstrated that trastuzumab-refractory breast cancer patients responded to lapatinib, and lapatinib in combination with trastuzumab significantly improve PFS compared with lapatinib alone.[77]

In recent years, there have been more developments of newer HER2-targeted therapies, such as pertuzumab and ado-trastuzumab emtansine. These new therapies have provided more treatment options for patients with metastatic breast cancer.

Pertuzumab is an HER2/neu receptor antagonist that augments the antitumor activity that complements the activity of trastuzumab (Herceptin). Pertuzumab binds to a different binding region of the HER2 receptor compared with trastuzumab, and thus pertuzumab needs to be used together with trastuzumab. Pertuzumab targets the extracellular dimerization domain (subdomain II) of the human epidermal growth factor receptor 2 protein (HER2) and, thereby, blocks ligand-dependent heterodimerization of HER2 with other HER family members, including EGFR, HER3, and HER4.[78] As a result, pertuzumab inhibits ligand-initiated intracellular signaling through two major signal pathways: mitogen-activated protein kinase (MAPK) and phosphoinositide 3-kinase (PI3K). Inhibition of MAPK and PI3K signaling pathways can result in cell growth arrest and apoptosis, respectively. In addition, pertuzumab mediates antibody-dependent cell-mediated cytotoxicity (ADCC). The drug's antitumor activity is significantly augmented when administered in combination with Herceptin.

In the CLEOPATRA trial, pertuzumab, trastuzumab, and taxane improve PFS and ORR compared with a placebo. Based on this result, currently the NCCN guidelines recommend pertuzumab as the first-line treatment of HER2 + metastatic breast cancer in combination with trastuzumab and docetaxel, and also as neoadjuvant therapy prior to surgery for breast cancer.[79]

PATIENT CASE 6

ML is a 57-year-old Hispanic female who was diagnosed with grade 3 ER/PR negative Her2/neu-positive breast cancer in March 2011. Her CT scan showed 10 lymph node involvements. Her social history includes heavy use of alcohol but she denies smoking.

ML's treatment history includes TCH × 6 cycles, bilateral mastectomy, and radiation. Trastuzumab was completed in August 2012 to complete one year of treatment. Her latest CT scan showed new lung nodules with bilateral lung nodules. ML was then treated with trastuzumab and capecitabine starting in October 2012.

Question: ML's CT scan in April 2013 indicated progression of bilateral pulmonary nodules/masses, suggesting progression of pulmonary metastases. ML was then treated with trastuzumab, pertuzumab, and docetaxel × 4 cycles. In August 2013, CT scan indicated further progression of bilateral pulmonary nodules with bone involvement. What therapy should be used now for ML?

Answer: With the new development of antibody drug conjugate, ML is eligible to receive ado-trastuzumab emtansine. ML's condition is indicated for the use of ado-trastuzumab emtansine for these reasons:

- ML has previously received a taxane and trastuzumab.
- ML's breast cancer is Her2/neu positive.

Ado-trastuzumab emtansine is a new, unique chemotherapy drug, because it is an antibody drug conjugate that targets the HER2 receptor. After binding to the HER2 receptor subdomain IV, ado-trastuzumab emtansine causes receptor-mediated internalization and subsequent lysosomal degradation, resulting in intracellular release of DM1, which disrupts microtubule networks in the tumor cell. Upon binding to the tubulin, DM-1 will disrupt the microtubule formation, ultimately resulting in cell-cycle arrest and apoptosis of the tumor cells. In addition to inhibiting HER2 receptor signaling and ADCC, ado-trastuzumab emtansine also inhibits the shedding of the HER2 extracellular domain in human breast cancer cells that overexpress HER2. Ado-trastuzumab emtansine is indicated for the treatment of HER2+ metastatic breast cancer patients who previously received trastuzumab and taxane in combination or separately. Ado-trastuzumab emtansine marks an advancement in breast cancer therapy as it improves PFS compared to capecitabine and lapatinib (9.6 vs. 6.4 months), and it improves OS compared to capecitabine and lapatinib (30.9 vs. 25.1 months).[80]

Both pertuzumab and ado-trastuzumab emtansine have shown marked improvements in previously targeted HER2/neu therapies. These two drugs have added more options for patients with metastatic breast cancer in previously untreated patients and for patients who have progressed after adjuvant trastuzumab therapies.

All HER2/neu targeted therapies can cause both symptomatic and asymptomatic reduction in left ventricular ejection fraction (LVEF) and heart failure. Evaluation of LVEF prior to and at least every three months during treatment is recommended for all HER2/neu therapies. Treatment interruption may be warranted for patients who develop clinically significant decrease in LVEF.[81]

ER/PR Receptors

ER and PR are nuclear hormone receptors that include androgen and retinoid receptors, and they operate as ligand-dependent transcription factors. Upon attachment of hormone, the DNA binding sites of the receptor will be unmasked, followed by migration to the nucleus, binding to specific hormone elements that may include transcription of mRNA and rRNA, and ultimately synthesis of new proteins.

ER and PR play an important role in determining therapy options in breast cancer management. ASCO recommends that both ER/PR analyses should be performed in all invasive breast cancers, and the information obtained will be used to select endocrine therapy for patients who have ER/PR-positive status.[82]

Quantification of ER/PR can be measured by several different assays such as ligand-binding assay and antibody-based assays, but IHC methods that recognize receptor protein have become the predominant method used to measure ER/PR in clinical practice.[83]

ASCO recommends that endocrine therapy be considered in patients whose breast tumors show at least 1% ER/PR-positive cells. Currently, there are two different types of hormonal therapies that are approved as adjuvant endocrine therapy for breast cancer, which are tamoxifen and aromatase inhibitors.[84]

Tamoxifen is a selective estrogen receptor modulator that inhibits growth of breast cancer cells through competitive antagonism of estrogen receptors. Aromatase inhibitors suppress estrogen levels by inhibiting the enzyme that is responsible for peripheral conversion of androgens to estrogens. There are currently three different types of aromatase inhibitors, which are exemestane, anastrozole, and letrozole.

Tamoxifen is the endocrine agent of choice for adjuvant treatment for premenopausal women and for postmenopausal women who are not candidates for the aromatase inhibitor. Aromatase inhibitors are the preferred adjuvant treatment for postmenopausal women as aromatase inhibitors are inactive in women with intact ovarian function.[75]

Endocrine therapy has been demonstrated to improve survival for women with non-metastatic hormone receptor-positive breast cancer and have generally favorable adverse effect profiles. Some examples of the adverse effects from endocrine therapy are as follows: tamoxifen is associated with hot flashes, vaginal discharge, and menstrual irregularities; whereas aromatase inhibitors are associated with increased osteoporosis, cardiovascular risk and hypercholesterolemia, hot flashes, and vaginal discharge.[85,86]

PATIENT CASE 7

HH, a 67-year-old man, presents with fatigue and a 10-lb weight loss over the past two weeks. Computed tomography (CT) scans of the chest, abdomen, and pelvis are obtained and demonstrate multiple bilateral pulmonary nodules up to 2 cm in size,

4 discrete liver metastases up to 3 cm in size, and bulky retroperitoneal adenopathy. Colonoscopy reveals a nonobstructing tumor 20 cm from the anal verge, biopsy of which is read as adenocarcinoma.

Question: What genetic test should be considered for this patient at this time before therapy is started?

Answer: Active K-RAS mutation test is recommended. Biopsy samples are sent for K-RAS testing, and results indicate wild-type K-RAS.

EGFR Antagonist and Targeted Therapy for Cancer

Epidermal growth factor receptor (EGFR, HER1, or c-ErbB-1) is activated by the homo- or heterodimerization that triggers autophosphorylation of the c-terminus region of the tyrosine kinase domain and initiates the cascade of downstream signal transduction leading to cell proliferation. Although EGFR is constitutively expressed in many normal epithelial tissues, it is commonly overexpressed in many human cancers, including those of the head, neck, colon, and rectum. Therefore, EGFR gene overexpression is a strong PGx indicator for selecting and predicting response to EGFR antagonists, such as EGFR blocking antibody cetuximab and panitumumab, and EGFR TKIs afatinib, erlotinib, and gefitinib.[87] Certain mutation of EGFR in cancer cells lead to enhanced EGFR tyrosine kinase activation even without ligand binding to EGFR. Generally, those cancer cells are resistant to treatments of any EGFR antagonists, such as cetuximab and panitumumab, which block EGFR binding to its ligands. In many cases, the EGFR TKIs, such as afatinib, erlotinib, and gefitinib, improve the clinical outcome in cancer patients with activating EGFR mutations (predominantly in never-before smokers, females, and tumors with adenocarcinoma histology) that are sensitive to these drugs.[8] Exon 19 deletions and the L858R point mutation are the two most common mutations sensitive to either erlotinib or gefitinib treatment. In contrast, T790M mutation and exon 20 insertion are commonly associated with lower response or resistance to erlotinib or gefitinib treatment.[88]

K-RAS is an oncogene that encodes a member of the small GTPase superfamily that plays an important role in the signal transduction of EGFR. The oncogenic mutations of the K-RAS gene lead to the accumulation of ras protein in the active GTP-bound state, which activates the downstream signal transduction pathway of EGFR without ligand binding. In addition to the mutations of the EGFR tyrosine kinase domain, mutated forms of the K-RAS have been shown to be present in many human tumors, including colon and lung cancers. Approximately 97% of K-RAS active mutations seen in NSCLC or colorectal cancer patients involve somatic mutations of codons 12 and 13, which are resistant to EGFR antagonists therapy, including both anti-EGFR monoclonal antibodies (cetuximab and panitumumab) and EGFR TKIs (afatinib, erlotinib, and gefitinib).[8,89] K-RAS RGQ kit is the US FDA-approved testing kit for K-RAS mutations. Generally, K-RAS mutations and EGFR mutations are mutually exclusive to each other. Unlike EGFR mutations that are predominant in never-before smokers, K-RAS mutations are more likely found in adenocarcinoma patients who are smokers, and usually it is an indicator of a poor prognosis for NSCLC.[8,88,89]

CETUXIMAB AND PANITUMUMAB

Cetuximab is a recombinant, human/mouse chimeric monoclonal antibody that binds specifically to the extracellular domain of the human (EGFR) and competitively inhibits the binding of EGF and other ligands, such as transforming growth factor-α.[8]

Panitumumab is a recombinant, human IgG2 kappa monoclonal antibody that also binds specifically to the human EGFR. Similarly, panitumumab (Vectibix®) also binds to the extracellular portion of the EGFR, which prevents ligand-induced EGFR receptor tyrosine kinase activation, resulting in inhibition of cell growth.[8] Binding of cetuximab and panitumumab to the EGFR on tumor cells has been shown to block receptor phosphorylation and activation of receptor-associated kinases, resulting in inhibition of cell growth, induction of apoptosis, and decreased activity of matrix metalloproteinase and production of vascular endothelial growth factor.[8]

Both cetuximab and panitumumab are indicated for the treatment of metastatic CRC in combination with FOLFOX or FOLFIRI regimen in patients who are K-RAS wild-type status. In the Crystal trial, FOLFIRI + cetuximab was compared with FOLFIRI alone, with the former having higher OS and PFS. In the PRIME study comparing FOLFOX + panitumumab vs. FOLFOX, the regimen FOLFOX + panitumumab yields higher OS and PFS, similar to cetuximab.[90,91]

Similar to other EGFR inhibitors, both cetuximab and panitumumab also share similar acneiform rash. Controversially, the rash might be a form of survival predictive markers. There are currently no prospective trials to confirm this, but the general consensus is to continue therapy and treat the side effects of the acneiform rash with both pharmacological and nonpharmacological measures.

Gefitinib (Iressa®) is another TKI that targets EGFR. It is indicated for patients with locally advanced or metastatic NSCLC after failure of platinum-based and docetaxel therapies.[92] Gefitinib is the first selective EGFR inhibitor that targets HER1 or ErbB-1. Clinically, when gefitinib was compared to carboplatin and paclitaxel, the PFS was higher with gefitinib in patients with EGFR mutation positive.[93]

Afatinib is an irreversible TKI and is indicated for the first-line treatment of patients with metastatic NSCLC. Afatinib covalently binds to the kinase domains of EGFR (ErbB1), HER2 (ErbB2), and HER4 (ErbB4), and irreversibly inhibits tyrosine kinase autophosphorylation, resulting in downregulation of ErbB signaling. Afatinib is indicated for the first-line treatment of patients with metastatic NSCLC whose tumors have EGFR exon 19 deletions or exon 21 mutation (L858R), and a secondary T790M mutation in some cases. In addition, afatinib inhibits in vitro proliferation of cell lines overexpressing HER2. Now afatinib can be added to a list of therapy options for patients with NSCLC.[94,95]

Erlotinib reversibly inhibits the kinase activity of EGFR, with higher binding affinity for EGFR exon 19 deletion or exon 21 mutation (L858R) than its affinity for the wild-type receptor. It is indicated as the first-line treatment for patients with metastatic NSCLC whose tumors have EGFR exon 19 deletion or exon 21 mutation (L858R) as detected by an FDA-approved test. Advanced NSCLC will be highly sensitive to EGFR-TKI, and TKIs are the first-line treatment for NSCLC with EGFR mutations.[96]

The presence of an EGFR-activating mutation in advanced stages of NSCLC treated with gefitinib or erlotinib increases the median survival from 10 up to 27 months. In the absence of such EGFR-activating mutations, gefitinib therapy is not superior to conventional chemotherapy. The presence of the T790M resistance mutation at presentation, together with an EGFR-activating mutation, predicts a shorter time to progression of the disease. Whereas the absence of EGFR-activating mutations is clearly associated with the nonresponse to gefitinib, it has been described that patients without EGFR-activating mutations seem to have a slightly better outcome with erlotinib compared with a placebo. In the EURTAC trial, erlotinib was compared to platinum doublet chemotherapy; in patients with EGFR mutation positive, PFS was higher in erlotinib compared with the platinum doublet chemotherapy. Additionally, when erlotinib was compared with gemcitabine and carboplatin in OPTIMAL trial, it yielded a higher PFS.[97]

All EGFR inhibitors share similar toxicities that are important to patients, which are diarrhea, uncommon interstitial pneumonitis, and acneiform rash. Acneiform rash is a characteristic of EGFR inhibitors as there are high levels of EGFR in the basal layer of epidermis.[96]

CRIZOTINIB AND CERITINIB

A small fraction of patients with NSCLC (4–8%) may have EML4-ALK mutation that can produce an inversion of chromosomes that results in fusion of oncogenes. Tumors that contain these EML4-ALK fusion oncogenes are associated with specific clinical features, such as little or no smoking history, younger age of patients, and adenocarcinoma subtype.[98]

EML4-ALK fusion oncogene provides a potential target for therapeutic intervention. Small molecule inhibitors that can target the ALK tyrosine kinase may be able to block ALK receptor, thus producing an anticancer effect on these tumor cells.

Crizotinib (Xalkori®) is a TKI that targets the ALK. Crizotinib is approved for the treatment of locally advanced or metastatic NSCLC with ALK-positive mutation that is detected by an FDA-approved test.[99] Crizotinib targets the ALK gene, which results in expression of oncogenic fusion proteins. These proteins function in activating and dysregulating a gene's expression, ultimately causing increased cell proliferation and survivals of the tumor cells that express these proteins. Crizotinib inhibits the activities from these proteins, thus reducing the cell proliferation and promoting apoptosis of the tumor cells that express these ALK mutations.[100]

Ceritinib is the newly approved ALK inhibitor. It is indicated for the treatment of patients with anaplastic lymphoma kinase (ALK)–positive metastatic NSCLC who have progressed on or are intolerant to crizotinib. This indication is approved under accelerated approval based on tumor response rate and duration of response.[101] There are currently two phase III trials studying ceritinib compared to single-agent chemotherapy and comparing ceritinib with platinum-based doublet in the first-line setting.[102,103] With the success of its first ALK inhibitors, several second-generation ALK inhibitors are currently under development to provide more potent and more selective inhibitors that will be paved for the future of NSCLC therapy (LDK378, alectinib).[104]

A NEW GENERATION OF THE BCR-ABL TKIs

As the first member of targeted therapy for cancers, imatinib (Gleevec; www.gleevec.com) has revolutionized the therapy of CML for over a decade by binding to the inactive configuration domain of bcr-abl kinase and competitively inhibiting the ATP-binding site of bcr-abl kinase. The bcr-abl kinase is a constitutively active product of BCR-ABL fusion gene resulted from reciprocal translocation between chromosomes 9 and 22, which is also known as Philadelphia translocation. BCR-ABL is usually associated with CML (95% of CML patients have this abnormality) and a subset of ALL. Imatinib is also a potent inhibitor of tyrosine kinase derived from c-Kit. Other than Philadelphia chromosome-positive CML, FDA also approved imatinib for the c-Kit positive (CD117) GIST and aggressive systemic mastocytosis (ASM) without D816V c-Kit mutation (or c-Kit mutation status unknown) based on the PGx studies.[8]

Although imatinib is effective in treating CML, about one-third of CML patients developed to be imatinib resistant due to primary resistance or secondary resistance. Primary resistance is defined as initial lacking of response due to low imatinib plasma concentration resulting from low activity of the human OCT1 and/or overexpression of multidrug resistance–associated protein 1 (MRP1, encoded by ABCC1). Secondary resistance to imatinib is most commonly caused by acquired bcr-abl kinase domain mutations, which account for 40–90% of patients with imatinib resistance. More than 100 point mutations in the BCR-ABLE fusion gene lead to single amino acid substitution and also result in resistance through a variety of mechanisms, including ATP phosphate binding loop (P-loop), catalytic domain, ATP-binding site, or activation loop conformational change.[105] The second-generation bcr-abl TKIs, such as dasatinib, nilotinib, bosutinib, and ponatinib, are able to overcome the majority of the secondary resistance to imatinib. Like imatinib, nilotinib also binds to the inactive conformation domain of bcr-abl kinase. Therefore, nilotinib is not the best alternative treatment for most imatinib-resistant mutations. On the other hand, the novel small molecular TKI dasatinib binds to the kinase domain in the open conformation, which is different from the mechanism of imatinib and nilotinib. Thus, dasatinib can be reserved for CML patients who are resistant to imatinib and nilotinib. However, T315I mutation confers resistance to imatinib, dasatinib, nilotinib, and bosutinib. Fortunately, the T315I mutation is sensitive to ponatinib treatment.[106,107]

In recent years, there have been more developments of therapies that target the Philadelphia chromosome. These two new therapies are indicated for CML patients who are Ph+. Bosutinib (Bosulif®) is a TKI that targets Philadelphia-positive CML patients, and TKIs' function as a switch for many cellular activities. Bosutinib inhibits the BCR-Abl kinase and also the Src-family kinases. Bosutinib is indicated for the treatment of patients with chronic, accelerated, or blast phase Philadelphia-positive CML with resistance or intolerance to prior therapy. Prior therapy is defined as therapy with imatinib, dasatinib, and nilotinib. Bosutinib has no activities against the T315I and V299L mutant cells.[108] Ponatinib (Iclusig®) is a protein inhibitor of Bcr-Abl and Src. Ponatinib is indicated for the treatment of patients with chronic phase, accelerated phase, or blast phase CML that is resistant or intolerant to prior TKI therapy or Philadelphia-positive ALL that is resistant or intolerant to prior tyrosine kinase therapy. Ponatinib inhibits tyrosine kinase activity of ABL and T315I mutant. Ponatinib offers advantages compared to other TKIs as it has activity of T315I mutant.[109]

BRAF AND MEK INHIBITION IN MELANOMA WITH BRAF V600 MUTATIONS

Metastatic melanoma is one of the most difficult advanced malignancies to treat. Most of the standard chemotherapy, radiotherapy, and even biological therapies, such as interleukin and interferon, have failed to significantly improve the OS for patients with advanced melanoma. One of the breakthroughs in the treatment of advanced melanoma is to block MAPK signaling targeting at BRAF V600E mutations. Until recently, there were no targeted therapies for BRAF mutations for advanced melanoma. Three therapies have emerged in recent years: vemurafenib (Zelboraf®), dabrafenib (Tafinlar®), and trametinib (Mekinist®). The MAPK pathway plays an important role in stimulating the growth of melanoma cells by activating the downstream RAF kinase that causes phosphorylation of the MEK kinases. Activated MEK kinases phosphorylate the ERK kinases and regulate cyclin D1 expression, leading to tumor cell proliferation. Certain BRAF serine–threonine kinase mutations lead to constitutive activation of BRAF protein in metastatic melanoma in the absence of growth factors, which activates the cascade of the RAS–RAF–MEK–ERK signaling pathway and significantly contributes to the growth and progress of melanoma tumors. BRAF V600E mutation is the most common mutation in melanoma and accounts for about 80% of all BRAF mutations. Other BRAF mutations include V600K, V600D/V600R, and G469A, which account for about 16%, 3%, and 1% of BRAF mutations, respectively.

Vemurafenib is the first drug in its class that is a BRAF inhibitor of the activated BRAF V600E gene. Vemurafenib is indicated for the treatment of unresectable or metastatic melanoma with BRAF V600E as detected by an FDA-approved test. Vemurafenib inhibits tumor growth by inhibiting kinase activities of BRAF and other kinases such as CRAF, ARAF, and FGR. Vemurafenib has antitumor effects in cellular and animal models of melanomas with mutated BRAF V600E and has been associated with prolonged OS and PFS. In a phase III study of vemurafenib vs. dacarbazine, the PFS and ORR were improved with vemurafenib.[110] However, after about seven months of treatment with vemurafenib, melanoma tumors acquire resistance to vemurafenib by bypassing BRAF signaling through stimulating other downstream or producing BRAF transcriptional splice variants.

Dabrafenib is an inhibitor of some mutated forms of BRAF kinases, as well as wild-type BRAF and CRAF kinases. Some mutations in the BRAF gene, including those that result in BRAF V600E, can result in constitutively activated BRAF kinases that may stimulate tumor cell growth. Dabrafenib is specifically indicated for the treatment of patients with unresectable or metastatic melanoma with BRAF V600E mutation. Dabrafenib is not indicated for the treatment of patients with wild-type BRAF melanoma.[67] In a phase III trial, dabrafenib was compared to dacarbazine, and the results shows that dabrafenib significantly increase PFS compared with dacarbazine.[111] Dabrafenib is also used in combination with trametinib for the treatment of patients with unresectable or metastatic melanoma with BRAF V600E or V600K mutations, although there is no evidence of improvement in disease-related symptoms or evidence of OS.

MEK INHIBITOR TRAMETINIB

Trametinib is an orally bioavailable inhibitor of MEK with potential antineoplastic activity. Trametinib specifically binds to and inhibits MEK 1 and 2, resulting in an inhibition of growth factor-mediated cell signaling and cellular proliferation in various cancers. MEK 1 and 2, dual specificity threonine/tyrosine kinases, are often upregulated in various cancer cell types, and play a key role in the activation of the RAS/RAF/MEK/ERK signaling pathway that regulates cell growth. Trametinib is specifically indicated for the treatment of patients with unresectable or metastatic melanoma with BRAF V600E or V600K mutations. It is not indicated for patients who have received prior BRAF-inhibitor therapy.[112] In a phase III trial, trametinib (package insert) was compared to chemotherapy; both PFS and OS were improved with trametinib compared to chemotherapy. In patients who previously used dabrafenib or other BRAF inhibitors, they did not appear to have benefited from trametinib.[113]

BRAF inhibitors are associated with dermatologic complications such as rash, photosensitivity reactions, and hyperkeratosis. Other toxicities associated with the use of BRAF inhibitors are alopecia, fatigue, nausea, and diarrhea. In addition to the dermatologic complications, trametinib is also associated with less common toxicities such as decreased cardiac ejection fraction, interstitial lung disease, and visual problems.[112]

PD-1 INHIBITORS

Recently, newer therapies have emerged for metastatic melanoma that target the programmed death 1 protein (PD-1), which helps tumor cells to evade the host immune system. By blocking the PD-1, the immune function can be enhanced and thus mediate antitumor activity in cells.[114] There are currently two approved therapies associated with the PD-1 inhibition: pembrolizumab (Keytruda®) and nivolumab (Opdivo®). Both pembrolizumab and nivolumab share the same indication for treatment of patients with unresectable or metastatic melanoma and disease progression following ipilimumab and if BRAF V600 mutation positive, a BRAF inhibitor.[115,116] Both pembrolizumab and nivolumab have shown promising data during their clinical trials, which led to accelerated approval by the USFDA.[117,118]

OTHERS

Cyclophosphamide is a prodrug that needs to be activated by cytochrome P450. The metabolite is then eliminated through detoxification by the aldehyde dehydrogenase (ALDH) and also GST.

Polymorphisms exist for the ALDH enzyme. It was found in one study by Ekhart et al. that carriers of the *ALDH3A1**2 allele are at an increased risk of hemorrhagic cystitis (11.95-fold) and carriers of the *ALDH1A1**2 allele are at increased risk of liver toxicity (5.13-fold) compared with noncarriers when treated with cyclophosphamide.[119]

The *ALDH3A1**2 variant allele has been observed with higher frequency in whites. An explanation of why patients carrying these two alleles would experience higher toxicities may be that the *ALDH3A1**2 and *ALDH1A1**2 variants cause decreased activity of ALDH, thus leading to decreased detoxification of the active metabolites and increased risk of hemorrhagic cystitis and liver toxicities in patients.[119]

CONCLUSIONS AND FUTURE PERSPECTIVES

One of the biggest challenges for cancer therapy is to manage the drug toxicities associated with cancer treatments. Accrued evidence demonstrates that PGx biomarkers related to DMEs, drug transporters, drug targets, and drug-detoxifying mechanisms play critical roles in predicting the safety, toxicity, and efficacy of cancer therapy in individuals or groups of patients. With clinically important biomarker polymorphisms delineated increasingly, the need for clinicians to understand the association of these clinically important polymorphisms with potential changes of pharmacokinetics and pharmacodynamics of certain cancer therapies has become crucial. Some of the tumor genetic variations are associated with genetic mutations (e.g., K-RAS and BRAF), molecular marker gene overexpression (CD20, ER, HER2, C-Kit, and CD25), chromosomal translocation (PML/RAR α, Ph chromosome, EML4-Alk), and germline variations (CYP 2D6). They play an important role in the management of cancer treatment efficacy and toxicities. The integration of PGx in cancer therapy offers the ability for clinicians to individualize and maximize the efficacy and safety of the cancer therapy. However, more pervasive clinical evidence is urgently needed for the FDA, NCCN, and ASCO to provide more practical guidance for clinicians to select the cancer-specific treatments and personalize the dose according to patients' PGx evidence.

STUDY QUESTIONS

1. Which of the following statement is false?
   a. A patient who is homozygous for mutant *DPD* should not be treated with 5-FU.
   b. Prior to starting rasburicase, a patient should be tested for *G6PD* deficiency.
   c. There are three types of biomarkers that are associated with a high risk of developing drug toxicity: drug-metabolizing enzymes, drug transports, and drug targets.

d. Carboplatin and cisplatin are activated through the CYP2B6 enzyme.

e. Higher levels of ERCC1 protein will cause less oxaliplatin ototoxicity.

2. Match the following chemotherapies with its PGx biomarkers

Chemotherapy	Biomarkers
Oxaliplatin	a. MTHFR
Cyclophosphamide	b. G6PD
Dabrafenib	c. UGT1A1
Irinotecan	d. DPD
Tamoxifen	e. CYP2B6
6-MP	f. CYP2D6
Methotrexate	g. TPMT
Capecitabine	h. ERCC1

3. What is the current recommendation for a patient who is treated with FOLFIRI (5-FU, leucovorin, and irinotecan) who is homozygous for *UGT1A1*28*?

4. What is the clinical recommendation for a patient who is being treated with 6-MP, and is heterozygous for *TPMT*?

a. Start with the normal dose, then allow 2 weeks to reach steady state before adjusting dose.

b. Start with 25% of normal dose, then allow 4 weeks to reach steady state before adjusting dose.

c. Start with 25% of normal dose, then allow 6 weeks to reach steady state before adjusting dose.

d. Start with 50% of normal dose, then allow 4 weeks to reach steady state before adjusting dose.

e. Start with 50% of normal dose, then allow 6 weeks to reach steady state before adjusting dose.

5. Consider the following statements:

I. Deficiency of CYP2D6 activities causes less toxicities in tamoxifen, such as hot flashes.

II. Majority of the biomarkers for drug toxicity are drug transporters.

III. Patients with heterozygote DPD activity treated with 5-FU should be treated with 50% dose reduction.

IV. Decreased GST activity is associated with decrease in cyclophosphamide toxicity.

V. *ALDH3A1*2* allele variant causes increased risk of hemorrhagic cystitis in patients.

Which of the above statements are true?

a. I, II, III, IV

b. I. III, V

c. I, IV, V

d. II, III, V

e. II and IV

Answer Key

1. d
2.

	Chemotherapy	Biomarkers
H	Oxaliplatin	a. MTHFR
E	Cyclophosphamide	b. G6PD
B	Dabrafenib	c. UGT1A1
C	Irinotecan	d. DPD
F	Tamoxifen	e. CYP2B6
G	6-MP	f. CYP2D6
A	Methotrexate	g. TPMT
D	Capecitabine	h. ERCC1

3. A reduction in the starting dose of irinotecan for any patient homozygous for the *UGT1A1*28* allele
4. d
5. b

REFERENCES

1. Jemal A, Center MM, DeSantis C, Ward EM. Global patterns of cancer incidence and mortality rates and trends. *Cancer Epidemiol Biomarkers Prev.* 2010;19(8):1893–907.
2. Albertini L, Siest G, Jeannesson E, Visvikis-Siest S. Availability of pharmacogenetic and pharmacogenomic information in anticancer drug monographs in France: Personalized cancer therapy. *Pharmacogenomics.* 2011;12:681–91.
3. McLeord HL, Krynetski EY, Relling MV, et al. Genetic polymorphism of thiopurinemethyltransferase and its clinical relevance for childhood acute lymphoblastic leukemia. *Leukemia.* 2000;14:567–72.
4. Feng X, Pearson D, Listiawan M, Cheung C. Pharmacogenomic biomarkers for toxicities associated with cancer therapy. *US Pharmacist.* 2012;37(1):2–7.
5. Feng X, Vyas D, Guan B. Novel immune and target therapy for skin cancer. *US Pharmacist.* 2012;37(11):7–11.
6. Lazarou J, Pomeranz BH, Corey PN. Incidence of adverse drug reactions in hospitalized patients: A meta-analysis of prospective studies. *JAMA.* 1998;279:1200– 205.
7. Weng L, Zhang L, Peng Y, Huang RS. Pharmacogenetics and pharmacogenomics: A bridge to individualized cancer therapy. *Pharmacogenomics.* 2013;14(3):15–24.
8. Feng X, Brazill B, Pearson D. Therapeutic application of pharmacogenomics in oncology: Selective biomarkers for cancer treatment. *US Pharmacist.* 2011;36(11):5–12.
9. Hoskins JM, Goldberg RM, Qu P, Ibrahim JG, McLeod HL. UGT1A1*28 genotype and irinotecan-induced neutropenia: Dose matters. *J. Natl Cancer Inst.* 2007;99:1290–95.
10. Gamelin E, Delva R, Jacob J, et al. Individual fluorouracil dose adjustment based on pharmacokinetic follow-up compared with conventional dosage: Results of a multicenter randomized trial of patients with metastatic colorectal cancer. *J Clin Oncol.* 2008;26:2099–105.
11. Schwab M, Zanger UM, Marx C, et al. Role of genetic and nongenetic factors for fluorouracil treatment-related severe toxicity: A prospective clinical trial by the German 5-FU toxicity study group. *J Clin Oncol.* 2008;26:2131–38.
12. Evans WE, Hon YY, Bomgaars L, et al. Preponderance of thiopurine S-methyltransferase deficiency and heterozygosity among patients intolerant to mercaptopurine or azathioprine. *J Clin Oncol.* 2001;19:2293–301.

13. Gor PP, Su HI, Gray RJ, et al. Cyclophosphamide-metabolizing enzyme polymorphisms and survival outcomes after adjuvant chemotherapy for node-positive breast cancer: A retrospective cohort study. *Breast Cancer Res.* 2010;12:2–10.

14. Ekhart C, Doodeman VD, Rodenhuis S, et al. Influence of polymorphisms of drug metabolizing enzymes (CYP2B6, CYP2C9, CYP2C19, CYP3A4, CYP3A5, GSTA1, GSTP1, ALDH1A1 and ALDH3A1) on the pharmacokinetics of cyclophosphamide and 4-hydroxycyclophosphamide. *Pharmacogenet Genomics* 2008;18:515–23.

15. Rodriguez-Antona C, Ingelman-Sundberg M. Cytochrome P450 pharmacogenetics and cancer. *Oncogene.* 2006;25:1679–91.

16. Aebi S, Davidson T, Gruber G, et al. Cardoso primary breast cancer: ESMO clinical practice guidelines for diagnosis, treatment and follow-up. *Ann Oncol.* 2011; 22(Suppl. 6):vi12–vi24.

17. Ansari M, Lauzon-Joset JF, Vachon MF, et al. Influence of GST gene polymorphisms on busulfan pharmacokinetics in children. *Bone Marrow Transplant.* 2010;45:261–7.

18. Hoffmann AC, Wild P, Leicht C, et al. MDR1 and ERCC1 expression predict outcome of patients with locally advanced bladder cancer receiving adjuvant chemotherapy. *Neoplasia.* 2010;12:628–36.

19. Gréen H, Söderkvist P, Rosenberg P, et al. Pharmacogenetic studies of paclitaxel in the treatment of ovarian cancer. *Basic Clin Pharmacol Toxicol.* 2008;104:130–7.

20. Desmarais JE, Looper KJ. Interactions between tamoxifen and antidepressants via cytochrome P450 2D6. *J Clin Psychiatry.* 2009;70:1688–97.

21. Gasche Y, Daali Y, Fathi M, et al. Codeine intoxication associated with ultrarapid CYP2D6 metabolism. *N Engl J Med.* 2004;351:2827–31.

22. Tomalik-Scharte D, Lazar A, Fuhr U, et al. The clinical role of genetic polymorphisms in drug-metabolizing enzymes. *Pharmacogenomics J.* 2008;8:4–15.

23. Ulrich CM, Robien K, McLeod HL. Cancer pharmacogenetics: Polymorphisms, pathways and beyond. *Nat Rev Cancer.* 2003;3:912–20.

24. "Camptosar (Irinotecan) Safety Labeling" FDA Medwatch. Available from: http://www. fda.gov/Safety/MedWatch/SafetyInformation/ucm215480.htm (accessed November 10, 2011).

25. Bhushan S, Howard M, Walko CM. Role of pharmacogenetics as a predictive biomarkers response and/or toxicity in the treatment of colorectal cancer. *Clin Colorectal Cancer.* 2009;8(1):15–21.

26. Bosma PJ, Chowdhury JR, Bakker C, et al. The genetic basis of the reduced expression of bilirubin UDP-glucuronosyltransferase 1 in Gilbert's syndrome. *N Engl J Med.* 1995;333:1171–75.

27. "Nilotinib" PharmGKB. Available from: http://www.pharmgkb.org/drug/PA162372877 (accessed November 10, 2011).

28. Omura K. Clinical implications of dihydropyrimidine dehydrogenase (DPD) activity in 5-FU based chemotherapy: Mutations in DPD gene, and DPD inhibitory fluoropyrimidines. *Int J Clin Oncol.* 2003;8:132–8.

29. Collie-Duguid ES, Etienne MC, Milano G, et al. Known variant DPYD alleles do not explain DPD deficiency in cancer patients. *Pharmacogenetics.* 2000;10:217–23.

30. Innocenti F, Iyer L, Ratain MJ. Pharmacogenetics: A tool for individualizing antineoplastic therapy. *Clin Pharmacokinet.* 2000;39:315–25.

31. Relling MV, Hancock ML, Boyett JM, et al. Prognostic importance of 6-mercaptopurine dose intensity in acute lymphoblastic leukemia. *Blood.* 1999;93:2817–23.

32. Sachse C, Brockmoller J, Hildebrand M, et al. Correctness of prediction of the CYP2D6 phenotype confirmed by genotyping 47 intermediate and poor metabolizer of debrisoquine. *Pharmacogenetics.* 1998;8:181–5.

33. Tan SH, Lee SC, Goh BC. Pharmacogenetics in breast cancer therapy. *Clin Cancer Res.* 2008;14:8027.

34. Technology Evaluation Center. CYP2D6 pharmacogenomic of tamoxifen treatment. *Assess Prog.* 2008;23(1):1–32.

35. Rohrbacher M, Kirchhof A, Geisslinger G, et al. Pyrosequencing-based screening for genetic polymorphisms in cytochrome P450 2B6 of potential clinical relevance. *Pharmacogenomics.* 2006;7(7):995–1002.

36. Wang H, Tompkins LM. CYP2B6: New insights into a historically overlooked cytochrome P450 isozyme. *Curr Drug Metab.* 2008;9:598–610.

37. Schwahn B, Rozen R. Polymorphisms in the methylenetetrahydrofolate reductase gene: Clinical consequences. *Am J Pharmacogenomics.* 2001;1:189–201.

38. Ulrich CM, Yasui Y, Storb R, et al. Pharmacogenetics of methotrexate: Toxicity among marrow transplantation patients varies with the methylenetetrahydrofolate reductase C677T polymorphism. *Blood.* 2001;98(1):231–4.

39. Cizmarikova M, Wagnerova M, Schonova L, et al. MDR1 (C3435T) polymorphism: Relation to the risk of breast cancer and therapeutic outcome. *Pharmacogenomics J.* 2010;10:62–69.

40. Keedy VL, Temin S, Somerfield MS, et al. American Society of Clinical Oncology provisional clinical opinion: Epidermal growth factor receptor (EGFR) mutation testing for patients with advanced non–small-cell lung cancer considering first-line EGFR tyrosine kinase inhibitor therapy. *J Clin Oncol.* 2011;29(15):2121–27.

41. Wolff AC, Hammond EHM, Schwartz JN, et al. American Society of Clinical Oncology/College of American Pathologists guideline recommendations for human epidermal growth factor receptor 2 testing in breast cancer. *J Clin Oncol.* 2008; 25(1):118–45.

42. Allegra CJ, Jessup JM, Somerfield MR, et al. American Society of Clinical Oncology provisional clinical opinion: Testing for KRAS gene mutations in patients with metastatic colorectal carcinoma to predict response to anti-epidermal growth factor receptor monoclonal antibody therapy. *J Clin Oncol.* 2009;27(12):2091–96.

43. NCCN. Clinical practice guideline in oncology: Chronic myelogenic leukemia (version 1.2015). Available from: http://www.nccn.org/professionals/physician_gls/pdf/cml.pdf (accessed April 20, 2015).

44. Demetri GD, Mehren MV, Antonescu CR, et al. NCCN Task Force report: Update on the management of patients with gastrointestinal stromal tumors. *J Natl Compr Canc Netw.* 2010;8(Suppl 2):S1–S40.

45. NCCN. Clinical practice guidelines in oncology™: Acute myeloid leukemia [v.1.2015]. Available from: http://www.nccn.org/professionals/physician_gls/pdf/aml.pdf (accessed April 2, 2015).

46. Lara PN Jr, Natale R, Crowley J, et al. Phase III trial of irinotecan/cisplatin compared with etoposide/cisplatin in extensive-stage small-cell lung cancer: Clinical and pharmacogenomic results from SWOG S0124. *J Clin Oncol.* 2009;27:2530–35.

47. Hoskins JM, Goldberg RM, Qu P, et al. UGT1A1*28 genotype and irinotecan-induced neutropenia: Dose matters. *J Natl Cancer Inst.* 2007;99:1290–95.

48. Toffoli G, Cecchin E, Corona G, et al. The role of UGT1A1*28 polymorphism in the pharmacodynamics and pharmacokinetics of irinotecan in patients with metastatic colorectal cancer. *J Clin Oncol.* 2006;24:3061–68.

49. Singer JB, Shou Y, Giles F, et al. UGT1A1 promoter polymorphism increases risk of nilotinib-induced hyperbilirubinemia. *Leukemia.* 2007;21:2311–15.

50. Lee W, Lockhart AG, Kim RB, et al. Cancer pharmacogenomics: Powerful tools in cancer chemotherapy and drug development. *Oncologist.* 2005;10:104–111.

51. Purinethol [package insert]. Horsham, PA: TEVA Biologics and Specialty Products; 2011.

52. Donnan JR, Ungar WJ, Mathews M, Hancock-Howard RL, Rahman P. A cost effectiveness analysis of thiopurinemethyltransferase testing for guiding 6-mercaptopurine dosing in children with acute lymphoblastic leukemia. *Pediatr Blood Cancer.* 2011;57(2):231–9.

53. Relling MV, Gardner EE, Sandborn WJ, Schmiegelow K, Pui CH, Yee SW, Stein CM, et al. Clinical pharmacogenetics implementation consortium guidelines for thiopurine methyltransferase genotype and thiopurine dosing: 2013 update. *Clin Pharmacol Ther.* 2013;93(4):324–5.

54. Yang JJ, Lim JYS, Huang J, et al. The role of inherited TPMT and COMT genetic variation in cisplatin-induced ototoxicity in children with cancer. *Nature.* 2013; 94(2):252–9.

55. Ross CJD, Katzov-Eckert H, Dube MP. Genetic variants in TPMT and COMT are associated with hearing loss in children receiving cisplatin chemotherapy. *Nature.* 2009; 41(12):1345–49.

56. TheraGuide 5-FU, Myriad Genetics Laboratories. Available from: http://www.myriadtests.com/hcp/theraguide_testing_process.htm (accessed August 18, 2014).

57. Caudle KE, Thorn CF, Klein TE, Swen JJ, McLeod HL, Diasio RB, Schwab M. Clinical Pharmacogenetics Implementation Consortium guidelines for dihydropyrimidine dehydrogenase genotype and fluoropyrimidine dosing. *Clin Pharmacol Ther.* 2013;94(6):640.

58. Takada K, Arefayene M, Desta Z, et al. Cytochrome P450 pharmacogenetics as a predictor of toxicity and clinical response to pulse cyclophosphamide in lupus nephritis. *Arthritis Rheumat.* 2004;50:2202–10.

59. Goetz MP, Murdter TE, Winter S. Tamoxifen use in postmenopausal breast cancer: CYP2D6 matters. *J Clin Oncol.* 2013;31(2):176–80.

60. Goetz MP, Suman V, Ames M, et al. Tamoxifen pharmacogenetics of CYP2D6, CYP2C19, and SULT1A1: Long term follow-up of the North Central Cancer Treatment Group 89-30-52 adjuvant trial. [Abstract] *31st Annual San Antonio Breast Cancer Symposium*, December 10–14, 2008, San Antonio, TX, A-6037, 2008.

61. Schroth W, Goetz MP, Hamann U, et al. Association between CYP2D6 polymorphisms and outcomes among women with early stage breast cancer treated with tamoxifen. *JAMA.* 2009;302(13):1429–36.

62. Brauch H, Murdter TE, Eichelbaum M, et al. Pharmacogenomics of tamoxifen therapy. *Clin Chem.* 2009;55(10):1770–82.

63. Ferradlceschi R, Newman WG. The impact of CYP2D6 genotyping on tamoxifen treatment. *Pharmaceuticals.* 2010;3:1122–38.

64. Elitek [package insert]. Bridgewater, NJ: Sanofi-Aventis U.S. LLC; 2011.

65. Browning LA, Kruse JA. Hemolysis and methemoglobinemia secondary to rasburicase administration. *Ann Pharmacother.* 2005;39:1932–35.

66. Beutler E. G6PD deficiency. *Blood.* 1994;84:3613–36.

67. Tafinlar [package insert]. Research Triangle Park, NC: Glaxo Smith Kline; 2014.

68. Sharom F. ABC multidrug transporters: Structure, function and role in chemoresistance. *Pharmacogenomics.* 2008;9:105–27.

69. Hoffmeyer S, Burk O, von Richter O, et al. Functional polymorphisms of the human multidrug-resistance gene: Multiple sequence variations and correlation of one allele with P-glycoprotein expression and activity in vivo. *Proc Natl Acad Sci U S A.* 2000; 97:3473–78.

70. Illmer T, Schuler US, Thiede C, et al. MDR1 gene polymorphisms affect therapy outcome in acute myeloid leukemia patients. *Cancer Res.* 2002;62(17):4955–62.

71. Kao CH, Hsieh JF, Tsai SC, et al. Quickly predicting chemotherapy response to paclitaxel-based therapy in non-small cell lung cancer by early technetium-99m methoxyisobutylisonitrile chest single-photon-emission computed tomography. *Clin Cancer Res.* 2000;6:820–4.

72. Robey RW, Honjo Y, Morisaki K, et al. Mutations at amino-acid 482 in the ABCG2 gene affect substrate and antagonist specificity. *Br J Cancer.* 2003;89:1971–78.

73. Herceptin [package insert]. South San Francisco, CA: Genentech; 2015.

74. Bang YJ, Van Cutsem E, Feyereislova A, et al. Trastuzumab in combination with chemotherapy versus chemotherapy alone for treatment of HER2-positive advanced gastric or gastro-oesophageal junction cancer (ToGA): A phase 3, open-label, randomised controlled trial. *Lancet.* 2010;376(9742):687–97.
75. NCCN. Clinical practice guidelines in oncology™: Breast cancer [v.2.2015]. http://www.nccn.org/professionals/physician_gls/pdf/aml.pdf (accessed April 23, 2015).
76. Tykerb [package insert]. Research Triangle Park, NC: GlaxoSmithKline; 2014.
77. Blackwell KL, Burnstein HJ, Storniolo AM, et al. Overall survival benefit with lapatinib in combination with trastuzumab for patients with human epidermal growth factor receptor 2–positive metastatic breast cancer: Final results from the EGF104900 study. *J Clin Oncol.* 2012;30(21):2585–92.
78. Perjeta [package insert]. South San Francisco, CA: Genentech; 2015.
79. Baselga J, Cortes J, Kim SB, et al. Pertuzumab plus trastuzumab plus docetaxel for metastatic breast cancer. *N Engl J Med.* 2012;366:109–19.
80. Verma S, Miles D, Gianni L, et al. Trastuzumab emtansine for HER2-positive advanced breast cancer. *N Engl J Med.* 2012;367:1783–91.
81. Kadcyla [package insert]. South San Francisco, CA; 2015.
82. Adjuvant endocrine therapy for women with hormone receptor–positive breast cancer: American Society of Clinical Oncology clinical practice guideline focused update 9/3/2014. http://www.asco.org/quality-guidelines/adjuvant-endocrine-therapy-women-hormone-receptor%E2%80%93positive-breast-cancer-american (accessed May 23, 2015).
83. Habel LA, Shak S, Jacobs MK, et al. A population-based study of tumor gene expression and risk of breast cancer death among lymph node-negative patients. *Breast Cancer Res.* 2006;8(3):R25.
84. Burnstein HJ, Temin S, Anderson H, et al. Adjuvant endocrine therapy for women with hormone receptor–positive breast cancer: American society of clinical oncology clinical practice guideline focused update. *J Clin Oncol.* 2014;32(21):2255–69.
85. Tamoxifen [package insert]. Wilmington, DE: Astra Zeneca Pharmaceuticals LP; 2004.
86. Arimidex [package insert]. Wilmington, DE: Astra Zeneca Pharmaceuticals LP; 2014.
87. Vlahovic G, Crawford J. Activation of tyrosine kinases in cancer. *Oncologist.* 2003; 8:531–8.
88. Jackman DM, Miller VA, Cioffredi LA, et al. Impact of epidermal growth factor receptor and KRAS mutations on clinical outcomes in previously untreated non-small cell lung cancer patients: Results of an online tumor registry of clinical trials. *Clin Cancer Res.* 2009;15:5267–73.
89. Van Krieken JH, Jung A, Kirchner T, et al. KRAS mutation testing for predicting response to anti-EGFR therapy for colorectal carcinoma: Proposal for an European quality assurance program. *Virchows Arch.* 2008;453:417–31.
90. Van Cutsem E, Kphne CH, Lang I, et al. Cetuximab plus irinotecan, fluorouracil, and leucovorin as first-line treatment for metastatic colorectal cancer: Updated analysis of overall survival according to tumor KRAS and BRAF mutation status. *J Clin Oncol.* 2011;29(15):2011–19.
91. Douillard JY, Siena S, Cassidy J, et al. Randomized, phase III trial of panitumumab with infusional fluorouracil, leucovorin, and oxaliplatin (FOLFOX4) versus FOLFOX4 alone as first-line treatment in patients with previously untreated metastatic colorectal cancer: The PRIME study. *J Clin Oncol.* 2010;28(31):4697–705.
92. Iressa [package insert]. Wilmington, DE: AstraZeneca; 2010.
93. Mok TS, Wu YL, Thongprasert S, et al. Gefitinib or carboplatin–paclitaxel in pulmonary adenocarcinoma. *N Engl J Med.* 2009;361(10):947–57.
94. Sequist LV, Yang JCH, Yamamoto N, Phase III study of afatinib or cisplatin plus pemetrexed in patients with metastatic lung adenocarcinoma with EGFR mutations. *J Clin Oncol.* 2013;31(27):3327–34.

95. Gilotrif [package insert]. Ridgefield, CT: Boehringer Ingelheim Pharmaceuticals Inc; 2014.
96. Tarceva [package insert]. South San Francisco, CA: Genentech, Inc; 2015.
97. Rosell R, Cancereny E, Gervais R, et al. Erlotinib versus standard chemotherapy as first-line treatment for European patients with advanced EGFR mutation-positive non-small-cell lung cancer (EURTAC): A multicentre, open-label, randomised phase 3 trial. *Lancet Oncol.* 2012;13(3):239.
98. Takahashi T, Sonobe M, Kobayashi M, et al. Clinicopathologic features of non-small-cell lung cancer with EML4-ALK fusion gene. *Ann Surg Oncol.* 2010;17(3):889.
99. Xalkori [package insert]. New York, NY: Pfizer; 2013.
100. Shaw AT, Kim DW, Nakagawa K, et al. Crizotinib versus chemotherapy in advanced ALK-positive lung cancer. *N Engl J Med.* 2013;368(25):2385.
101. Zykadia [package insert]. East Hanover, NJ: Novartis Pharmaceutical; 2014.
102. K378 Versus Chemotherapy in Previously Untreated Patients With ALK Rearranged Non-small Cell Lung Cancer. https://clinicaltrials.gov/ct2/show/NCT01828099?term=NCT01828099&rank=1 (accessed April 27, 2015).
103. LDK378 Versus Chemotherapy in ALK Rearranged (ALK Positive) Patients Previously Treated With Chemotherapy (Platinum Doublet) and Crizotinib. https://clinicaltrials.gov/ct2/show/NCT01828112?term=NCT01828112&rank=1 (accessed April 27, 2015).
104. Seto T, Kiura K, Nishio M, et al. CH5424802 (RO5424802) for patients with ALK-rearranged advanced non-small-cell lung cancer (AF-001JP study): A single-arm, open-label, phase 1-2 study. *Lancet Oncol.* 2013;14(7):590.
105. De Braekeleer E, Douet-Guilbert N, Rowe D, et al. ABL1 fusion genes in hematological malignancies: A review. *Eur J Haemotol.* 2011;86(5):361–71.
106. Terasawa T, Dahabreh I, Trikalinos T. BCR-ABL mutation testing to predict response to tyrosine kinase inhibitors in patients with chronic myeloid leukemia. *PLoS Curr Evid Genomic Tests.* 2011;2:RRN1204. doi: 10.1371/currents.RRN1204.
107. Tanaka R, Kimura S. Abl tyrosine kinase inhibitors for overriding Bcr-Abl/T315I: From the second to third generation. *Expert Rev Anticancer Ther.* 2008;8(9):1387–98.
108. Bosulif [package insert]. New York, NY: Pfizer Labs; 2014.
109. Iclusig [package insert]. Cambridge, MA: Ariad Pharmaceuticals; 2015.
110. Chapman PB, Hauschild A, Robert C, et al. Improved survival with vemurafenib in melanoma with BRAF V600E mutation. *N Engl J Med.* 2011;364(26):2507.
111. Hauschild A, Grob JJ, Demidov LV, et al. Dabrafenib in BRAF-mutated metastatic melanoma: A multicentre, open-label, phase 3 randomised controlled trial. *Lancet.* 2012; 380(9839):358.
112. Mekinist [package insert]. Research Triangle Park, NC: GlaxoSmithKline; 2014.
113. Flaherty KT, Robert C, Hersey P, et al. Improved survival with MEK inhibition in BRAF-mutated melanoma. *N Engl J Med.* 2012;367(2):107–14.
114. Brahmer JR, Tykodi SS, Chow LQ, et al. Safety and activity of anti-PD-L1 antibody in patients with advanced cancer. *N Engl J Med.* 2012;366(26):2455–65.
115. Keytruda [package insert]. Whitehouse Station, NJ: Merck & Co. Inc; 2015.
116. Opdivo [package insert]. Princeton, NJ: Bristol-Myers Squibb Company; 2015.
117. Robert C, Ribas A, Wolchok JD. Anti-programmed-death-receptor-1 treatment with pembrolizumab in ipilimumab-refractory advanced melanoma: A randomised dose-comparison cohort of a phase 1 trial. *Lancet.* 2014;384:1109–17.
118. Robert C, Long GV, Brady B, et al. Nivolumab in previously untreated melanoma without BRAF mutation. *N Engl J Med.* 2015;372:320–30.
119. Ekhart C, Rodenhuis S, Smits PHM, et al. Relations between polymorphisms in drug-metabolising enzymes and toxicity of chemotherapy with cyclophosphamide, thiotepa, and carboplatin. *Pharmacogenet Genomics.* 2008;18(11):1009–15.

7 Clinical Applications of Pharmacogenomics in Drug Therapy of Cardiovascular Diseases

Tibebe Woldemariam

CONTENTS

KEY CONCEPTS

- Recommendations for genetic testing prior to prescribing certain drugs in product labeling including warfarin, whose metabolism is affected by a patient's *CYP2C9* and *VKORC1* genotypes.
- The genetic polymorphisms of a drug transporter have been strongly implicated in statin-induced myopathy. Although this may not be predictive enough to use clinically, it has enhanced our understanding of this potentially serious adverse drug event.
- β-Blocker pharmacogenomic research in heart failure has noticeably associated with various components of the adrenergic signaling pathway, and these data have the potential to influence future drug development in heart failure.
- The clinical utilization of pharmacogenomics in cardiovascular disease is promising, and clinical implementation has begun in certain centers, and the findings will provide important insight into the challenges and future directions for cardiovascular pharmacogenomics.
- Evaluation of the ultimate effects of a combination of polymorphisms in drug-metabolizing enzymes, drug transporters, drug targets, and/or disease progression genes on cardiovascular drug response.
- The potential for improvement in cardiovascular disease management may arise from the applications of pharmacogenomics.

INTRODUCTION

Cardiovascular diseases remain the number one cause of death in the United States and will likely soon become the number one cause of death globally. At the same time, new understandings of individual characteristics including genetic variations are reforming our ability to treat various aspects of cardiovascular disorders. Although large randomized clinical trials clearly demonstrate population benefits with many cardiovascular drugs, individual patients exhibit remarkable variability in efficacy and adverse effects. There are many sources of variability in response to drug therapy, such as noncompliance and unknown drug interactions. This chapter focuses on genetic mechanisms contributing to variability in response to cardiovascular drugs used to treat hypertension, hyperlipidemia, arrhythmia, and heart failure (Figure 7.1).

Since its inception, the field of pharmacogenetics has extended to study a wide range of cardiovascular drugs and has become a typical research discipline. Cardiovascular pharmacogenomics offers improved prevention of adverse drug events and treatment outcomes. Variants common in the population have been shown to modify targets, metabolism, and transport of drugs and could be utilized to predict an individual's treatment response. Evidence suggests that the common polymorphic variants of modifier genes could influence drug response in cardiovascular disease (CVD) in a variety of areas, including heart failure, arrhythmias, dyslipidemia, and hypertension.[1]

In the realm of cardiovascular pharmacogenetics, major advances have identified markers in each class for a variety of therapeutics, some with a potential to refining patient outcomes. Included in this chapter are the major pharmacogenetic variants

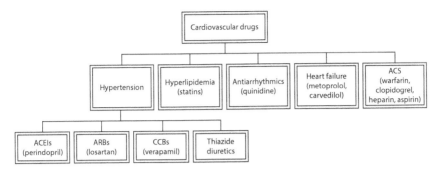

FIGURE 7.1 Cardiovascular disorders that reflect sources of pharmacogenomics variability. ACS, acute coronary syndrome; ACEIs, angiotensin-converting enzyme inhibitors; ARBs, angiotensin receptor blockers; CCBs, calcium channel blockers.

associated with commonly used cardiovascular medications including warfarin, clopidogrel, and simvastatin. The data on warfarin, clopidogrel, and simvastatin were found to be sufficient, well replicated, and clinically important. There are now examples of clinical application of pharmacogenetic data of these drugs to guide therapy. Other cardiovascular drug classes covered in this chapter that may be closest to clinical application of pharmacogenetics are the β-blockers, angiotensin-converting enzyme inhibitors (ACEIs), angiotensin II receptor blockers (ARBs), diuretics, and antiarrhythmic drugs. Examples of cardiovascular drugs with evidence of relationship between genetics and efficacy or toxicity are summarized in Table 7.1.

WARFARIN

The oral anticoagulant, warfarin, is prescribed for the long-term treatment and prevention of thromboembolic events. It has a very narrow and highly variable therapeutic range. The dose requirement and risk of bleeding are influenced by intake of vitamin K, illness, age, gender, concurrent medication, body surface, and genetics. In addition to the possible or demonstrated influence of a large number of genes,[2] warfarin's effect is influenced by two major genes: one involved in its biotransformation (CYP2C9) and the other involved in its mechanism of action (VKORC1).

Warfarin is administered as a racemic mixture of the *R* and *S* stereoisomers. (*S*)-warfarin is two to five times more potent than (*R*)-warfarin and is mainly metabolized by CYP2C9. (*R*)-warfarin is mainly metabolized via CYP3A4, with involvement of several other cytochrome P450 enzymes.[3] An investigation of the pharmacodynamics and pharmacokinetic properties of warfarin showed the additive involvement of two genes to determine its dosage. One of these genes encodes CYP2C9, which is responsible for approximately 80% of the metabolic clearance of the pharmacologically potent *S*-enantiomer of warfarin. There are three allele types: CYP2C9*1, *2, and *3, and both CYP2C*2 and *3 cause a reduction in warfarin clearance. A 10-fold difference in warfarin clearance was observed between groups of individuals having the genotype of the highest metabolizer (CYPC9*1 homozygote) and lowest metabolizer (CYP2C9*3/*3).[4]

TABLE 7.1

Examples of Cardiovascular Drugs with Evidence of Relationship between Genetics and Efficacy or Toxicity

Drug/Drug Class	Gene(s) Associated with Efficacy or Toxicity[a]
β-Agonists	ABRB2
β-Blockers	ADRB1
	ACE
	Gs protein α subunit
	CYP2D6
ACEIs	ACE
	AGT
	AT₁ receptor
	Bradykinin B₂
Antiarrhythmics	Congenital LQTS-associated genes
	NAT2 (procainamide)
	CYP2D6
P2Y12 blockers	Platelet P2Y12 receptor
Abciximab	Platelet glycoprotein IIIa
Aspirin	COX1
Heparin	Platelet Fc receptor
Warfarin	CYP2C9, VKORC1
ARBs	AT₁ receptor
Digoxin	P-glycoprotein
Diuretics	G protein B₃ subunit
	ADD1
Hydralazine	NAT2
Lipid-lowering drugs	ApoE
Statins	Cholesteryl ester transfer protein
	Stromelysin-1
	β-fibrinogen LDL receptor
	Lipoprotein lipase
	ACE
Gemfibrozil	ApoE
	Stromelysin-1

Source: From Johnson J, et al., *Clin Pharmacol Therap*, 90, 519–531, 2011.

[a] The reader is referred to focused reviews for detailed information as described elsewhere.

The VKORC1 gene encodes the vitamin K epoxide reductase enzyme. Warfarin exerts its anticoagulant effect by inhibiting vitamin K epoxide reductase, which catalyzes the conversion of vitamin K epoxide to vitamin K. Vitamin K is an essential cofactor in the synthesis of several clotting factors. A common noncoding variant, −1639G>A, is associated with an increased sensitivity to warfarin.[5] Patients who carry the −1639G>A polymorphism in the promoter region of the VKORC1 gene are more sensitive to warfarin and require lower doses.

Polymorphisms in CYP2C9 and VKORC1 account for approximately 40% of the variance in warfarin dose.[6] Current warfarin labeling suggests lower doses for patients with certain genetic variations. Patients with the CYP 2C9*2 and CYP 2C9*3 alleles need lower warfarin maintenance doses than patients with the wild-type allele in order to achieve their desired international normalized ratio (INR). The two allelic variants will result in slower metabolism of warfarin. Additionally, VKORC1 (vitamin K epoxide reductase A1) haplotype necessitates a lower warfarin maintenance dose due to decreased expression of the messenger RNA responsible for production of proteins necessary for VKORC1 enzyme formation.[7]

Prospective randomized clinical trials are currently underway utilizing dosing algorithms that incorporate genetic polymorphisms in CYP2C9 and VKORC1 to determine warfarin dosages.[6] FDA-approved testing for *VKORC1* and *CYP2C9* variants is available for patients on warfarin. Variations, such as missense mutations in the gene encoding for VKORC1, cause warfarin resistance. Hence, the patient's genotypic variance in the two enzymes predicts response to warfarin therapy and the amount of warfarin dose needed.[8]

CLINICAL APPLICATION

Prospective randomized clinical trials are currently underway employing dosing algorithms that incorporate genetic polymorphisms in CYP2C9 and VKORC1 to determine warfarin dosages. Variations of the *VKORC1* gene that might require an adjustment in the dose of warfarin occur in up to 89% of Asians and in about one-third of Caucasians and African Americans. More than one in ten Caucasians has a relevant variation in CYP2C9, and the variation is less common in non-Caucasians.[9] When these genetic variations are known, warfarin can be initiated at a higher or lower dose as appropriate. The warfarin dose calculator at www.warfarindosing. org can also take these variations into consideration when they are known. In addition, the dosing calculator uses clinical factors, such as drug–drug interactions, age, weight, for determining dose.

At present, available genotype-guided warfarin initiation dosing algorithms show limited clinical validity and utility compared with standard therapy, especially for patients who are dosed based on their clinical information and close INR monitoring.[10] A study with high dropout rate, which used CYP2C9 only to determine dosing, has shown some benefit on anticoagulation control with minor bleeding events.[11] An observational study reported a relationship between providing warfarin genetic information to clinicians and reduced hospitalizations.[12]

Other anticoagulants of minor pharmacogenomic interest include heparin, dabigatran, rivaroxaban, and apixaban. Dabigatran was approved by the FDA in October 2010 for prevention of stroke and blood clots in people with atrial fibrillation. Rivaroxaban was approved in November 2011 to treat atrial fibrillation and lower the risk of blood clots after hip and knee replacements. Apixaban was approved in December 2012 to lower the risk of stroke and dangerous blood clots in patients with atrial fibrillation.

Dabigatran is administered as a prodrug that is rapidly biotransformed after absorption. Both dabigatran and rivaroxaban are substrates for the drug efflux pump P-glycoprotein, coded by the *ABCB1* gene; and dabigatran (but not rivaroxaban) undergoes metabolism by CYP450 enzymes, predominantly CYP3A4 and CYP2J2.[13-15] To date, no data regarding pharmacogenomic studies of these agents have been published, and it is currently unknown whether genetic variants modulate their effectiveness.

Most pharmacogenetic studies on heparin[16] have been related to heparin-induced thrombocytopenia (HIT) and its thromboembolic complications. Recently, it has been proposed that the PlA2 polymorphism of the GPIIIa gene may modulate the pro-thrombotic effects of HIT.[17] However, data on this association are not consistent.[18] There are also controversial data regarding the association between platelet FcγRIIA H131R and platelet factor 4 polymorphisms with HIT,[19-21] whereas the association of hemostatic polymorphisms with thromboembolic complications is certainly weak.[18] It was shown that the homozygous 131Arg/Arg genotype resulted in significantly more common HIT events.[20] However, other studies did not support this finding.[21] These preliminary results demonstrated that more evidence is needed before the patient's genotype can be used for prevention of HIT events.[20,21]

CLOPIDOGREL

Antiplatelet agents such as aspirin and ADP receptor antagonists are effective in reducing recurrent ischemic events. Considerable interindividual variability in the platelet inhibition obtained with these drugs has initiated a search for explanatory mechanisms and ways to improve treatment. In recent years, numerous genetic polymorphisms have been linked to reduced platelet inhibition and lack of clinical efficacy of antiplatelet drugs, particularly clopidogrel and aspirin.

The mechanism of action of clopidogrel, a thienopyridine, is to inhibit platelet function. This will result in the prevention of cardiovascular events in patients with acute coronary syndrome (ACS). This prodrug when enzymatically modified produces bioactive thiol metabolite (SR 26334). This metabolite irreversibly binds to the platelet P2Y12 receptor, inhibiting ADP-mediated platelet aggregation. Interpatient genetic variability affects the response to clinical outcomes of clopidogrel therapy.[22]

SNPs in certain genes involved in clopidogrel metabolism, transport, and signaling could affect the pharmacokinetics and pharmacodynamics of clopidogrel, which include CYP1A2, CYP2C19, CYP3A4, CYP3A5, P-glycoprotein (ABCB1), paraoxonase 1 (PON1), and P2Y12. Persuasive evidence suggests that genetic variability in CYP2C19 affects the efficacy of clopidogrel and has a role in preventing cardiovascular events.[22]

CYP2C19 genetic variations comprise one-third of all patients with a loss-of-function allele, resulting in reduced conversion of clopidogrel to its active metabolite, leading to more cardiovascular events.[23] Genetic variation is commonly seen in Asians. It is estimated that 50% of Asians have one loss-of-function allele in CYP2C19, resulting in impaired bioactive conversion of clopidogrel to its active metabolite.[24]

It is vital that for clopidogrel, there is no specific guidance for responding to genotype test results. For poor CYP2C19 metabolizers, one of the options is to increase the dose of clopidogrel to a 600 mg loading dose followed by 150 mg once daily. Platelet response is improved in poor metabolizers, but there are no data on outcomes. Experts argue that there are no definitive data on other choices, such as switching to prasugrel (Effient), specifically in clopidogrel nonresponders. Like clopidogrel, prasugrel is a prodrug that is converted to an active metabolite. However, it is converted mainly by CYP3A4 and CYP2B6 instead of CYP2C19.[25]

CLINICAL APPLICATION

Guided by the replicated findings of CYP2C19*2 as a determinant of clopidogrel response in patients undergoing percutaneous coronary intervention for atherosclerotic heart disease, the FDA approved new labeling of clopidogrel in March 2010. This includes a boxed warning alerting physicians to the genetic findings and suggests alternative antiplatelet therapy in CYP2C19*2 homozygotes.[26] These may include prasugrel and ticagrelor, which are not markedly affected by the CYP2C19 genotype.[27] The idea of increasing the clopidogrel dose has also been evaluated, but not specifically in CYP2C19*2 homozygotes.[28]

ASPIRIN

Aspirin is widely used for the prophylaxis of cardiovascular events in patients with cardiovascular risk factors or established atherosclerotic disease. Aspirin undergoes polymorphic metabolism. Among the enzymes involved in aspirin biodisposition, a major role is played by the enzymes UDP-glucuronosyltransferase (UGT) 1A6, CYP2C9, and the xenobiotic/medium chain fatty acid:CoA ligase ACSM2, although other UGTs and ACSMs enzymes may significantly contribute to aspirin metabolism. UGT1A6, CYP2C9, and ACSM2 are polymorphic, as well as PTGS1 and PTGS2, the genes coding for the enzymes cyclooxygenases COX1 and COX2, respectively. The genes associated with response to aspirin also include several platelet glycoproteins (GPIIb-IIIa, GPVI, GPIa, and GPIb).[29,30]

The mechanism of action of aspirin is by irreversibly inactivating cyclooxygenase-1 (COX-1), so that the conversion of arachidonic acid to prostaglandin G2/H2, along with TXA2, is inhibited. The inhibition of TXA2 production leads to a lower expression of glycoprotein (GP) IIb/IIIa, causing an inhibition of platelet activation and aggregation. Aspirin has been proven to have efficacy in the secondary and primary prevention of CVD. However, serious vascular events still occur in patients despite the use of aspirin. A reduced ability of aspirin to inhibit platelet aggregation has been associated with an increased risk of adverse events.[31] A substantial number of patients experience recurrent events. Such aspirin resistance is generally defined as failure of aspirin to produce an expected biological response: for example, inhibition of platelet aggregation or of thromboxane A2 synthesis. While its etiology is not evident, genetic factors are likely to play their part. Aspirin's ability to suppress platelet function varies widely among individuals, and lesser suppression of platelet function is associated with increased risk of myocardial infarction, stroke, and cardiovascular death.[31] Platelet response to aspirin is a complex phenotype involving multiple genes and molecular pathways. Aspirin response phenotypes can be categorized as directly or indirectly related to COX-1 activity, with phenotypic variation indirectly related to COX-1 being much more prominent. Recent data indicate that variability in platelet response to aspirin is genetically determined, but the specific gene variants that contribute to phenotypic variation are not known.

STATINS

Hydroxymethylglutaryl-CoA (HMG-CoA) reductase inhibitors competitively inhibit the enzyme HMG-CoA reductase. HMG-CoA reductase is the rate-limiting hepatic enzyme responsible for converting HMG-CoA to mevalonate, which is a precursor of cholesterol. HMG-CoA reductase inhibitors decrease LDL concentrations by reducing hepatic cholesterol production and increasing LDL clearance from the blood. The persistent inhibition of cholesterol synthesis in the liver also decreases concentrations of very low density lipoproteins.[32] In addition, HMG-CoA reductase inhibitors increase HDL concentrations and decrease triglyceride concentrations.[32] The reduction in triglyceride concentrations may be through hepatic synthesis inhibition and lipoprotein lipase activity enhancement in adipocytes. In addition to specific lipoprotein effects, HMG-CoA reductase inhibitors have been shown to halt the progression and facilitate the regression of atherosclerotic lesions.

The potential value of statin pharmacogenetic information include identification of genes and genetic variants that influence statin responsiveness, which holds promise for identifying molecular components of physiological lipid and inflammatory pathways that mediate statin effects. However, the most important clinical benefits of statin pharmacogenetic knowledge would be based on identifying a set of genotypes that aid in predicting the outcomes of statin treatment in terms of reduced risk for cardiovascular events, together with reduced risk for serious ADRs.

Statins are among the most widely used medications and are largely safe and well tolerated. Nevertheless, myopathy is the most common side effect of statins, with symptoms ranging from mild myalgias without creatine kinase (CK) elevation to life-threatening rhabdomyolysis with markedly elevated CK levels, muscle damage, and acute renal injury. The risk for statin-induced myopathy is greater with higher statin doses or inhibition of statin metabolism or clearance due to drug interactions or decreased hepatic or renal function.[32] In addition, there is a heritable component to the risk for statin-induced myopathy. The strongest data in this regard exist for the solute carrier organic anion transporter family, member *1B1 (SLCO1B1)* gene. This gene encodes the organic anion transporting polypeptide (OATP) 1B1, which transports most statins, with the exception of fluvastatin, to the liver. SLCO1B1 genotype exerts a large effect on risk for simvastatin myotoxicity.

Current studies that identify genomics of LDL response are also a key step in addressing the broader question of whether genomic markers identify subjects who develop coronary events during statin therapy.

A number of large randomized controlled trials have demonstrated significant reductions in coronary events and stroke with statin therapy for both primary and secondary prevention of CVD.[33]

β-BLOCKERS

Beta-adrenergic receptor antagonists (β-blockers) are an important class of cardiovascular drugs used for a range of conditions including cardiac arrhythmias, ACS, stable angina, hypertension, and heart failure. β-Blockers antagonize endogenous catecholamines at β-adrenergic receptors, of which two subtypes, β1 and β2, are most important for cardiovascular pharmacology.[34] Important autoregulatory mechanisms include G-protein-coupled receptor kinases (GRKs); enzymes that moderate signaling through phosphorylation of activated β-receptors; and presynaptic α2C-adrenergic receptors (ADRA2C), which regulate norepinephrine release via a negative feedback pathway.[35]

Two of the most commonly used β-blockers in heart failure, metoprolol and carvedilol, both undergo substantial metabolism by the highly polymorphic CYP2D6 enzyme, whose gene contains loss of function, deletion, and duplication polymorphisms. The pharmacokinetics of these drugs is affected based on CYP2D6 genotype; however, there is less evidence for differences in efficacy or side effects.[36]

There are also a number of studies suggesting functional polymorphisms in adrenergic receptor signaling genes are associated with differential response to β-blockers, particularly in hypertension and heart failure. The genes with the strongest data are *ADRB1, ADRA2C, GRK5*, and *GRK43*.[37] These data suggest differential responses to β-blockers by genotype that includes blood pressure response, improvement in left ventricular ejection fraction (LVEF), and survival differences in hypertension and heart failure.

A large prospective randomized pharmacogenomics trial PEAR (Pharmacogenomic Evaluation of Antihypertensive Responses) used genome-wide association studies (GWASs) genotyping to determine the relationship between genetic polymorphisms and antihypertensive medications. Atenolol, hydrochlorothiazide (HCTZ) as well as a combination of the two were studied. It was found that participants who had the T/T and T/C genotypes of the intergenic SNP rs1458038 near FGF5 had a better response to atenolol compared to those who had the C/C genotype. The study concluded that white Caucasian hypertensive individuals with the risk allele for HTN(T) might have a better response to atenolol compared to HCTZ. This gene might be used in the future as a marker for high sympathetic nervous system and renin–angiotensin activity.[38,39]

Even though the pharmacogenetic data for β-blockers have not yet been significant to warrant clinical utility, they are the next closest examples among the cardiovascular drugs that solidify clinical application of pharmacogenomics. Currently, investigations are looking at the benefits of β-blocker therapy for some subjects with heart failure.

ANGIOTENSIN-CONVERTING ENZYME INHIBITORS

The renin–angiotensin–aldosterone system (RAAS) is important for the development of hypertension, and several antihypertensive drugs target this system.

ACEIs block the conversion of angiotensin I to angiotensin II through competitive inhibition of the angiotensin-converting enzyme (ACE). Angiotensin is formed via the RAAS, an enzymatic cascade that leads to the proteolytic cleavage of angiotensin I by ACEs to angiotensin II. RAAS impacts cardiovascular, renal, and adrenal functions via the regulation of systemic blood pressure (BP) and electrolyte and fluid balance. Reduction in plasma levels of angiotensin II, a potent vasoconstrictor and negative feedback mediator for renin activity, by ACEIs leads to increased plasma renin activity and decreased BP, vasopressin secretion, sympathetic activation, and cell growth. Decreases in plasma angiotensin II levels also results in a reduction in aldosterone secretion, with a subsequent decrease in sodium and water retention.[30]

ACEIs also inhibit the breakdown of bradykinin, a potent vasodilator, by kininase II, an enzyme identical to ACE, which may increase levels of nitric oxide. Bradykinin-induced vasodilation is thought to be of secondary importance in the BP lowering effect of ACEIs.

Most data regarding genomic modulators of response refer to a common insertion/deletion (I/D) polymorphism in the *ACE* gene that is strongly correlated with plasma enzyme levels.[40,41] In an observational study, carriers of the DD genotype were found to have significantly higher mortality than those with the I/I genotype, with the risk for heterozygotes being intermediate.[42] However, prospective trials have failed to validate these findings,[43,44] and, at present, there is no strong evidence that this variant influences response to ACEIs. A number of other candidates in the renin–angiotensin–aldosterone pathway have been analyzed, including angiotensinogen (AGT) and angiotensin II receptor types I and II (AGTR1 and AGTR2). Signals for higher risk of adverse cardiovascular outcomes while taking ACEIs have emerged with variants in AGT[45]; however, the data are not conclusive.

Recently, the Perindopril Genetic Association study (PERGENE)[46] evaluated 12 genes from the pharmacodynamic pathway of ACEIs in 8907 patients with stable CAD treated with perindopril or placebo. Two polymorphisms in AGTR1 and a third in the bradykinin type I receptor were significantly associated with the combined primary outcome of cardiovascular mortality, nonfatal myocardial infarction, and resuscitated cardiac arrest during 4.2 years of follow-up. A pharmacogenetic score combining these SNPs demonstrated a stepwise decrease in treatment benefit of perindopril with an increasing score and identified a subgroup of 26.5% of the population that did not benefit from therapy.[46] This finding requires independent replication but represents a potential advance toward understanding variable response to ACEIs.

CLINICAL APPLICATION

Although currently there are no treatment guidelines for the RAAS-acting drugs based on genomics, large multigene haplotype and genome-wide pharmacogenomic studies such as PERGENE may provide clearer evidence to improve therapeutic choices, reduce side effects, improve BP control, and prevent organ damage.

ANGIOTENSIN II RECEPTOR BLOCKERS

ARBs inhibit activation of the AT1 (Angiotensin I) receptor. The role of ARBs in pharmacogenomics still appears to be inconclusive. The Swedish Irbesartan Left Ventricular Hypertrophy Investigation versus Atenolol (SILVHIA) trial found that the ACE I/D6 and CYP11B2 C-344T polymorphisms were associated with modified BP response. This study consisted of 50 subjects who took irbesartan.[7] Subjects carrying the ACE I/I genotype and those carrying the CYP11B2-344 T/T variant had greater BP response to irbesartan. However, Redon et al.[47] conducted a similar study with 206 subjects treated with telmisartan and found no significant association with the RAAS gene polymorphisms.[8] Ortlepp et al. had results that differed from the SILVHIA trial. These results concluded that the C allele of the CYP11B2 C-344T polymorphism is associated with a better response to candesartan.[48]

CLINICAL APPLICATION

Genetic determinants of BP and long-term outcomes in hypertensive patients are being identified. Although the results to date are promising, broader studies would help clarify the role of the RAAS polymorphisms in the use of ARBs.

THIAZIDE DIURETICS

Varied BP response to diuretics is observed in hypertensive patients. It has been proposed that genetic polymorphisms in several candidate genes such as the ACE, alpha-adducin (ADD1), G protein b3-subunit (GNB3) gene, angiotensinogen (AGT), and angiotensin II receptor 1 (AGTR1) may influence BP response to diuretic therapy.[49] ACE insertion/deletion (ACE I/D) polymorphism has been extensively studied for association with BP lowering response to diuretics[50]; however contradictory results have been reported in different studies. Both association and lack of association of ACE genotypes with BP lowering response to diuretics have been reported in different studies.

Carriers of polymorphisms in the ADD1 alpha-adducin Trp460Trp (homozygous patients for the Trp460Trp allele) gene have shown a reduction in the effective renal plasma flow, effective renal blood flow, and glomerular filtration rate compared to noncarriers of the Trp460 allele patients with the *Gly460Gly* gene.[51]

The GenHAT multicenter randomized clinical trial studied the pharmacogenetic association of the atrial natriuretic precursor A (NPPA) T2238C genetic variant with CVD outcomes in patients with hypertension. The study concluded that (NPPA) 2238 T>C polymorphism was associated with better cardiovascular outcomes in patients taking chlorthalidone compared to amlodipine. Carriers of the T/T allele carriers on amlodipine had better outcomes when taking amlodipine.[52]

CALCIUM CHANNEL BLOCKERS

Drugs in this class block voltage-gated calcium channels in the heart and vascular smooth muscle, thereby reducing intracellular calcium.[53] Very few studies have described the pharmacogenomic associations of calcium channel blockers (CCBs).

One study showed that three SNPs in the CACNA1C gene had a significant impact in lowering BP with CCBs.[54] In addition, the effects of CYP3A5*3 and *6 variants on verapamil treatment for hypertension risk outcomes in blacks and Hispanics were studied.[55] Individuals that are homozygous for the T allele of NPPA T2238C had more favorable clinical outcomes when treated with a CCB, whereas C carriers responded better to a diuretic.[52] Carriers of the β-adrenergic receptor 1 (BADRB1) Ser49–Arg389 haplotype carriers had higher death rates compared to those with other haplotypes when treated with verapamil.[56]

The majority of studies looking at CCBs have focused on calcium-signaling genes, such as *CACNA1C*, *CACNB2*, and *KCNMB1*. Although *CACNB2* has been reported as a hypertension gene in GWASs, it is evident that further studies are still needed to confirm genetic associations in CCB response.[57]

ANTIARRHYTHMICS

During the past two decades, pharmaceutical companies have been faced with the withdrawal of some of their marketed drugs because of rare, yet lethal, post marketing reports associated with ventricular arrhythmias. The implicated drugs include antiarrhythmics, noncardiac drugs, such as antibiotics, histamine blockers, and antipsychotics. DNA variants underlie not only variability in cardiac rhythm but also the response of normal and abnormal cardiac rhythms to drug exposure. These undesired effects include prolongation of the QT interval, which may lead to characteristic ventricular tachyarrhythmias, known as torsades de pointes (TdP). These clinical symptoms of the acquired long QT syndrome (LQTS) are also found in an inherited form of the disease, called congenital LQTS.

Currently, a number of environmental (nongenetic) and genetic risk factors for acquired LQTS have been described. Nongenetic factors include female gender, hypokalemia, and other heart diseases. The knowledge of genetic risk factors is emerging rapidly. During the last decade, mutations in several genes encoding ion channels have been shown to cause congenital LQTS. In acquired LQTS, a number of "silent" mutations in these LQTS genes have been identified, and functional polymorphisms in the same genes have been found to be associated with an increased vulnerability for the disease. Furthermore, there is also evidence that interindividual differences in drug metabolism, caused by functional polymorphisms in drug-metabolizing enzyme genes, may be a risk factor for acquired LQTS, especially if multiple drugs are involved. This review evaluates the current knowledge on these risk factors for acquired LQTS, with an emphasis on the genetic risk factors. It also assesses the potential to develop pharmacogenetic tests that will enable clinicians and pharmaceutical companies to identify at an early stage patients or individuals in the general population who are at risk of acquired LQTS.[58]

Congenital LQTS is a rare genetic disease with discernible prolongation of the QT interval. Other signs and symptoms seen with LQTS are recurrent syncope, palpitations, seizures, and the development of TdP that can lead to sudden death. TdP is morphologically distinctive since affected individuals may develop polymorphic ventricular tachycardia.[59] On the other hand, similar LQTS features are present due to certain drug's adverse effects, termed drug-induced LQTS (diLQTS). Antiarrhythmic drugs that commonly cause LQTS are sotalol, dofetilide, or quinidine.

diLQTS is more likely associated with lower, rather than high doses of quinidine. This could be in part due to mechanism of quinidine where at low doses it produces arrhythmogenic effects on cardiac repolarization. However, quinidine inhibits repolarization-related arrhythmias at higher doses.[58] The genes that cause patients to be at a higher risk for diLQTS have not been fully identified. Some studies have associated allele variants of IKs (slow rectifier potassium channels)[45] with an increased susceptibility to drug-induced TdP.[39] Others have found women to be at a greater risk for developing diLQTS due to the difference in modulation of IKr (rapid rectifier potassium channels).[39]

CLINICAL APPLICATION

Identifying patients at risk for long QT-related arrhythmias during drug therapy may become possible as platforms to identify both common and rare variants are increasingly deployed. Early studies suggest that available atrial fibrillation therapies may be less effective in some genetically defined subsets, and these patients thus may be candidates for alternate approaches.

N-ACETYLTRANSFERASE

The acetylation polymorphism elucidates another genetic polymorphism of a drug-metabolizing enzyme studied in the early period of pharmacogenetics. NAT, a phase II conjugating liver enzyme, catalyzes the N-acetylation and O-acetylation of arylamine carcinogens and heterocyclic amines. The slow acetylator phenotype often experiences toxicity from drugs such as procainamide and hydralazine, whereas the fast acetylator phenotype may not respond to hydralazine. Slow acetylators are also at risk for sulfonamide-induced toxicity and can suffer from idiopathic lupus erythematosus while taking procainamide.[60] The slow acetylator phenotype is an autosomal recessive trait. Studies have shown large variations of the slow acetylator phenotype among ethnic groups. Allelic variation at the *NAT2* gene locus accounts for the polymorphism seen with acetylation of substrate drugs.

CONCLUSION AND FUTURE PERSPECTIVES

Translating fundamental discovery in genome science to individual patients and populations is a challenge for the field, and genotyping for selection of appropriate drug therapy is one of the first ways this is being accomplished. In oncology, tumor genotyping is rapidly becoming standard of care to identify specific mutations that then dictate selection of therapy.[61] Pre-prescription germline genotyping is becoming standard of care to reduce the risk of serious adverse reactions to carbamazepine and abacavir.[62] In cardiovascular therapy, a number of centers are now deploying programs to use *CYP2C19* genotypes to guide clopidogrel therapy. The National Institutes of Health is making major investments in discovery of new pharmacogenomic pathways and in how that information can be used to improve healthcare. These include the warfarin trials and efforts such as the Pharmacogenomics Research Network and the Electronic Medical Records and Genomics Network. Identifying outcomes and analyzing costs are other challenges. The questions surrounding genotype testing for warfarin and clopidogrel dosing include the following: Will testing help improve outcomes? Will the costs of testing be offset by the benefits? For warfarin, dosing recommendations for different genotypes are derived from the results of clinical studies. And there are data to suggest that hospitalizations could be reduced with genotype testing. However, there is no evidence that testing is cost-effective, and it is not clear that dosing according to genotype is better than careful INR monitoring and dose adjustment. For clopidogrel, there are no data on dosing or treatment strategies to improve outcomes based on genotype. Medical practitioners also question the value of adding genetic testing while INRs are being routinely monitored. Critics of the black box warning for clopidogrel argue that clinicians have no specific guidance for responding to the warning. Despite extensive research, much of the research in cardiovascular pharmacogenomics remains in the discovery phase, with researchers struggling to demonstrate clinical utility and validity because of poor study design, inadequate sample sizes, lack of replication, and heterogeneity across the patient populations and phenotypes. In order to progress pharmacogenetics in cardiovascular therapies, researchers need to utilize next-generation sequencing technologies, develop clear phenotype definitions, and engage in multicenter collaborations. These efforts do not need larger sample sizes but have to replicate associations and confirm results across different ethnic groups.

STUDY QUESTION

DV is a 65-year-old man who recently had an acute myocardial infarction (MI). To prevent subsequent ischemic events, DV's physician recommends antiplatelet therapy and prescribes clopidogrel. His current medications included amlodipine 10 mg daily, hydrochlorothiazide 25 mg daily, aspirin 81 mg daily, pravastatin 40 mg daily, metformin ER 850 mg daily, and clopidogrel 75 mg daily. Four months later, DV suffered another acute MI, and his physician suspects that the patient has not been taking his medications as directed, or alternatively, that clopidogrel therapy may have been unsuccessful. How would genetic testing benefit DV and his physician?

Answer

Clopidogrel is a prodrug, and in order to be active in vivo, it must be transformed to a more active metabolite. CYP2C19 is responsible for its metabolic activation, and *CYP2C19* loss of activity alleles appear to be associated with higher rates of recurrent cardiovascular events in patients receiving clopidogrel. CYP2C19 poor metabolizer status is associated with diminished antiplatelet response to clopidogrel. At least one loss-of-function allele is carried by 24% of the white non-Hispanic population, 18% of Mexicans, 33% of African Americans, and 50% of Asians.[54] The liver enzyme CYP2C19 is primarily responsible for the formation of the active metabolite of clopidogrel. Pharmacokinetic and antiplatelet tests of the active metabolite of clopidogrel show that the drug levels and antiplatelet effects differ depending on the genotype of the *CYP2C19* enzyme.

The following represent the different alleles of CYP2C19 that make up a patient's genotype:

- The CYP2C19*1 allele has fully functional metabolism of clopidogrel.
- The CYP2C19*2 and *3 alleles have no functional metabolism of clopidogrel. These two alleles account for most of the reduced function alleles in patients of Caucasian (85%) and Asian (99%) descent classified as poor metabolizers.
- The CYP2C19*4, *5, *6, *7, and *8 and other alleles may be associated with absent or reduced metabolism of clopidogrel but are less frequent than the CYP2C19*2 and *3 alleles.[55]

Noticing DV's CYP2C19 genotype may reveal that he carries a variant that diminishes the antiplatelet effect of clopidogrel. If this were the case, alternative antiplatelet therapies may have been considered, reducing the chance that DV would suffer a second cardiac event. Homozygous carriers, who are poor CYP2C19 metabolizers, make up 3–4% of the population. A current FDA-boxed warning states that poor CYP2C19 metabolizers may not benefit from clopidogrel and recommends that prescribers consider alternative treatment for patients in this category. However, routine CYP2C19 testing is not recommended, and no firm recommendations have been established regarding dose adjustments for CYP2C19 status.

Clinicians should be aware that the low exposure seen in poor metabolizers also occurs in patients taking drugs that inhibit CYP2C19.[54] Although a higher dose regimen (600 mg loading dose followed by 150 mg once daily) in poor metabolizers increases antiplatelet response, an appropriate dose regimen for this patient population has not been established in a clinical outcome trial. Switching to prasugrel (Effient) is also another alternative to this problem.

ACKNOWLEDGMENT

The author acknowledge the contributions of Drs Anahita Malekakhlagh and Caroline Forrester.

REFERENCES

1. Wilkinson G. (2005). Drug metabolism and variability among patients in drug response. *New England Journal of Medicine*, 352: 2211–2221.
2. Wadelius M, Pirmohamed M. (2007). Pharmacogenetics of warfarin: Current status and future challenges. *Pharmacogenomics Journal*, 7: 99–111.
3. PharmGKB. *Warfarin pathway pharmacokinetics: Representation of the candidate genes involved in transport, metabolism and clearance of warfarin.* Palo Alto, CA: Stanford University. Available from: http://www.pharmgkb.org/pathway/PA145011113 (accessed February 24, 2012).
4. Voora D, McLeod HL, Elby C, Gage BF. (2005). The pharmacogenetics of coumarin therapy. *Pharmacogenomics*, 6: 503–513.

5. Tantisira KG, Lasky-Su J, Harada M, Murphy A, Litonjua AA, Himes BE, Lange C, et al. (2011). Genomewide association between GLCCI1 and response to glucocorticoid therapy in asthma. *New England Journal of Medicine*, 365: 1173–1183.

6. Yip V, Pirmohamed M. (2013). Expanding role of pharmacogenomics in the management of cardiovascular disorders. *American Journal of Cardiovascular Drugs*, 13: 151–162.

7. Jiayi L, Wang S, Barone J, Malone B. (2009). Warfarin pharmacogenomics. *Pharmacy and Therapeutics*, 34: 422–427.

8. Dean L. (2013). *Warfarin therapy and the genotypes CYP2C9 and VKORC1.* Medical Genetics Summaries [Internet]. Bethesda (MD): National Center for Biotechnology Information (US); 2012. Bookshelf ID: NBK84174.

9. Henry I. Bussey, Ann K Wittkowsky, Elaine M. Hylek, Marie B. Walker. (2007). Genetic testing to aid in warfarin (Coumadin) dosing? – Not Yet Ready for Prime Time. ClotCare Online Resource.

10. Tang HL, Shi WL, Li XG, Zhang T, Zhai SD, Xie HG. (2015). Limited clinical utility of genotype-guided warfarin initiation dosing algorithms versus standard therapy: A meta-analysis and trial sequential analysis of 11 randomized controlled trials. *The Pharmacogenomics Journal*, 1–9. Apr 14. doi: 10.1038/tpj.2015.16. [Epub ahead of print].

11. Caraco Y, Blotnick S, Muszkat M. (2008). CYP2C9 genotype-guided warfarin prescribing enhances the efficacy and safety of anticoagulation: A prospective randomized controlled study. *Clinical Pharmacology & Therapeutics*, 83: 460–470.

12. Epstein RS, Moyer TP, Aubert RE, O'Kane DJ, Xia F, Verbrugge RR, Gage BF, Teagarden JR. (2010). Warfarin genotyping reduces hospitalization rates results from the MM-WES (Medco-Mayo Warfarin Effectiveness study). *Journal of the American College of Cardiology*, 55: 2804–2812.

13. Gnoth MJ, Buetehorn U, Muenster U, Schwarz T, Sandmann S. (2011). In vitro and in vivo P-glycoprotein transport characteristics of rivaroxaban. *Journal of Pharmacology and Experimental Therapeutics*, 338: 372–380.

14. Blech S, Ebner T, Ludwig-Schwellinger E, Stangier J, Roth W. (2008). The metabolism and disposition of the oral direct thrombin inhibitor, dabigatran, in humans. *Drug Metabolism and Disposition*, 36: 386–399.

15. Stangier J, Clemens A. (2009). Pharmacology, pharmacokinetics, and pharmacodynamics of dabigatran etexilate, an oral direct thrombin inhibitor. *Clinical and Applied Thrombosis/Hemostasis*, 15(Suppl 1): 9S–16S.

16. Warkentin TE, Greinacher A. (2004). Heparin-induced thrombocytopenia: Recognition, treatment, and prevention: The Seventh ACCP Conference on Antithrombotic and Thrombolytic Therapy. *Chest*, 1263(Suppl): 311S–337S.

17. Harris K, Nguyen P, Van Cott EM. (2008). Platelet PlA2 polymorphism and the risk for thrombosis in heparin-induced thrombocytopenia. *American Journal of Clinical Pathology*, 129: 282–286.

18. Carlsson LE, Lubenow N, Blumentritt C, Kempf R, Papenberg S, Schröder W, Eichler P, Herrmann FH, Santoso S, Greinacher A. (2003). Platelet receptor and clotting factor polymorphisms as genetic risk factors for thromboembolic complications in heparin-induced thrombocytopenia. *Pharmacogenetics*, 13: 253–258.

19. Ahmed I, Majeed A, Powell R. (2007). Heparin induced thrombocytopenia: Diagnosis and management update. *Postgraduate Medical Journal*, 83: 575–582.

20. Burgess J, Lindeman R, Chesterman C, Chong B. (1995). Single amino acid mutation of Fc gamma receptor is associated with the development of heparin-induced thrombocytopenia. *British Journal of Haematology*, 91: 761–766.

21. Arepally G, McKenzie S, Jiang X, Poncz M, Cines D. (1997). Fc gamma RIIA H/R 131 polymorphism, subclass-specific IgG anti-heparin/platelet factor 4 antibodies and clinical course in patients with heparin-induced thrombocytopenia and thrombosis. *Blood*, 89: 370–375.

22. Johnson J, Cavallari L, Beitelshees A, Lewis J, Shuldiner A, Roden D. (2011). Pharmacogenomics: Application to the management of cardiovascular disease. *Clinical Pharmacology & Therapeutics*, 90: 519–531.

23. Pare G, Eikelboom J. (2011). CYP2C19 genetic testing should not be done in all patients treated with clopidogrel who are undergoing percutaneous coronary intervention. *Circulation: Cardiovascular Interventions*, 4: 514–521.

24. Kitzmiller J, Groen D, Phelps M, Sadee W. (2011). Pharmacogenomic testing: Relevance in medical practice: Why drugs work in some patients but not in others. *Cleveland Clinic Journal of Medicine*, 78: 243–257.

25. Eli Lilly. (July 2009). Product information for Effient. Eli Lilly: Indianapolis, IN.

26. Mega JL, Close SL, Wiviott SD, Shen L, Hockett RD, Brandt JT, Walker JR, et al. (2009). Cytochrome P450 genetic polymorphisms and the response to prasugrel: Relationship to pharmacokinetic, pharmacodynamic, and clinical outcomes. *Circulation*, 119: 2553–2560.

27. Wallentin L, James S, Storey RF, Armstrong M, Barratt BJ, Horrow J, Husted S, et al. (2010). Effect of CYP2C19 and ABCB1 single nucleotide polymorphisms on outcomes of treatment with ticagrelor versus clopidogrel for acute coronary syndromes: A genetic substudy of the PLATO trial. *The Lancet*, 376: 1320–1328.

28. Gladding P, White H, Voss J, Ormiston J, Stewart J, Ruygrok P, Bvaldivia B, Baak R, White C, Webster M. (2009). Pharmacogenetic testing for clopidogrel using the rapid INFINITI analyzer: A dose-escalation study. *Journal of the American College of Cardiology & Cardiovascular Intervention*, 2: 1095–1101.

29. Brodde OE, Bruck H, Leineweber K, Seyfarth T. (2001). Presence, distribution and physiological function of adrenergic and muscarinic receptor subtypes in the human heart. *Basic Research in Cardiology*, 96: 528–538.

30. Brugts JJ, Ferrari R, Simoons ML. (2009). Angiotensin-converting enzyme inhibition by perindopril in the treatment of cardiovascular disease. *Expert Review in Cardiovascular*, 7: 345–360.

31. Awidi A, Saleh A, Dweik M, Kailani B, Abu-Fara M, Nabulsi R, Bener A. (2011). Measurement of platelet reactivity of patients with cardiovascular disease on-treatment with acetyl salicylic acid: A prospective study. *Heart Vessels*, 26(5): 516–522.

32. Thompson PD, Clarkson P, Karas RH. (2003). Statin-associated myopathy. *Journal of American Medical Association*, 289: 1681–1690.

33. Baigent C, Keech A, Kearney PM, Blackwell L, Buck G, Pollicino C, Kirby A, et al. (2005). Efficacy and safety of cholesterol-lowering treatment: Prospective meta-analysis of data from 90,056 participants in 14 randomised trials of statins. *The Lancet*, 366: 1267–1278.

34. Johnson JA, Cavallari LH. (2013). Pharmacogenetics and cardiovascular disease—Implications for personalized medicine. *Pharmacology Review*, 65(3): 987–1009.

35. Reiter E, Lefkowitz RJ. (2006). GRKs and beta-arrestins: Roles in receptor silencing, trafficking and signaling. *Trends Endocrinology & Metabolism*, 17: 159–165.

36. Chan SW, Hu M, Tomlinson B. (2012). The pharmacogenetics of β-adrenergic receptor antagonists in the treatment of hypertension and heart failure. *Expert Opinion on Drug Metabolism and Toxicology*, 8: 767–790.

37. Vandell AG, Lobmeyer MT, Gawronski BE, Langaee TY, Gong Y, Gums JG, Beitelshees AL, et al. (2012). G protein receptor kinase 4 polymorphisms: β-blocker pharmacogenetics and treatment-related outcomes in hypertension. *Hypertension*, 60: 957–964.

38. Cordeiro J, Brugada R, Wu Y, Hong K, Dumaine R. (2005). Modulation of IKr inactivation by mutation N588K in KCNH2: A link to arrhythmogenesis in short QT syndrome. *Cardiovascular Research*, 67: 498–509.

39. Kannankeril P, Roden D, Darbar D. (2010). Drug-induced long QT syndrome. *Pharmacological Reviews*, 62: 760–781.

40. Rigat B, Hubert C, Alhenc-Gelas F, Cambien F, Corvol P, Soubrier F. (1990). An insertion/deletion polymorphism in the angiotensin I-converting enzyme gene accounting for half the variance of serum enzyme levels. *Journal of Clinical Investigation*, 86: 1343–1346.

41. Tiret L, Rigat B, Visvikis S, Breda C, Corvol P, Cambien F, Soubrier F. (1992). Evidence, from combined segregation and linkage analysis, that a variant of the angiotensin I-converting enzyme (ACE) gene controls plasma ACE levels. *American Journal of Human Genetics*, 51: 197–205.

42. Bleumink GS, Schut AF, Sturkenboom MC, van Duijn CM, Deckers JW, Hofman A, Kingma JH, Witteman JC, Stricker BH. (2005). Mortality in patients with hypertension on angiotensin-I converting enzyme (ACE)-inhibitor treatment is influenced by the ACE insertion/deletion polymorphism. *Pharmacogenetics & Genomics*, 15: 75–81.

43. Arnett DK, Boerwinkle E, Davis BR, Eckfeldt J, Ford CE, Black H. (2002). Pharmacogenetic approaches to hypertension therapy: Design and rationale for the Genetics of Hypertension Associated Treatment (GenHAT) study. *Pharmacogenomics Journal*, 2: 309–317.

44. Harrap SB, Tzourio C, Cambien F, Poirier O, Raoux S, Chalmers J, Chapman N, et al. (2003). The ACE gene I/D polymorphism is not associated with the blood pressure and cardiovascular benefits of ACE inhibition. *Hypertension*, 42: 297–303.

45. Verschuren JJ, Trompet S, Wessels JA, Guchelaar HJ, de Maat MP, Simoons ML, Jukema JW. (2011). A systematic review on pharmacogenetics in cardiovascular disease: Is it ready for clinical application? *European Heart Journal*, 33: 165–175.

46. Brugts JJ, Isaacs A, Boersma E, van Duijn CM, Uitterlinden AG, Remme W, Bertrand M, et al. (2010). Genetic determinants of treatment benefit of the angiotensin-converting enzyme-inhibitor perindopril in patients with stable coronary artery disease. *European Heart Journal*, 31: 1854–1864.

47. Redon J, Luque-Otero M, Martell N, Chaves F. (2004). Renin–angiotensin system gene polymorphisms: Relationship with blood pressure and microalbuminuria in telmisartan-treated hypertensive patients. *The Pharmacogenomics Journal*, 5: 14–20.

48. Ortlepp J, Hanrath P, Mevissen V, Kiel G, Borggrefe M, Hoffmann R. (2002). Variants of the CYP11B2 gene predict response to therapy with candesartan. *European Journal of Pharmacology*, 445: 151–152.

49. Huang CC, Chung CM, Hung SI, Leu HB, Wu TC, Huang PH, Lin SJ, Pan WH, Chen JW. (2011). Genetic variation in renin predicts the effects of thiazide diuretics. *European Journal of Clinical Investigation*, 41(8): 828–835. doi:10.1111/j.1365-2362.2011.02472.x.

50. Nordestgaard BG, Kontula K, Benn M, Dahlöf B, de Faire U, Edelman JM, Eliasson E, et al. (2010). Effect of ACE insertion/deletion and 12 other polymorphisms on clinical outcomes and response to treatment in the LIFE study. *Pharmacogenetics and Genomics*, 20(2): 77–85. doi: 10.1097/FPC.0b013e328333f70b.

51. Beeks E. (2004). Adducin Gly460Trp polymorphism and renal hemodynamics in essential hypertension. *Hypertension*, 44: 419–423.

52. Lynch A, Boerwinkle E, Davis B, Ford C, Eckfeldt J, Leiendecker-Foster C, Arnett D. (2008). Pharmacogenetic association of the NPPA T2238C genetic variant with cardiovascular disease outcomes in patients with hypertension. *JAMA*, 299: 296–307.

53. Katz A. (1986). Pharmacology and mechanisms of action of calcium-channel blockers. *Journal of Clinical Hypertension*, 2: 28S–37S.

54. Bremer T, Man A, Kask K, Diamond C. (2006). CACNA1C polymorphisms are associated with the efficacy of calcium channel blockers in the treatment of hypertension. *Pharmacogenomics*, 7: 271–279.

55. Langaee T, Gong Y, Yarandi H, Katz D, Cooper-Dehoff R, Pepine C, Johnson J. (2007). Association of CYP3A5 polymorphisms with hypertension and antihypertensive response to verapamil. *Clinical Pharmacology & Therapeutics*, 81: 386–391.

56. Pacanowski M, Gong Y, Cooper-Dehoff R, Schork N, Shriver M, Langaee T, Johnson J. (2008). Beta-adrenergic receptor gene polymorphisms and beta-blocker treatment outcomes in hypertension. *Clinical Pharmacology & Therapeutics*, 84: 715–721.

57. Beitelshees A, Gong Y, Wang D, Schork N, Cooper-Dehoff R, Langaee T, Johnson J. (2007). KCNMB1 genotype influences response to verapamil SR and adverse outcomes in the International Verapamil SR/Trandolapril Study (INVEST). *Pharmacogenetics and Genomics*, 17, 719–729.

58. Aerssens J, Paulussen AD. (2005). Pharmacogenomics and acquired long QT syndrome. *Pharmacogenomics*, 6(3): 259–270.

59. Li G, Cheng G, Wu J, Zhou X, Liu P, Sun C. (2013). Drug-induced long QT syndrome in women. *Advances in Therapy*, 30: 793–802.

60. Wolkenstein P, Carriere V, Charue D, Bastuji-Garin S, Revuz J, Roujeau JC, Beaune P, Bagot M. (1995). A slow acetylator genotype is a risk factor for sulphonamide-induced toxic epidermal necrolysis and Stevens-Johnson syndrome. *Pharmacogenetics*, 5: 255–258.

61. Mega JL, Simon T, Collet JP, Anderson JL, Antman EM, Bliden K, Cannon CP, et al. (2010) Reduced function CYP2C19 genotype and risk of adverse clinical outcomes among patients treated with clopidogrel predominantly for PCI: A meta-analysis. *JAMA* 304: 1821–1830.

62. Roden DM, Stein CM. (2009). Clopidogrel and the concept of high-risk pharmacokinetics. *Circulation* 119: 2127–2130.

8 Clinical Applications of Pharmacogenomics in Drug Therapy of Neurologic and Psychiatric Disorders

Megan J. Ehret

CONTENTS

KEY CONCEPTS

- Genetic variations contribute to individual differences in pharmacokinetics and pharmacodynamics of neurologic and psychiatric drugs.
- The translation of pharmacodynamic testing into clinical practice has multiple limitations.
- Challenges exist in determining specific genetic components of SSRI response to test pharmacogenomically.
- Larger population-based studies are needed to determine the relationship between polymorphisms and response to drug treatment in neurology and psychiatry.

INTRODUCTION

The human brain is one of the most complex organs in the body. This complexity makes the treatment of central nervous system (CNS) disorders very challenging. CNS medications are the second largest class of drugs ($n \geq 25$) whose labels contain pharmacogenomic information approved by the US Food and Drug Administration (FDA) (Table 8.1).[1] Genetic variables influencing the pharmacokinetics and pharmacodynamics of these medications have been extensively described in the literature, but large-scale randomized controlled clinical studies specifically designed to assess whether clinical testing is better than standard of care are still limited or even lacking.[2] These studies are difficult to design and perform because of the multidimensional biological causes of neurologic and psychiatric diseases and limited knowledge of the mechanism of action of their treatments.[3] Currently, drug labeling and consensus guidelines offer some guidance on the potential clinical utility of pharmacogenetic testing with the use of these medications.

PHARMACOGENETIC STUDIES OF THE RESPONSE TO NEUROLOGIC/PSYCHOTROPIC MEDICATIONS

Pharmacogenomic studies have been completed with multiple classes of neuropsychiatric medications, with the major focus on interindividual variation in drug efficacy. The vast majority of these studies have used a candidate gene approach, which is usually based on the receptor pharmacology of the neuropsychiatric medications. Although some studies have shown statistically significant effect sizes for several of these candidate genes (dopamine 2 receptor and serotonin transporter), the studies do not yield adequate sensitivity and specificity to reasonably guide clinical practice.[2,4,5] Additionally, alternative treatment strategies for the genetic carriers who may not respond well to medications have not been empirically tested, because most prescribed neurologic or psychotropic drugs have similar primary targets in each major class.[2]

Another area of research where functional significance is important, but one in which fewer studies have focused, is the cytochrome P450 (CYP450) system in neurologic or psychiatric pharmacogenomics.[6] This may be due to the lack of compelling empirical support for a relationship between plasma drug levels and

TABLE 8.1
FDA Product Labeling for CNS Medications

Drug Name	Pharmacogenetic Biomarker	Label Information
Amitriptyline	CYP2D6	• PM: May have higher than expected plasma concentrations of TCAs when given usual doses, and the increase in plasma concentration range can be large.
Aripiprazole	CYP2D6	• PMs: Initial dose should be reduced to ½ of the usual dose. • PMs: 80% increase in aripiprazole exposure and about a 30% decrease in exposure to the active metabolite compared to EMs. • Mean elimination half-lives are 75 h for EMs and 146 h for PMs.
Aripiprazole (Abilify Maintena™)	CYP2D6	• PMs and taking a CYP3A4 inhibitor for greater than 14 days: adjusted dose of 200 mg. • PMs: Adjusted dose of 300 mg.
Atomoxetine	CYP2D6	• PMs: 10-fold higher AUC and a 5-fold higher C_{max} to a given dose of atomoxetine compared with EMs; higher rates of some adverse effects. • Children/adolescents up to 70 kg body weight: Initiation dose should be 0.5 mg/kg/day and only increased to the usual start dose of 1.2 mg/kg/day if symptoms fail to be improved after 4 weeks and the initial dose is well tolerated. • Children/adolescents >70 kg body weight: Initial dose should be 40 mg/day and only increased to the usual target dose of 80 mg/day if symptoms fail to be improved after 4 weeks and the initial dose is well tolerated.
Carbamazepine	HLA-B*1502	• Chinese ancestry: Strong association between risk of developing SJS/TEN and the presence of HLA-B*1502 • Patients with ancestry in genetically at-risk populations should be screened for the presence of HLA-B*1502 prior to initiating treatment with carbamazepine. • Patients testing positive should not be treated with carbamazepine unless the benefit clearly outweighs the risk.
Citalopram	CYP2C19	• Maximum dose should be limited to 20 mg/day in patients who are CYP2C19 PMs due to risk of QT prolongation.
Clomipramine	CYP2D6	• PMs: Higher than expected plasma concentrations of TCAs when given usual doses. Depending on the fraction of drug metabolized by CYP2D6, the increase in plasma concentration may be small or quite large (eightfold increase in plasma AUC of the TCAs).
Clozapine	CYP2D6	• PMs: May develop higher than expected plasma concentrations of clozapine when given usual doses.

(Continued)

TABLE 8.1 *(Continued)*
FDA Product Labeling for CNS Medications

Drug Name	Pharmacogenetic Biomarker	Label Information
Desipramine	*CYP2D6*	• PMs: Higher than expected plasma concentrations of TCAs when given usual doses, the increase in plasma concentration may be small or quite large (eightfold increase in plasma AUC of the TCAs).
Diazepam (rectal gel formulation only, not the oral tablet)	*CYP2C19*	• Marked interindividual variability in the clearance of diazepam reported, likely attributable to variability of CYP2C19.
Doxepin	*CYP2D6*	• PMs: 2D6 may have higher plasma levels than normal subjects.
Fluoxetine	*CYP2D6*	• PMs: Metabolized *S*-fluoxetine at a slower rate and achieved higher concentrations of *S*-fluoxetine. Compared to normal metabolizers, the total sum at steady state of the plasma concentrations of the 4 active enantiomers was not significantly greater among poor metabolizers. The net pharmacodynamic activities were essentially the same for PMs and normal metabolizers.
Fluoxetine and olanzapine combination	*CYP2D6*	• Same as fluoxetine.
Fluvoxamine	*CYP2D6*	• In vivo study of fluvoxamine single-dose pharmacokinetics in 13 PMs demonstrated altered pharmacokinetic properties compared to 16 EMs: mean C_{max}, AUC, and $t_{1/2}$ were increased by 52%, 200%, and 62%, respectively
Iloperidone	*CYP2D6*	• PMs: Dose should be reduced by ½. • Observed mean elimination half-lives for iloperidone, P88, and P95 in EMs are 18, 26, and 23 h, respectively; and in PMs are 33, 37, and 31 h, respectively.
Imipramine	*CYP2D6*	• PMs: Higher than expected plasma concentrations of TCAs when given usual doses. Depending on the fraction of drug metabolized by CYP2D6, the increase in plasma concentration may be small, or quite large (eightfold increase in plasma AUC of the TCAs).
Modafinil	*CYP2D6*	• PMs: Levels of CYP2D6 substrates may be increased by coadministration of modafinil. • Dose adjustments may be necessary for patients being treated with CYP2D6 substrates or inhibitors.
Nortriptyline	*CYP2D6*	• PMs: Higher than expected plasma concentrations of TCAs when given usual doses. Depending on the fraction of drug metabolized by CYP2D6, the increase in plasma concentration may be small or quite large (eightfold increase in plasma AUC of the TCAs).

(Continued)

TABLE 8.1 *(Continued)*
FDA Product Labeling for CNS Medications

Drug Name	Pharmacogenetic Biomarker	Label Information
Paroxetine	*CYP2D6*	• Only information about inhibition of CYP2D6.
Perphenazine	*CYP2D6*	• PMs: Will metabolize perphenazine more slowly and will experience higher concentrations compared with EMs.
Phenytoin	*HLA*-B*1502	• Chinese ancestry: Strong association between the risk of developing SJS/TEN and the presence of *HLA*-B*1502 in patients using carbamazepine. • Limited evidence suggests the polymorphism is also a risk for the development of SJS/TEN in patients of Asian ancestry taking other antiepileptic drugs associated with SJS/TEN, including phenytoin. • Phenytoin should be avoided as an alternative for carbamazepine in patients positive for *HLA*-B*1502
Pimozide	*CYP2D6*	• PMs: Exhibit higher pimozide concentrations than EMs. Time to achieve steady-state concentrations is expected to be longer in PMs because of the prolonged $t_{1/2}$. Alternate dosing strategies are recommended in PMs. • Children: At doses >0.05 mg/kg/day, *CYP2D6* genotyping should be performed. • In PMs, doses should not exceed 0.05 mg/kg/day, and doses should not be increased earlier than 14 days. Adult dosing: doses above 4 mg/day, *CYP2D6* genotyping should be performed. In PMs, doses should not exceed 4 mg/day, and doses should not be increased earlier than 14 days.
Protriptyline	*CYP2D6*	• PMs: Higher than expected plasma concentrations of TCAs when given usual doses; the increase in plasma concentration may be small or quite large (eightfold increase in plasma AUC of the TCAs).
Risperidone	*CYP2D6*	• EMs: Convert risperidone rapidly to 9-hydroxyrisperidone, whereas PMs convert it much more slowly. • After single and multiple dose studies, the pharmacokinetics are similar in EMs and PMs.
Tetrabenazine	*CYP2D6*	• Doses above 50 mg should not be given without *CYP2D6* genotyping. • PMs: Will have substantially higher levels of the primary drug metabolites than EMs. • PMs: Maximum recommended total daily dose is 50 mg and the maximum recommended single dose is 25 mg. • EMs: Maximum recommended total daily dose is 100 mg and the maximum recommended single dose is 37.5 mg.

(Continued)

TABLE 8.1 *(Continued)*
FDA Product Labeling for CNS Medications

Drug Name	Pharmacogenetic Biomarker	Label Information
Thioridazine	*CYP2D6*	• Reduced CYP2D6 activity: Would be expected to augment the prolongation of the QTc interval associated with thioridazine and may increase the risk of serious, potentially fatal cardiac arrhythmias. • Thioridazine is contraindicated in individuals with a known genetic defect leading to reduced levels of activity of CYP2D6.
Trimipramine	*CYP2D6*	• PMs: Higher than expected plasma concentrations of TCAs when given usual doses. Depending on the fraction of drug metabolized by CYP2D6, the increase in plasma concentration may be small, or quite large (eightfold increase in plasma AUC of the TCAs).

Valproic acid	UCD	• Contraindicated in patients with known urea cycle disorders.
		• Prior to the initiation of valproate therapy, evaluation for UCD should be considered for patients with:
		  • a history of unexplained encephalopathy or coma, encephalopathy associated with a protein load, pregnancy-related or postpartum encephalopathy, unexplained mental retardation, or history of elevated plasma ammonia or glutamine;
		  • cyclical vomiting and lethargy, episodic extreme irritability, ataxia, low BUN, or protein avoidance;
		  • a family history of UCD or a family history of unexplained infant deaths (particularly males); or
		  • other signs or symptoms of UCD.

Venlafaxine	*CYP2D6*	• Total concentrations of active metabolites were similar between PMs and EMs. • No dosage adjustment is required when venlafaxine is coadministered with a CYP2D6 inhibitor.

Note: AUC, area under the curve; C_{max}, maximum drug plasma concentration; CYP, cytochrome 450; EM, extensive metabolizers; PM, poor metabolizers; SJS, Steven–Johnson syndrome; $t_{1/2}$, half-life; TCA, tricyclic antidepressant; TEN, toxic epidermal necrolysis; UCD, urea cycle disorder.

neuropsychiatric drug efficacy. Using pharmacogenomics in determining the pharmacokinetics of medications has been useful in the potential prediction of adverse effects, demonstrated by the development of clinical guidelines for dosing of medications based on *CYP* genotypes.[7]

A final approach to the study of pharmacogenomics in neurologic and psychiatric research is the utilization of genome-wide association studies (GWASs). A major limitation to the use of GWASs is the large sample sizes assumed to be necessary to overcome the statistical penalty acquired by the genotyping of hundreds of thousands of single nucleotide polymorphisms (SNPs).[8] To date, GWASs have not yet been useful in predicting genome-wide significant or clinically useful predictors of antidepressant response, lithium treatment response, or antipsychotic drug response.[9–16] Other limitations to the use of GWASs include the use of chronic patients, lack of medication adherence monitoring, and ambiguity of phenotype definition.[2]

Key Points: The translation of pharmacogenomic testing into clinical practice has several limitations in CNS disorders as below:

1. Lack of clear relationships between the serum concentrations of many neuropsychiatric drugs and response to treatment
2. Wide therapeutic ranges for many neuropsychiatric drugs
3. Lack of clinical guidelines and limited number of well-designed trials investigating the use of genomic testing in methods that could be translated into clinical practice
4. Use of multiple drugs to treat neuropsychiatric illnesses, which complicates the understanding of how and when to utilize available testing options
5. Diverse ethnicities of the patients with CNS disorders

The Clinical Pharmacogenomics Implementation Consortium established in 2009, was formed to establish evidence-based pharmacogenetic guidelines and disseminate them to clinicians in the field. The guidelines created are designed to assist and guide drug therapy in situations when genetic information is available, but the guidelines do not specifically state if or for whom pharmacogenetic testing should be obtained (www.pharmgkb.org).[17] Currently, the group has established or is in the process of establishing guidelines for the use of tricyclic antidepressants, selective serotonin reuptake inhibitors (SSRIs), carbamazepine, phenytoin, and valproic acid. The website offers a convenient interface to enter the patient's genotype for specific dosing recommendations given to the drugs they are prescribed.

PATIENT CASE

RJ, a 55-year-old female widowed school teacher, has been admitted to the inpatient psychiatric unit after a suicide attempt. RJ has a history of mild cognitive impairment and has been experiencing symptoms of depression after the passing of her husband six months ago. RJ describes her days as being spent sleeping and not eating. She is unable to concentrate on things, rereading the same newspaper article several times. She feels guilty about her husband's passing, stating how she should have died first. At admission, RJ is receiving donepezil 10 mg daily, lisinopril 10 mg daily, metformin 500 mg twice daily, omeprazole 20 mg daily, and a multivitamin daily.

The physician would like to start RJ on citalopram due to the low cost of the medication and the low incidence of adverse effects. The physician initiates the medication at 20 mg daily and titrates the dose to 40 mg over the course of a week while RJ is an inpatient.

Question: What pharmacogenomic issue could be useful in understanding the potential concerns with increasing citalopram to 40 mg daily?

Answer: In August 2011, the US FDA issued a Drug Safety Communication stating that citalopram should no longer be used at doses greater than 40 mg per day. Doses greater than 40 mg per day could cause potentially dangerous abnormalities in the electrical activity of the heart. The maximum dose recommended is 20 mg per day for patients older than 60 years of age, CYP2C19 poor metabolizers, and patients taking CYP2C19 inhibitors. These patients are likely to have increased plasma levels of citalopram and an increased QT interval prolongation and Torsade de Pointes. Patients should notify clinicians if they experience dizziness, palpitations, or syncope.

Given that citalopram is a frequently used drug for the treatment of depression, with approximately 31.5 million prescriptions dispensed at US outpatient retail pharmacies in 2011, CYP2C19 genotyping would be beneficial to do prior to dispensing to determine metabolizer status, which could alter dosing parameters. The risk of QTc prolongation could be minimized with proper dosing and understanding of metabolizer status prior to initiation of citalopram, especially in an aging population.

Additionally, the patient is receiving omeprazole 20 mg daily, which is a CYP2C19 inhibitor. This could further inhibit the metabolism of citalopram, resulting in accumulation of the drug in the body.

CYTOCHROME P450 ENZYME SYSTEM

To date, there is only one FDA-approved pharmacogenetic test for use in psychiatry, the Roche Diagnostics AmpliChip CYP450 Test (www.roche.com), which assesses 27 alleles in *CYP2D6* and 3 alleles in *CYP2C19*, although other lab-developed tests are available at select laboratories.[18] The clinical uptake of the AmpliChip has been modest at best, given several limitations. There are concerns about the interpretation of tests, the paucity of prospective data suggesting that test utilization influences clinical outcomes, and the lack of reimbursement for an expensive test.[2,6,19] Additionally, there is rapid progression in genotyping technology, with several companies producing genotyping platforms specifically designed for pharmacogenomics, focusing on genetic markers associated with absorption, distribution, metabolism, and enzymes.[2] These companies are marketing these different platforms either to prescribing clinicians or directly to the consumer. It is important to consider that not all pharmacogenomic platforms are the same. Each platform may not examine the same variants in the same gene, and these variants may be more or less important, depending on the ethnicity of the patient being tested. If a particular platform does not test for a particular variant, it may code the patient as a wild-type or no variant, when in fact he or she has a variant, which the platform cannot detect. The use of these additional platforms for clinical practice has been slow to enter all areas of clinical practice due to the high cost and long turnaround time, although this is changing as more pharmacogenomic testing is being covered by US insurance companies.

TARGET GENES INVOLVED IN NEUROLOGY AND PSYCHIATRY

There are a large number of genes involved in both the efficacy and safety of medications used for the treatment of neurologic and psychiatric medical conditions. The following sections contain detailed information on many genes that are currently being researched.

DEPRESSION–SEROTONIN TRANSPORTER GENE AND SEROTONIN RECEPTOR GENE

One of the most extensively studied genes affecting the treatment of depression is the serotonin transporter gene (SLC6A4), which is located at 17q.[20,21] SLC6A4 is a protein structure made up of 12 transmembrane helices with an extracellular loop between helices 3 and 4. This transporter is responsible for the reuptake of serotonin (5HT) into the presynaptic neuron. Variations in SLC6A4 allele frequencies occur across ancestral populations.

The most widely studied variant of SLC6A4 is the indel promoter polymorphism, which is frequently referred to as 5-HTTLPR. The polymorphism consists of a variant that is either 43 or 44 bp in size. There are many variations of both the long and short alleles of the polymorphism. Additionally, there has been some evidence that there is an interaction between the SNP (irs25531) located immediately upstream of the indel polymorphism and the activity of the long allele of the transporter protein.[22]

Extensive research, including several meta-analyses, has focused on the pharmacogenomic variability of SLC6A4 on antidepressant response to SSRIs. One meta-analysis of 15 studies concluded that patients who are homozygous for the long allele and are of European ancestry has a more consistent therapeutic response to SSRI treatment, while an additional meta-analysis ($n = 28$ various ethnicities) concluded that there is no significant effect of the transporter promoter length polymorphism on the rates of antidepressant response. The authors stated that there was substantial unexplained heterogeneity of effect sizes across the studies, eluding additional interacting factors that could contribute to an association in some cases.[23,24]

Key Points: SSRIs have a broad therapeutic index, and the use of genetic testing for dose-related outcomes is controversial. Despite the optimism in using pharmacogenomic testing in determining SSRI response, challenges still exist in determining the specific genetic components of SSRI response to testing.

Polymorphisms in the genes that code for various serotonin receptors have also been studied in regard to their roles in altering the efficacy of various antidepressants. The 5-HT1A and 5-HT2A receptors have also been studied with varying results. Additional areas of interest, which have been studied to a lesser extent, to determine an association regarding efficacy of the antidepressants include the following: G-protein coupled receptors, tryptophan hydroxylase (TPH) I, monoamine oxidase, dopamine receptor, noradrenergic receptor, nitric oxide, angiotensin-converting enzyme, interleukin-1β, stress hormone system, and phosphodiesterase. A recent meta-analysis concluded that evidence suggests that 5-HTTLPR, 5-HT1A, 5-HT2A, TPH1, and brain-derived neurotropic factor may modulate antidepressant response, despite the presence of some heterogeneity across the studies.[5]

SCHIZOPHRENIA

DRUG EFFICACY

A substantial number of trials have been conducted to predict the efficacy of antipsychotics in the treatment of schizophrenia with both the first- and second-generation antipsychotics. Studies have supported the importance of both the dopaminergic and serotonergic systems in mediating efficacy, but less is known about the substantiated roles of other neurotransmitter systems at this time.[5] These studies are difficult to complete, however, as there are differing sample and illness characteristics, antipsychotic medication types, treatment durations, adherence, adjuvant treatments, and outcome phenotypes.

ADVERSE EFFECTS

The C allele of rs381929 polymorphism (–759C/T) has been consistently associated with weight gain in patients taking antipsychotic drugs. A meta-analysis of the association between this polymorphism and weight gain concluded that the –759T allele is associated with less weight gain.[25] For clinical practice, if testing is available, antipsychotic drugs that have been demonstrated to stimulate weight gain should be avoided in carriers of the protective T allele.[25]

Clozapine-induced agranulocytosis (CIA) has a poorly understood mechanism. The relative rarity of CIA significantly limits the collection of large samples required to conduct robust genomic samples needed to determine the predictability of the adverse event. Several mechanisms have been studied, and one resulted in a commercially available product, clozapine metabolite-induced neutrophil toxicity and immune system mediation.

From 2002 to 2007, PGxHealth (pgxhealth.com) completed a GWAS that demonstrated that HLA-DQB1 6672G>C was a risk for CIA, and a 16.9-fold increased risk for carriers of this marker. In 2007, a commercial test was marketed to determine whether a patient was "lower risk" or "higher risk" for CIA. The problem of the test was the 21.5% sensitivity, which has been a limiting factor in its application in clinical settings.[26,27]

Another association that has been demonstrated in a small trial is the reduction on dihydronicotinamide riboside quinone oxidoreductase 2 mRNA levels seen in patients with CIA compared with controls. The complexity of possible genetic mechanisms in CIA may be very challenging, but adding to the complexity is the additional need for studies with varying ethnicities and environmental factors, which may also play a role in the rate of development of CIA.[28,29]

Tardive dyskinesia (TD), typically experienced not only by those who receive first-generation antipsychotics but also by those who receive second generation antipsychotics, can cause potentially permanent abnormal motor movements. It has been associated with genes in the dopamine system, including dopamine 2 (D2) and dopamine 3 (D3) receptors and the catechol-O-methyltransferase (COMT) enzyme. Several studies have demonstrated that the minor T allele of the taq 1A polymorphism, rs1800497, appears to be protective against TD, while the rs6280 SNP of the DRD3 receptor has shown inconsistent results.[30–32] Additionally, a functional polymorphism that codes for a substitution of methionine (met) for valine (val) at codon 158 has been reported to have an association with TD.[33] Other pathways that have been investigated with varying results include the following: accumulation of free radicals or oxidative stress, 5-HT2A, and CYP2D6.[34–36]

Key Points: Antipsychotic-associated weight gain is a serious consequence for morbidity and mortality for patients. Pharmacogenomic approaches have allowed for detection of more than 300 possible candidate genes for this adverse effect. Given the variable histories of prior drug exposure and medication adherence of patients treated with these drugs, it has confounded attempts to identify genetic effects on this complex phenotype. Clinicians should monitor body mass index, total fat mass, blood glucose, and insulin levels in patients for the development of obesity.[37]

PATIENT CASE

TH, a 45-year-old male who was resistant to schizophrenia treatment, is currently being treated with haloperidol 10 mg twice daily. He has been treated with several different antipsychotics in the past with no efficacy. His physician explains the risk of TD to his family and what to watch for as TH starts on the haloperidol.

Question: Are there any pharmacogenomic tests that can be completed to determine if TH is at risk for developing TD? If there are, do clinicians have an ethical obligation to complete the testing prior to prescribing antipsychotics?

Answer: TD is a long-term, potentially irreversible muscular side effect associated with the use of antipsychotics. It is characterized by random movements in the tongue, lips, or jaw as well as facial grimacing, movements of arms, lips, fingers, and toes. It can also be characterized by swaying movements of the trunk or hips. Extensive research has been conducted to determine if genetics plays a role in determining who is at risk for developing TD when prescribed antipsychotics. Currently, the DRD2 polymorphism has mixed results, although a meta-analysis demonstrates that DRD2 taq A2/A2 genotype increases the risk of TD.[38] The serine-9-glycine polymorphism in the DRD3 receptor has been studied with mixed results.[39] There have been consistent results with the G-protein signaling 2 (*RGS 2*) gene in identifying those at risk for TD.[39] The evidence currently states that there is a major role of dopamine and serotonin receptors and the broader dopamine and serotonin systems in the development of TD and extrapyramidal symptoms during antipsychotic treatment.

One area of concern with pharmacogenomic testing and TD is the uncertainty that comes with the results. The sensitivity, specificity, and predictive value of testing have not been established to date. Another challenge we face as clinicians is consent to perform the pharmacogenomic test. Given the nature of the illness we are treating, this can be quite challenging. Additionally, explaining the future applications of genetic information is controversial in this field. All of these areas must be considered prior to deciding whether pharmacogenomic testing will be appropriate to the patient population you are treating.[39]

BIPOLAR DISORDER

Lithium-responsive bipolar disorder is linked to several unique characteristics, including euphoric manias, positive family history, few comorbidities, and symptom-free intervals between episodes.[40,41] Pharmacogenetic association studies have focused on selected genes. Studies that have investigated the mechanism of action of lithium and predication via SNP pharmacogenomic testing have only modest success, with few replications.[40,41] GWASs may prove to be a better design in determining the combination of genetic variations, which may detect lithium response.[40,41]

ALZHEIMER DISEASE

Numerous studies have been conducted to determine the influence of the *apoE* genotype on drug response in Alzheimer disease (AD). In the monogenic-related studies, the *apoE*-4/4 carriers are the poorest responders to medication.[42–45] In trigenic-related studies, the *apoE*-4/4, presenilin 1 and 2, the best responders are those patients carrying the 331222-, 341222-, and 441112-genomic profiles. The worst responders in all genomic clusters were those patients with the 441122+ genotypes.[42–45] These results demonstrate the deleterious effect of the *apoE*-4/4 genotype on AD, in sporadic and familial late-onset AD, therapeutics in combination with the other AD-related genes.[42–45]

The current treatments for AD, the cholinesterase inhibitors, are metabolized via the CYP450 pathway. Poor metabolizers and ultrarapid metabolizers are the poorest responders to drug treatment, while the extensive and intermediate metabolizers are the best responders.[42–45] In light of the emerging data, it seems very plausible that the determination of response to drug treatment of AD could depend on the interaction of genes involved in the drug metabolism and those genes associated with AD pathogenesis.[42–45]

Additionally, the gene encoding for choline transferase, which encodes the major catalytic enzyme of the cholinergic pathway, is associated with response to acetylcholinesterase (AChE) inhibitors (AChEIs). A SNP in the promoter region of choline *O*-acetyltransferase (CHAT), rs733722, accounts for 6% of the variance in response to AChEIs.[46]

Key Points: Larger population confirmation studies are still needed on polymorphisms in genes of the cholinergic markers that have been found to predict better response to AChEIs: AChE, butyrylcholinesterase, choline AChE, and paraoxonase.[47]

EPILEPSY

Little evidence currently exists to correlate any CYP450 polymorphisms with clinical response, serum drug concentrations, or other measurable outcomes in those taking antiepileptic medications. Phenytoin metabolism is the only anticonvulsant with evidence to suggest that variation in based on various *CYP2C9* alleles can be predictive of toxicity.[48,49]

Recent research has shown that the human leukocyte antigen (HLA) *HLA*-B*1502 shows a strong association with carbamazepine-induced Stevens–Johnson syndrome (SJS) in Han Chinese.[50] However, this association was not found in any Caucasian populations. These findings did prompt the US FDA to change the labeling information of carbamazepine to include information about *HLA*-B*1502, and the US FDA now recommends genotyping individuals of Asian ancestry for the allele prior to using the drug.[51]

Key Points: *HLA*-B*1502 should be genotyped prior to initiation of carbamazepine in the Asian ancestry population.

MULTIPLE SCLEROSIS

Candidate genes for *HLA* and interferon receptor polymorphisms have been studied to determine response to therapy with inconsistent definitions of response and markers studied. The HLA studies demonstrated that allelic variation has a correlation with antibody production, but not with treatment response.[52,53] Interferon receptor studies have demonstrated conflicting results.[53–55] One completed GWAS found some promising results but was limited by only examining SNPs detected by the author's microarray.[55,56]

PARKINSON DISEASE

Genetic variability, to some extent, is involved in the interindividual variability seen with the response to drug treatments with Parkinson disease (PD). Although clinical relevance has not been conclusively determined, there are interesting associations between pharmacotherapy and genetic polymorphisms for the following pairs:

- L-Dopa with COMT and dopamine receptors (a SNP at nucleotide 1947 that encodes for low or high activity genotype is associated with changes in response to both tolcapone and L-Dopa; *DRD2* gene may play a pivotal role in more frequent motor complications of long-term L-Dopa use; associations of late-onset hallucinations with the C allele of the taq1A polymorphism 10.5 kb of *DRD2*; and similar results obtained in *DRD1*, *DRD2*, *DRD3*, and *DRD4* in patients with PD with or without chronic visual hallucinations).

- COMT inhibitors with COMT and glucuronosyltransferase (UGT)1A9 (*COMT* haplotype appears to have little influence on the development of L-Dopa-induced dyskinesias).
- Selegiline with *CYP2B6* (*CYP2B6**18 and other *CYP2B6* defect variant alleles may be potential biomarkers for altered selegiline biodisposition; currently, there are no recommendations for selegiline use and *CYP2B6* pharmacogenetics).
- Pramipexole with *DRD3* [no significant associations between *DRD3* (MscI polymorphism) and *DRD4* (120 bp tandem duplication polymorphism in the promoter region)] polymorphisms and the phenomenon of "sleep attacks."[57]

PHARMACOGENOMICS OF ALCOHOL, NICOTINE, AND DRUG ADDICTION TREATMENTS

One of the most promising areas of research in pharmacogenomics and addiction treatment is the μ opioid receptor gene, *OPRM1* A118G (rs561720) polymorphism. Three randomized, placebo-controlled studies have shown that individuals with a G allele display longer abstinence and greater reduction in the positive effects of alcohol when taking naltrexone.[58–60] Additional studies are needed to test and develop adaptive approaches to treatment based on genotype information.[61]

The most encouraging pharmacogenetic data for smoking cessation treatment relates to *CYP2A6* polymorphisms. Two open-label studies of approximately 500 European subjects each found a link between nicotine metabolism and the effectiveness of nicotine replacement therapy. *CYP2A6* rapid metabolizers were less likely to achieve abstinence with transdermal nicotine replacement.[62,63]

Key Points: Further research is needed to determine how genetic information in the treatment of addiction disorders will be incorporated into clinical practice given the controversy regarding the implementation of genetic testing for complex phenotypes. Prospective studies and pragmatic clinical trials evaluating the use of genetic testing in a clinical setting and the effect on treatment outcome are warranted to further evaluate prospective testing in addiction medicine prior to implementation.[64]

CONCLUSIONS AND FUTURE PERSPECTIVES

With the increasing number of medications used to treat all of the varying disorders of the CNS, prescribing the correct medications for each patient will become increasingly more difficult. There are numerous genetic factors that contribute to the CNS disorders and many more which may ultimately lead to response or failure of a medication. In addition, the number of adverse drug events caused by these medications is staggering.

There are several factors affecting the implementation of neurologic and psychiatric pharmacogenomics. Initially, cost was the largest barrier to implementation of pharmacogenomics into mainstream practice, but as genotyping technologies decrease in cost, it may be a cost-effective strategy to incorporating pharmacotherapy. The concern regarding implementation is the lack of understanding of the implications of testing.[65] Although pharmacogenetic markers in the treatment of neurology and psychiatry are unlikely to attain perfect sensitivity and specificity, they are still beneficial and can be meaningful in informing clinical decisions, potentially clarifying prognosis, and guiding the development of a clinical treatment plan.[66,67]

Clinical Implications

1. Poor metabolizers: Examples of extreme adverse effects can be found in the literature; with the ability to predict poor metabolizers of *CYP2D6*, the clinician would be able to minimize adverse effects by either choosing a medication that is metabolized by an alternate enzyme or by modifying the dosing strategy for the drug metabolized via CYP2D6.
2. CYP1A2: This is an inducible enzyme. When patients who smoke cigarettes are treated with olanzapine and clozapine on an inpatient unit where smoking is restricted, blood levels can decrease upon release when smoking is reinitiated. If patient's CYP1A2 activity can be determined prior to discharge, proper dosing of these drugs could be determined to prevent relapse.
3. Appropriate antidepressant selection for patients: The trial-and-error method of selecting antidepressants can take a toll on a patient and clinician trying to determine the appropriate drug and dose for a patient. Determination of genetic susceptibility to response prior to initiation of drugs could speed the process of recovery for patients.
4. *HLA-*B-1502: Determination of susceptibility of patients of Asian ancestry to develop SJS with the initiation of carbamazepine. Determination of *HLA-*B-1502 allele status prior to the initiation of carbamazepine allows for a decrease in the potentially life-threatening adverse effects.

STUDY QUESTIONS

1. Which of the following is a limitation to the translation of pharmacogenomic testing in CNS clinical practice?
   a. Narrow therapeutic range for many neuropsychiatric drugs
   b. Use of monotherapy to treat neuropsychiatric illnesses
   c. Lack of clinical guidelines for pharmacogenomic testing use
   d. Limited ethnicities of the patients with CNS disorders

2. Which of the following genes has been extensively studied in the treatment response to depression?
   a. *SLC6A4*
   b. MAO
   c. Stress hormone system
   d. Nitric oxide

3. Possession of which of the following genotypes, according to current literature, appears to be protective against the development of tardive dyskinesia?
   a. *DRD3* rx6280
   b. T allele of rs1800497
   c. *HLA*-DQB1 6672 G>C
   d. C allele of rx381929

4. Which of the following genes should be genotyped prior to the initiation of carbamazepine in the population of Asian ancestry?
   a. *SLC6A4*
   b. *DRD3*
   c. *HLA*-B*1502
   d. *CYP3A4*

5. Which of the following drugs would show an increase in blood levels if smoking was restricted in a person who smokes?
   a. Olanzapine
   b. Quetiapine
   c. Lurasidone
   d. Aripiprazole

Answer Key

   1. c
   2. a
   3. b
   4. c
   5. a

REFERENCES

1. FDA. Table of pharmacogenomic biomarkers in drug labels. Available from: http://www.fda.gov/drugs/scienceresearch/researchareas/pharmacogenetics/ucm083378.htm (accessed February 24, 2015).
2. Malhota AK, Zhang JP, Lencz T. Pharmacogenetics in psychiatry: Translating research into clinical practice. *Mol Psychiatry* 2012;8:760–769.
3. Arranz MJ, Kapur S. Pharmacogenetics in psychiatry: Are we ready for widespread clinical use? *Schizophr Bull* 2008;6:1130–1144.
4. Zhang JP, Lencz T, Malhotra AK. D2 receptor genetic variation and clinical response to antipsychotic drug treatment: A meta-analysis. *Am J Psychiatry* 2010;167:763–772.
5. Serretti A, Kato M, De Ronchi D, Kinoshita T. Meta-analysis of serotonin transporter gene promoter polymorphism (5-HTTLPR) association with selective serotonin reuptake inhibitor efficacy in depressed patients. *Mol Psychiatry* 2007;12:247–257.

6. Fleeman N, Dundar Y, Dickson R, Jorgensen A, Pushpakom S, McLeod C, et al. Cytochrome P450 testing for prescribing antipsychotics in adults with schizophrenia: Systematic review and meta-analyses. *Pharmacogenomics J* 2011;11:1–14.

7. de Leon J, Armstrong SC, Cozza KL. Clinical guidelines for psychiatrists for the use of pharmacogenetic testing for CYP450 2D6 and CYP450 2C19. *Psychosomatics* 2006;47:75–85.

8. Cichon S, Craddock N, Daly M, Farone SV, Gejman PV, Kelose J, et al. Genome-wide association studies: History, rationale, and prospects for psychiatric disorders. *Am J Psychiatry* 2009;166:540–556.

9. Uher R, Perround N, Ng MY, Hauser J, Henigsberg N, Maier W, et al. Genome-wide pharmacogenetics of antidepressant response in the GENDEP project. *Am J Psychiatry* 2010;167:555–564.

10. Garriock HA, Kraft JB, Shyn SI, Peters EJ, Yokoyama JS, Jenkins GD, et al. A genome-wide association study of citalopram response in major depressive disorder. *Biol Psychiatry* 2011;67:133–138.

11. Ising M, Lucae S, Binder EB, Bettecken T, Uhr M, Ripke S, et al. A genome-wide association study points to multiple loci that predict antidepressant drug treatment outcome in depression. *Arch Gen Psychiatry* 2009;66:966–975.

12. Perlis RH, Smoller JW, Ferreira MA, McQuillin A, Bass N, Lawrence J, et al. A genome-wide association study of response to lithium for prevention of recurrence in bipolar disorder. *Am J Psychiatry* 2009;166:718–725.

13. Aberg K, Adkins DE, Bukszar J, Webb BT, Caroff SN, Miller DD, et al. Genome-wide association study of movement-related adverse antipsychotic effects. *Biol Psychiatry* 2010;67:279–282.

14. Adkins DE, Aberg K, McClay JL, Bukszar J, Zhao Z, Jia P, et al. Genome-wide pharmacogenomics study of metabolic side effects to antipsychotic drugs. *Mol Psychiatry* 2011;16:321–332.

15. Alkelai A, Greenbaum L, Rigbi A, Kanyas K, Lerer B. Genome-wide association study of antipsychotic-induced parkinsonism severity among schizophrenia patients. *Psychopharmacology (Berl)* 2009;206:491–499.

16. McClay JL, Adkins DE, Aberg K, Stroup S, Perkins DO, Vladimirov VI, et al. Genome-wide pharmacogenomics analysis of response to treatment with antipsychotics. *Mol Psychiatry* 2011;16:76–85.

17. The Pharmacogenomics Knowlegebase. Available from: www.pharmgkb.org (accessed February 24, 2015).

18. Amplichip CYP450 Test. Available from: http://www.roche.com/products/product-details.htm?type=product&id=17 (accessed February 24, 2015).

19. Evaluation of Genomic Applications in Practice and Prevention (EGAPP) Working Group. Recommendations for the EGAPP Working Group: Testing for cytochrome P450 polymorphisms in adults with nonpsychotic depression treated with selective serotonin reuptake inhibitors. *Genet Med* 2007;9:819–825.

20. Frueh FW. Regulation, reimbursement, and the long road of implementation of personalized medicine—A perspective from the United States. *Value Health* 2013;16:S27–S31.

21. Rammamoorthy S, Cool DR, Mahesh VB, Leibach FH, Melikian HE, Blakely RD, et al. Regulation of the human serotonin transporter. Cholera toxin-stimulation of serotonin uptake in human placental chorioarcinoma cells is accompanied by increased serotonin transporter mRNA levels and serotonin transporter-specific ligand binding. *J Biol Chem* 1993;268:21626–21631.

22. Lesch KP, Aulakh CS, Wolozin BL, Tolliver TJ, Hill JL, Murphy DL, et al. Regional brain expression of serotonin transporter mRNA and its regulation by reuptake inhibiting antidepressants. *Brain Res Mol Brain Res* 1993;17:31–35.

23. Serretti A, Cusin C, Rossini D, Artioli P, Dotoli D, Zanardi R, et al. Further evidence of a combined effect of SERTPR and TPH on SSRIs response in mood disorders. *Am J Med Genet B Neuropsychiatr Genet* 2004;15:36–40.

24. Taylor MJ, Sen J, Bhagwagar Z. Antidepressant response and the serotonin transporter gene-linked polymorphic region. *Biol Psychiatry* 2010;68:536–543.

25. Kato M, Serretti A. Review and meta-analysis of antidepressant pharmacogenetic findings in major depressive disorder. *Mol Psychiatry* 2010;15:473–500.

26. Dettling M, Cascorbi I, Opgen-Rhein C, Schaub R. Clozapine-induced agranulocytosis in schizophrenic Caucasians: Confirming clues for associations with human leukocyte class I and II antigens. *Pharmacogenomics J* 2007;7(5):325–332.

27. Athanasiou MC, Dettling MC, Cascorbi I, Mosyagin I, Salisbury BA, Pierz KA, et al. Candidate gene analysis identifies a polymorphism in HLA-DQB1 associated with clozapine-induced agranulocytosis. *J Clin Psychiatry* 2011;72(4):458–463.

28. Ostrousky O, Meged S, Loewenthal R, Valevski A, Weizman A, Carp H, et al. NQO2 gene is associated with clozapine-induced agranulocytosis. *Tissue Antigens* 2003;62(6):483–491.

29. Chowdhury NI, Remington G, Kennedy JL. Genetics of antipsychotic-induced side effects and agranulocytosis. *Curr Psychiatry Rep* 2011;13:156–165.

30. Neville MJ, Johnstone EC, Walton RT. Identification and characterization of ANKK1: A novel kinase gene closely linked to DRD2 on chromosome band 11q23.1. *Hum Mutat* 2004; 23(6):540–545.

31. Zai CC, Romano-Silva MA, Hwang R, Zai GC, Deluca V, Muller DJ, et al. Genetic study of eight AKT1 gene polymorphisms and their interaction with DRD2 gene polymorphisms in tardive dyskinesia. *Schizophr Res* 2008;106(2):248–252.

32. Tsai HT, North KE, West SL, et al. The DRD3 rs6280 polymorphism and prevalence of tardive dyskinesia: a meta-analysis. *Am J Med Genet B Neuropsychiatr Genet* 2010;153B(1):57–66.

33. Mannisto PT, Ulnianen I, Lundstrom K, et al. Characteristics of catechol-O-methyltransferase (COMT) and properties of selective COMT inhibitors. *Prog Drug Res* 1992;39:291–350.

34. Hori H, Ohmori O, Shinkai T, Kojima J, Okano C, Suzuki T, et al. Manganese superoxide dismutase gene polymorphism and schizophrenia: Relation to tardive dyskinesia. *J Neuropsychopharmacol* 2000;23(2):170–177.

35. Lerer B, Segman RH, Tan EC, Basile VS, Cavallaro R, Aschauer HN, et al. Combined analysis of 635 patients confirms an age-related association of the serotonin 2A receptor gene with tardive dyskinesia and specificity for the non-orofacial subtype. *Int J Neuropsychopharmacol* 2005;8(3):411–425.

36. Patsopoulos NA, Ntzani EE, Zintzaras E, et al. CYP2D6 polymorphisms and the risk of tardive dyskinesia in schizophrenia: A meta-analysis. *Pharmacogenet Genomics* 2005;15(3):151–158.

37. Panariello F, De Luca V, de Bartolomeis A. Weight gain, schizophrenia and antipsychotics: New findings from animal model and pharmacogenomic studies. *Schizophr Res Treat* 2011;2011:459284, pp. 1–16.

38. De Luca V, Mueller DJ, de Bartolomeis A, Kennedy J. Association of the HTR2C gene and antipsychotic induced weight gain: A meta-analysis. *Int J Neuropsychopharmacol* 2007;10:697–704.

39. Zai CC, De Luca V, Hwang RW, et al. Meta-analysis of two dopamine D2 receptor gene polymorphisms with tardive dyskinesia in schizophrenic patients. *Mol Psychiatry* 2007;12:794–795.

40. Shamy MCF, Zai C, Basile VS, Kennedy J, Muller DJ, Masellis M. Ethical and policy considerations in the application of pharmacogenomics testing for tardive dyskinesia: Case study of the dopamine D3 receptor. *Curr Pharmacogenomics Pers Med* 2011;9:94–101.

41. McCarthy MJ, Leckband SG, Kelsoe JR. Pharmacogenetics of lithium response in bipolar disorder. *Pharmacogenomics* 2010;11:1439–1465.

42. Cacabelos R. Pharmacogenomics in Alzheimer's disease. *Mini Rev Med Chem* 2002; 2:59–84.

43. Cacabelos R. Pharmacogenomics for the treatment of dementia. *Ann Med* 2002;34: 357–379.

44. Cacabelos R. The application of functional genomics to Alzheimer's disease. *Pharmacogenomics* 2003;4:597–621.

45. Cacabelos R. Pharmacogenomics and therapeutic prospects in Alzheimer's disease. *Exp Opin Pharmacother* 2005;6:1967–1987.

46. Cacabelos R. Pharmacogenomics and therapeutic prospects in dementia. *Eur Arch Psychiatry Clin Neurosci* 2008;258:28–47.

47. Harold D, MacGregor S, Patterson CE, et al. A single nucleotide polymorphism in CHAT influences response to acetylcholinesterase inhibition in Alzheimer's disease. *Pharmacogenet Genomics* 2006;16:75.

48. Noetzi M, Eap CB. Pharmacodynamic, pharmacokinetic, and pharmacogenetic aspects of drugs used in the treatment of Alzheimer's disease. *Clin Pharmacokinet* 2013; 52:225–241.

49. Ferraro TN, Buono RJ. The relationship between the pharmacology of antiepileptic drugs and human gene variation: An overview. *Epilepsy Behav* 2005;7:18–36.

50. van der Weide J, Steijins LSW, van Weelden MJM, de Haan K. The effect of genetic polymorphism of cytochrome P450 CYP2C9 on phenytoin dose requirement. *Pharmacogenet Genomics* 2001;11:287.

51. Chung WH, Hung SI, Hong HS, et al. Medical genetics: A marker for Stevens-Johnson syndrome. *Nature* 2004;428:486.

52. Ferrell PB, McLeod HL. Carbamazepine, HLA-B*1502 and risk of Stevens-Johnson syndrome or toxic epidermal necrolysis. *N Engl J Med* 1995;33:1600–1608.

53. Comabella M, Fernandez-Arquero M, Rio J, et al. HLA class I and II alleles and response to treatment with interferon-beta in relapsing-remitting multiple sclerosis. *J Neuroimmunol* 2009;210:116–119.

54. Domanski P, Colamonici OR. The type-I interferon receptor. The long and short of it. *Cytokine Growth Factor Rev* 1996;7:143–151.

55. Sriam U, Barcellos LF, Villoslada P, et al. Pharmacogenomic analysis of interferon receptor polymorphisms in multiple sclerosis. *Genes Immun* 2003;4:147–152.

56. Byun E, Caillier SJ, Montalban X, et al. Genome-wide pharmacogenomic analysis of the response to interferon beta therapy in multiple sclerosis. *Arch Neurol* 2008; 65:337–344.

57. Gilgun-Sherki Y, Djadetti R, Melamed E, Offen D. Polymorphism in candidate genes: Implications for the risk and treatment of idiopathic Parkinson's disease. *Pharmacogenomics J* 2004;4:291–306.

58. Anton RF, Oroszi G, O'Malley S, Couper D, Swift R, Pettinati H, et al. An evaluation of mu-opioid receptor (OPRM1) as a predictor of naltrexone response in the treatment of alcohol dependence: Results from the combined pharmacotherapies and behavioral interventions for alcohol dependence (COMBINE) study. *Arch Gen Psychiatry* 2008;65:135–144.

59. Kim SG, Kim CM, Choi SW, Jae YM, Lee HG, Son BK, et al. A micro opioid receptor gene polymorphism (A118G) and naltrexone treatment response in adherent Korean alcohol-dependent patients. *Psychopharmacology* 2009;201:611–618.

60. Oslin DW, Berrettini W, Kranzler HR, Pettinati H, Gelernter J, Volpicelli JR, et al. A functional polymorphism of the mu-opioid receptor gene is associated with naltrexone response in alcohol-dependent patients. *Neuropsychopharmacology* 2003; 28:1546–1552.

61. Ray LA, Hutchison KE. Effects of naltrexone on alcohol sensitivity and genetic moderators of medication response: A double-blind placebo-controlled study. *Arch Gen Psychiatry* 2007;64:1069–1077.
62. Oslin DW, Berrettini WH, O'Brien CP. Targetting treatments for alcohol dependence: The pharmacogenetics of naltrexone. *Addict Biol* 2006;11:397–403.
63. Schnoll RA, Patterson F, Wileyto EP, Tyndale RF, Benowitz N, Lerman C. Nicotine metabolic rate predicts successful smoking cessation with transdermal nicotine: A validation study. *Pharmacol Biochem Behav* 2009;92:6–11.
64. Lerman C, Tyndale R, Patterson F, Wileyto EP, Shields PG, Pinto A, et al. Nicotine metabolite ratio predicts efficacy of transdermal nicotine for smoking cessation. *Clin Pharmacol Ther* 2006;79:600–608.
65. Sturgess JE, George TP, Kennedy JL, Heinz A, Muller DJ. Pharmacogenetics of alcohol, nicotine and drug addiction treatments. *Addict Biol* 2011;16:357–376.
66. Hresko A, Haga S. Insurance coverage policies for personalized medicine. *J Pers Med* 2012;2:201–216.
67. De Leon J. Pharmacogenomics: The promise of personalized medicine for CNS disorders. *Neuropsychopharmacology* 2009;34:159–172.

9 Applying Pharmacogenomics in the Therapeutics of Pulmonary Diseases

Wei Zhang and Jason X.-J. Yuan

CONTENTS

KEY CONCEPTS

- Pharmacogenetic and pharmacogenomic studies aim to elucidate genetic determinants for drug response, a complex phenotype.
- Responses to therapeutics may be affected by various genetic and non-genetic factors.
- Several pulmonary disorders, such as chronic obstructive pulmonary disease (COPD), asthma, and pulmonary hypertension, have been under

extensive genetic research, which provides the basis for pharmacogenetic and pharmacogenomic discovery.

- Pharmacogenetic and pharmacogenomic studies in pulmonary diseases are still in the early stage. Future integration of various "omics" data will provide a more comprehensive picture of therapeutic responses in patients with these diseases.

INTRODUCTION

The launch of the Human Genome Project in the 1990s and the completion of the initial human genome reference sequences[1,2] in the early 21st century opened an era of research efforts focusing on investigating genetic variations and their implications for common, multifactorial diseases and phenotypes. Along with our much-improved understanding of the genetic make-up of humans, advances in technologies of high-throughput profiling of genetic variations during the past decade have allowed extensive exploration of the genetic contributions to various complex traits, such as genetic susceptibility to human complex diseases, and naturally occurring variations in physiological traits (e.g., adult height, skin color, blood pressure) and in responses to drugs. The availability of the genotypes of common genetic variants, particularly those in the form of single nucleotide polymorphisms (SNPs), through microarray-based genotyping platforms and more recently the next-generation sequencing (NGS) technologies, has facilitated the widespread use of genome-wide association study (GWAS) to assess common SNPs for statistical associations with complex traits.[3] The unprecedented enhancement of our understanding of the genetic contributions to complex traits can be evidenced by the substantial expansion of the GWAS Catalog[4] maintained by the National Human Genome Research Institute (NHGRI), which currently covers >15,000 associated genetic variants for more than 400 human complex traits.

Physicians prescribe drugs to treat diseases on the basis of their pharmacological characteristics and based on the probability that a patient may respond with reliable and reproducible clinical outcomes. However, the variability in response of the patient to the drug, likely ranging from beneficial therapeutic effects to serious adverse effects—even fatality—has long been demonstrated from clinical observations. Furthermore, drug response differences are common among patients, therefore presenting challenges that require the right drug and right dose for the right patient. Of particular interest are those drugs with a narrow therapeutic window, such as the oral anticoagulant warfarin, the overdose of which may cause severe bleeding in 1–3% of treated patients.[5]

In patient care, variability in response to the drug is the result of the interaction of genetic and nongenetic factors. Of them, the common nongenetic (or environmental) factors include smoking status, food intake, concomitant drug therapy, alcohol use, compliance, psychological status, and pregnancy and lactation status for females. The genetic factors refer to the patient's genetic make-up, specifically the states of common SNPs in the human genome, which may contribute substantially to the variability of clinical outcomes from the drug. For example, a patient's response to a particular drug may be affected by the activity of drug-metabolizing enzymes, whose expression can be regulated by local (*cis*-acting) SNPs (expression quantitative trait loci or eQTL). A patient with a gain-of-function variant allele in the protein-coding region of a cytochrome P450 (CYP) enzyme may have higher catalytic activity to metabolize its substrate drug, thus rendering the patient to require higher doses for effective treatment. Specifically, taking advantage of the high-quality genotyping data that may be available through high-throughput genotyping and sequencing platforms, pharmacogenetic and pharmacogenomic studies aim to elucidate the contributions of genetic variants to interindividual variability of drug response.

GENERAL PHARMACOGENETIC AND PHARMACOGENOMIC APPROACHES

Depending on the hypothesis, there are two general strategies used to elucidate the genetic basis of drug response: the candidate gene approach and the whole-genome approach. For the former, the search for drug response–associated genetic variants is within a well-defined set of genes and/or pathways. Previous studies have implicated genetic variations related to many mechanisms that may be relevant to drug therapy, through their effects on the genes encoding drug-metabolizing enzymes, transporters, and receptors, as well as their effects on pharmacokinetics (affecting drug concentrations) and pharmacodynamics (affecting drug action) characteristics of a drug.[6] For example, P450 enzymes are a superfamily of heme-containing proteins expressed abundantly in the hepatocytes, enterocytes, and in the lung, kidney, and brain. P450 enzymes CYP1, CYP2, and CYP3 are among the major families responsible for the oxidative metabolism of drugs and environmental chemicals.

In contrast, the whole-genome approach or GWAS scans the entire human genome for genetic variants associated with a particular complex trait. The whole-genome approach is unbiased because it does not require prior knowledge of candidate genes or pathways. GWAS has been a powerful tool for elucidating genetic variants for complex traits, by assuming the common variant–common disease/trait hypothesis, which predicts that common disease-causing alleles, or variants, will be found in human populations that manifest a given trait. In population genetics, linkage disequilibrium (LD) is a general characteristic in the human genome.[7] The existence of LD suggests nonrandom association of two alleles at two or more loci in the human genome, thus allowing the possibility of "tagging" causal variants by other known or genotyped variants. The technical advances of the past decade have allowed various cost-efficient approaches, including microarray based and sequencing based, for genotyping tens of thousands or millions of SNPs in one experiment. Therefore, by taking advantages of the LD characteristics in the human genome and the genotyping technologies, GWAS-based statistical methods for testing associations facilitate the applications of the whole-genome approach in detecting variants associated with complex traits, including response phenotypes of drugs.

APPLYING PHARMACOGENOMICS TO PULMONARY DISORDERS

Similar to other complex diseases, pharmacogenetic and pharmacogenomic studies on pulmonary diseases aim to implicate genetic variants responsible for interindividual responses to drugs, with the ultimate goal of realizing personalized care of patients with pulmonary diseases. Given the strong associations of immune response genes with many pulmonary diseases, pharmacogenetic research in pulmonary diseases has often been focused on these candidate genes. With the availability of whole-genome approaches, pharmacogenomic studies have also been carried out in a whole-genome, unbiased fashion. Specifically, several pulmonary disorders or syndromes, specifically chronic obstructive pulmonary disease (COPD), asthma, and pulmonary artery hypertension (PAH), which have attracted abundant genetic, pharmacogenetic, and/or pharmacogenomic research efforts, are reviewed in the following sections with an emphasis on clinical applications. Table 9.1 summarizes the genetic and pharmacogenomic loci as reviewed for these pulmonary disorders.

Chronic Obstructive Pulmonary Disease

Brief Background

COPD is characterized by the progressive development of airflow limitation that is not fully reversible.[8] COPD is a heterogeneous disorder or syndrome with multiple disease subphenotypes, including two main conditions: emphysema and chronic bronchitis.[8] In emphysema, the walls between many air sacs in the lungs are damaged, resulting in narrowing of the small airways and breakdown of lung tissues.[9,10] In contrast, in chronic bronchitis, the lining of the airways is constantly irritated and inflamed, thus worsening airflow obstruction by luminal obstruction of small airways, epithelial remodeling, and alteration of airway surface tension, predisposing to collapse.[11–13] The results of these damages cause progressive loss of lung function,

---

TABLE 9.1

Summary of Genes Associated with Pulmonary Diseases and Pharmacogenomic Loci

Disorder	COPD	Asthma	PH
Genes associated with disease risk and pathogenesis	*TNF-α/TNF-β*	*IL-33*	*BMPR2*
	CHRNA3	*TSLP*	*ACVRL1*
	FAM13A	*IL1RL1*	*SMAD1, SMAD4, SMAD9*
	HHIP	*ORMDL3*	*BMPR1B*
	RIN3		*CAV1*
	MMP12		*KCNK3*
	TGFB2		*TRPC6*
Pharmacogenomic loci	*ADRB2*	*ADRB2*	*BMPR2*
		GSNOR	*ACVRL1*

Note: COPD, chronic obstructive pulmonary disease; PH, pulmonary hypertension.

---

leading to various symptoms such as coughing that produces large amounts of mucus, wheezing, shortness of breath, and chest tightness among other symptoms.[14]

The prevalence of COPD is substantial with >300 millions of estimated patients in 2010 worldwide.[15] Overall, the number of deaths from COPD has decreased slightly from 3.1 million to 2.9 million between 1990 and 2010,[16] making it the fourth leading cause of death globally.[17] COPD is also a major cause of disability and the third leading cause of death in the United States. The total number of global COPD is expected to continually increasing, as the population continues to get older. The World Health Organization (WHO) predicts that COPD will become the third leading cause of death worldwide by 2030.[18]

Independently of other risk factors, regular cigarette smoking is the leading cause of COPD, accounting for approximately 80% of cases.[19] Most COPD cases are those who used to smoke or are current smokers. In multivariate analysis, cigarette smokers, as compared with nonsmokers, were at higher risk for developing COPD with dose–response effects.[19] Total deaths from COPD are projected to increase by more than 30% in the next 10 years unless urgent actions are taken to reduce the underlying risk factors, especially tobacco use.[18] Long-term exposure to other lung irritants in the environment,[20,21] such as indoor and outdoor air pollution, occupational dusts and chemicals (vapors, irritants, and fumes), and frequent lower respiratory infections during childhood are also risk factors that contribute to the pathogenesis and development of COPD. In addition, genetic susceptibility of individuals could be used to explain why not all smokers would develop COPD.[22] For example, a small proportion of COPD cases (about 1–5%) have been known to have α1-antitrypsin deficiency.[23,24] This risk is even higher if an individual who is α1-antitrypsin deficient also smokes regularly.[23]

Therapy for COPD

Although COPD cannot be cured completely, drug treatment for COPD and subsequent exacerbations of respiratory symptoms must be included in an effective strategy for COPD management, in which the major therapeutic approaches include risk factor reduction, such as smoking cessation and treatment with bronchodilators and corticosteroid therapy. When selecting a treatment plan, the benefits and risks to the individual and to the community must be taken into account. To date, none of the existing medications for COPD have been shown to modify the long-term decline in lung function, which is the hallmark of this syndrome. Therefore, pharmacotherapy for COPD is primarily meant to decrease symptoms and complications. In particular, bronchodilator medications, which include β-agonists, anticholinergics, theophylline,[25–27] and a combination of one or more of these drugs, as well as glucocorticosteroids,[28] are central to the symptomatic management of stable COPD and control of exacerbations.

Bronchodilators

Bronchodilators help open the airways in the lungs by relaxing smooth muscle around the airways. Bronchodilators can be classified as short or long acting. The short-acting bronchodilators, sometimes called "quick-reliever" can quickly decrease shortness of breath for about 4 to 6 hours. Common short-acting bronchodilators include β-agonists, such as albuterol, levalbuterol, pirbuterol, and ipratropium.[29–31] In addition, albuterol and ipratropium[29] are a common combination of short-acting inhaled bronchodilators used in COPD patients. In contrast, the long-acting bronchodilators, such as salmeterol, formoterol, and arformoterol,[32–35] are characterized by a long maintenance period of effective symptom improvement (24 hours). Therefore, these drugs are not used for acute shortness of breath.

Anti-Inflammatory Drugs

Anti-inflammatory drugs help reduce and prevent inflammation inside the airways. Commonly used inhaled corticosteroids include mometasone, fluticasone, budesonide, and beclomethasone,[36–40] in addition to corticosteroid pills such as prednisone and methylprednisolone.[41–43] Combinations of long-acting bronchodilators and corticosteroids, such as Advair® (fluticasone and salmeterol),[36,44] are also used in some patients to prevent exacerbations of COPD.

Genetics and Pharmacogenetics of COPD

COPD is a heterogeneous syndrome, possibly caused by the interactions of both genetic susceptibility and environmental factors. Although cigarette smoking is the primary environmental risk factor for developing COPD,[19] only about 15% of smokers develop clinically significant COPD, suggesting that there are other factors and potential genetic factors that contribute to the pathogenesis of COPD.[45] Genomic research on the genetic basis for COPD has identified the genes that are associated with COPD pathogenesis and heterogeneity by scanning variations in the genes coding protease/anti-protease systems.[46] For example, it is generally accepted that COPD is associated with an abnormal inflammatory response.[47] Therefore, many different immune response mediators, such as TNF-α (tumor necrosis factor alpha) and TNF-β (tumor necrosis factor beta), have been implicated in the pathogenesis of COPD.[46,48,49] TNF-α mediated inflammation is thought to play a key role in the respiratory and systemic features of COPD.[46,50,51] An SNP in the promoter region of the TNF-α gene directly affects the regulation of gene expression, and thus is an eQTL. This SNP has been shown to be associated with COPD phenotypes in Asian individuals,[52–54] but not in European individuals,[55,56] potentially explaining the ethnic disparities of this disease.[57] More recently, a population-based GWAS has identified genetic risk loci for COPD, in which a susceptibility locus on chromosome 19q13 was identified in a total of 3499 cases and 1922 control subjects from four cohorts,[58] further extending previous GWAS findings in smaller cohorts.[59] Notably, a recent analysis of 6633 individuals with moderate-to-severe COPD and 5704 control individuals confirmed association at three known loci: CHRNA3 (encoding nicotine cholinergic receptor alpha 3), FAM13A (encoding family with sequence similarity 13, member A), and HHIP (encoding hedgehog interacting protein), in addition to novel loci near RIN3 (encoding Ras and Rab interactor 3) with significant evidence of association, as well as MMP12 (encoding matrix metallopeptidase 12) and TGFB2 (encoding transforming growth factor, beta 2).[60]

The goal of understanding the genetic defects in patients with COPD will be not only to redefine the disease phenotypes based on the genetic information, but also to alternatively approach patients based on the understanding of COPD pathogenesis, which may lead to improved clinical outcomes. To date, the majority of COPD pharmacogenetic and pharmacogenomic studies published have focused on immediate bronchodilator responsiveness (BDR) as a clinical outcome, given that short-acting β-agonists, acting as bronchodilators to quickly relieve acute shortness of breath, are the most commonly prescribed drugs for COPD.[9,29–31] In a genome-wide linkage analysis using short tandem repeat markers, regions on chromosomes 3 and 4 have revealed some evidence for linkage to BDR,[61] suggesting that BDR is likely an actionable phenotype of COPD[62] that may identify responder subgroups to bronchodilators with different clinical outcomes.

Specifically, most studies have focused on the roles of common genetic variants in the *ADRB2* gene (encoding β2-adrenergic receptor) on BDR, although the findings have been inconclusive so far.[63] In a Japanese COPD cohort study, *ADRB2* polymorphism was demonstrated to be a determinant of preferential BDR to either β2-agonists or anticholinergics in patients with COPD.[64] Furthermore, although a significant correlation between BDR measures to salbutamol and to oxitropium was observed, there were individuals who responded preferentially to one of the two agents.[64] The genetic effects of *ADRB2* polymorphisms may explain some of the variability in response to therapeutic doses of short-acting β2-agonists in patients with COPD.[65] The present findings suggest that the *ADBR2* gene haplotypes may affect the severity of obstructive ventilatory impairment but not the immediate response to salbutamol during acute exacerbations of COPD.[66] Yet in another recent pharmacogenomic study, there was little evidence for the association between *ADRB2* variants and response to indacaterol (an ultra-long-acting β2-agonist),[67] suggesting that *ADRB2* genetic variation is unlikely to play a significant role in differential response to indacaterol treatment in COPD patients.[68] These inconclusive results highlight the difficulties in translating these findings to clinically relevant outcomes of COPD.

ASTHMA

Brief Background

Asthma is a disorder that causes the airways of the lungs to swell and narrow (i.e., airway obstruction), leading to various phenotypes or groupings of clinical characteristics.[69] Common symptoms of asthma are wheezing, shortness of breath, chest tightness, and coughing.[70] When an asthma attack occurs, the lining of the air passages swells and the muscles surrounding the airways become tight, caused by some combination of airway smooth muscle constriction and inflammation of the bronchi,[70] which reduces the amount of air that can pass through the airway. Previous studies indicate that inflammatory processes associated with type 2 helper T-cell (Th2) immunity are present in approximately half of the population with asthma.[71] In particular, cytokines, interleukin-4 (IL-4), and interleukin-13 (IL-13) that signal through two different but overlapping receptors have been implicated in asthma.[72–75]

An estimated 300 million people have asthma worldwide.[70] In the United States, asthma prevalence increased from 7.3% in 2001 to 8.4% in 2010, with 25.7 million persons suffering from asthma.[76] Children aged 0–17 years had higher asthma prevalence (9.5%) than adults (7.7%) for the years 2008–2010,[76] while females had higher asthma prevalence than males (9.2% vs. 7.0%).[76] Common asthma triggers include animals (pet hair or dander); dust mites; certain medicines (e.g., aspirin); changes in weather (most often cold weather); chemicals in the air or in food; exercise; mold; pollen; respiratory infections, such as the common cold; strong emotions (stress); and tobacco smoke.[77-79] Childhood exposure to a wider range of microbes was found to explain a substantial fraction of the lower asthma incidence relative to reference.[80,81] For example, in a birth cohort study in rural areas, the amount and microbial content of house dust were inversely associated with asthma and wheezing.[81] Some people with asthma have a personal or family history of allergies, such as hay fever (allergic rhinitis) or eczema.[82,83] Other health conditions, including a runny nose, sinus infections, reflux disease, psychological stress, and sleep apnea, can worsen asthma and make it harder to manage.

Therapy for Asthma

The goals for asthma treatment are to control airway swelling, to stay away from substances that trigger symptoms, and to help patients to carry on normal activities without asthma symptoms. According to their respective effects, drugs used to treat asthma are generally categorized into two classes: bronchodilators, which relax airway smooth muscle, and anti-inflammatory drugs, which suppress airway inflammation. Also, newer drugs (leukotriene modifiers) and combinations of drugs (corticosteroids and long-acting β-adrenergic agonists) are sometimes used.[70] According to the onset of action, these treatments can also be grouped into two classes. The first class is long-term medicine (maintenance medicines) to help prevent asthma attacks, which are used to prevent symptoms for patients with moderate-to-severe asthma. The second class is quick relief (rescue medicines) for use during acute attacks, which are taken to relieve acute symptoms during asthma attacks such as coughing, wheezing, and trouble breathing. Notably, many of the therapies for asthma are also used in treating COPD, as described in the above section.

Bronchodilators

Bronchodilators relieve the symptoms of asthma by relaxing the smooth muscles that can tighten around the airways during asthma attacks. This helps to open up the airways. Short-acting bronchodilator inhalers (rescue inhalers) are used to quickly relieve acute symptoms, such as coughing, wheezing, chest tightness, and shortness of breath, all of which are caused by asthma. Commonly prescribed short-acting bronchodilators for asthma include albuterol, alupent, levalbuterol, and pirbuterol.[84-86] In contrast to their short-acting counterparts, long-acting bronchodilators are used to provide long-term control, not quick relief, of asthma symptoms. Commonly prescribed long-acting bronchodilators for asthma include salmeterol and formoterol.[44,87]

Anti-Inflammatory Drugs

Inhaled corticosteroids are the preferred medicine for long-term control of asthma. These therapies prevent asthma attacks and work by reducing swelling and mucus production in the airways. As a result, the airways are less sensitive and less likely to react to asthma triggers and cause serious asthma symptoms. The main types of anti-inflammatory drugs for asthma control are steroids or corticosteroids. Other anti-inflammatory drugs include mast cell stabilizers, leukotriene modifiers, and immunomodulators.[88–90] Inhaled steroids for better asthma control include flunisolide, mometasone, triamcinolone acetonide, fluticasone, budesonide, and beclomethasone.[91–94] There are also some combinations of a steroid and long-acting bronchodilator, such as Advair (fluticasone and salmeterol), Dulera® (mometasone and formoterol), and Symbicort® (budesonide and formoterol).[44,95,96]

Genetics and Pharmacogenetics of Asthma

Genetics is believed to partially contribute to the risk of asthma. Notably, recent advances in genotyping technology have allowed GWAS and meta-analyses of GWAS on asthma, which have begun to shed light on both common and distinct pathways that contribute to asthma.[97] For example, associations with genetic variants in *IL-33* (encoding interleukin-33),[98] *TSLP* (encoding thymic stromal lymphopoietin),[99,100] and *IL1RL1* (encoding interleukin 1 receptor-like 1)[101] highlight a central role of the innate immune response pathways that promote the activation and differentiation of T-helper 2 cells in the pathogenesis of both asthma and allergic diseases. In contrast, variation at the 17q21 asthma locus, encoding the *ORMDL3* (encoding ORM1-like 3), is specifically associated with risk for the onset of asthma in childhood.[102,103]

Despite the availability of several classes of asthma drugs and their overall effectiveness, a significant portion of patients fail to respond to these therapeutic agents. Evidence suggests that genetic factors may partly mediate the heterogeneity in asthma treatment response.[104,105] The responses to treatments such as β2-agonists, leukotriene modifiers, and inhaled corticosteroids have demonstrated substantial interindividual variability.[105] Although many studies are limited by small sample sizes and results remain to be confirmed in larger replication cohorts, several candidate genes have been identified to be associated with variable response to common asthma drugs.[106] High-throughput technologies are also allowing for large-scale genetic investigations. Thus, the future is promising for a personalized treatment of asthma, which will improve therapeutic outcomes, minimize side effects, and lead to more cost-effective care. Some major pharmacogenomic findings in the context of asthma are summarized.

Similar to what has been found in COPD, genetic variants in *ADRB2* were observed to be associated with the variability in response to β2-agonists in patients with asthma. In a meta-analysis examining the association between *ADRB2* polymorphisms and the response to inhaled β2-adrenergic agonists in children with asthma, a significant association was made between favorable therapeutic response in asthmatic children and the Arg/Arg genotype at position 16 of the *ADRB2* gene, compared with the genotype Arg/Gly or Gly/Gly.[107] Therefore, clinical genotyping for *ADRB2* polymorphisms may help determine whether asthmatic children are drug resistant to β2-adrenergic agonists. In contrast, in another cohort of pediatric asthmatics, children whose genotypes were homozygous for *ADRB2* Gly16Gly had a more rapid response to β2-agonist treatment.[108] However, the effect of the *ADRB2* Arg16Gly polymorphisms appeared to be drug specific, with apparently no effect for salmeterol and fluticasone propionate in a cohort of patients with persistent asthma,[109] the observation of which was also confirmed in a more recent clinical trial.[110]

Racial and ethnic disparities in the quality of asthma care have been well documented.[111] In the United States, African Americans (AAs) are disproportionately affected by asthma. Though socioeconomic factors play a role in this disparity, there is evidence that genetic factors may also influence the development of asthma and responses to drug treatments in AA children. Pharmacogenomic studies may identify genetic variants that contribute to these racial disparities. For example, in a cohort of 107 AA children with severe asthma, the genotype variation in *GSNOR* (encoding glutathione-dependent *S*-nitrosoglutathione reductase) was found be associated with a decreased response to albuterol in AA children.[112] Further analysis in this cohort suggested that a combination of four SNPs within *GSNOR*, *ADRB2*, and *CPS1* (encoding carbamoyl phosphate synthetase 1) gave a 70% predictive value for lack of response to this therapy, indicating that genetic variants could contribute to the observed population disparities in asthma.[112]

Pulmonary Hypertension

Brief Background

Pulmonary hypertension (PH) is a progressive disease characterized by increased pressure in pulmonary arteries. By definition, PH is characterized by an increase in mean pulmonary arterial pressure (PAP) to ≥25 mmHg at rest, and a mean primary capillary wedge pressure of ≤15 mmHg.[113,114] As PH develops, blood flow through the pulmonary arteries is restricted and the right side of the heart becomes enlarged due to the increased strain of pumping blood through the lungs, which in turn leads to common symptoms, such as breathlessness, fatigue, weakness, angina, syncope, and abdominal distension.[114]

PH begins with inflammation and changes in the cells that line the pulmonary arteries. These changes make it hard for the heart to push blood through the pulmonary arteries and into the lungs, thus causing increased pressure in arteries. PH is classified into five groups by the WHO according to the cause of the condition and treatment options. Pulmonary arterial hypertension (PAH) is the Group 1 PH. In PAH, the pulmonary arteries constrict abnormally, which forces the heart to work faster and causes increased blood pressure within the lungs. Group 1 PAH includes PAH that has no known cause (primary PAH or idiopathic PAH [IPAH]); PAH that is inherited; PAH that is caused by drugs or toxins; PAH that is caused by conditions such as connective tissue diseases, liver disease, sickle cell disease, and HIV infection; and PAH that is caused by conditions that affect the veins and small blood vessels of the lungs. Group 1 PAH that occurs with a known cause often is called associated PAH. Groups 2 through 5 are sometimes called secondary PH.

Epidemiologically, IPAH without a known cause is rare, with an estimated 15–50 cases per million.[115] IPAH has an annual incidence of 1–2 cases per million people in the United States and Europe, but is 2–4 times more common in women than in men.[116] PH that occurs with another disease or condition is more common. For example, the prevalence is 0.5% in HIV-infected patients[117] and around 3% in patients with sickle cell disease.[118,119] PH usually develops between the ages of 20 and 60, although it can occur at any age. People who are at increased risk for PH include those with a family history,[120] those with certain diseases or conditions,[121] those who use street drugs (e.g., cocaine) or certain diet medicines (e.g., fenfluramine),[122–127] and those who live at high altitudes.[128,129]

Therapy for Pulmonary Hypertension

Treatments, including medicines, procedures, and other therapies, as well as lifestyle changes have been targeted to relieve PH symptoms and slow the progress of the disease. Among the five PH groups, PAH is the most studied group and, therefore, all of the currently available drug classes, such as prostacyclin analogs, endothelin receptor antagonists, and phosphodiesterase (PDE) type 5 inhibitors have been developed to treat PAH. Notably, significant progress in the understanding of the pathogenesis of PAH has resulted in a shift from vasodilator therapy to the development of specific drugs targeting seminal molecular derangements of this disorder. However, limited treatment data exist for the non-PAH forms of PH that are currently less studied. The major classes of treatments for PAH are summarized as follows.

Prostanoids

Prostanoids, such as epoprostenol and treprostinil, have vasodilator, antiproliferative, and immunomodulatory effects and represent established therapies for severe cases of PAH. In particular, continuous intravenous epoprostenol is a standard care for individuals with serious or life-threatening PAH, and is also the most effective therapy to date.[130] Epoprostenol is a prostacyclin (also called prostaglandin I2 or PGI$_2$), a prostaglandin member of the family of lipid molecules known as eicosanoids. Prostacyclin prevents formation of the platelet plug involved in primary hemostasis by inhibiting platelet activation,[131] thus acting as an effective vasodilator.

Endothelin Receptor Antagonists

Endothelin receptor antagonists, such as bosentan, ambrisentan, and macitentan, have been used to treat PAH by improving exercise ability and decreasing clinical worsening. For example, bosentan is an effective, safe drug used for patients with PAH at a dose of 125 mg twice daily.[132] Recent meta-analysis further suggests that bosentan can treat PAH effectively with an increased incidence of abnormal liver function testing compared with the placebo.[133] In a recent multicenter, double-blinded, randomized, placebo-controlled, and event-driven phase III trial, macitentan, a new ETA/ETB antagonist, was shown to significantly reduce morbidity and mortality among patients with PAH.[134] Notably, macitentan represents the latest addition to the drug therapies for PAH.[135] Compared with other analogs, macitentan is characterized by fewer contraindications, use in hepatic impairment, and once-daily administration.[136]

Phosphodiesterase Type-5 Inhibitors

PDE5 inhibitors inhibit the degradation of cyclic guanosine monophosphate (cGMP), which is catalyzed by PDE5.[137] Randomized clinical trials in monotherapy or combination therapy have been conducted in PAH patients with PDE5 inhibitors sildenafil and tadalafil, both of which can significantly improve clinical status, exercise capacity, and hemodynamics of PAH patients,[137] acting by increasing the levels of nitric oxide (NO) and improving pulmonary haemodynamics.[138]

Guanylate Cyclase Activators

The NO-soluble guanylate cyclase–cGMP signal-transduction pathway is impaired in many cardiovascular diseases, including PAH. Riociguat, a guanylate cyclase activator, which works both in synergy with and independently of NO to increase levels of cGMP,[139] reduces right ventricular (RV) systolic pressure and RV hypertrophy (RVH), and improves RV function compared with vehicle. Riociguat has a greater effect on hemodynamics and RVH than sildenafil.[139]

Genetics and Pharmacogenetics of Pulmonary Hypertension

PAH is an uncommon disease in the general population. The prevalence of heritable PAH remains unknown. The reason for incomplete penetrance of heritable PAH is not well understood yet. During the past decade, genetic and genomic approaches have been applied to dissect genetic contributors to PAH, with major discoveries obtained in the field of hereditary predisposition to PAH.[140] Notably, *BMPR2* (encoding bone morphogenetic protein receptor type 2) was identified as the major predisposing gene, and *ACVRL1* (encoding activin A receptor type II-like 1) as the major gene when PAH is associated with hereditary hemorrhagic telangiectasia.[141] Over 300 independent *BMPR2* mutations have been identified, accounting for approximately 75% of patients with a known family history of PAH, and up to 25% of apparent sporadic cases have been associated with mutations in this gene as the major genetic determinant.[142] Taken together, these observations support a prominent role for TGF-β family members in the development of PAH. Consequently, a series of candidate gene–based studies have been carried out to delineate novel genetic variants by examining TGF-β receptors and effectors in patient cohorts. These studies have implicated genes such as *SMAD9* (encoding SMAD family member 9), *SMAD4* (encoding SMAD family member 4), *SMAD1* (encoding SMAD family member 1), *BMPR1B* (encoding bone morphogenetic protein receptor type IB), and *CAV1* (encoding caveolin 1, caveolae protein, 22 kDa) in PAH.[140] More recently, exome sequencing in a family with multiple affected family members without identifiable heritable PAH mutations was found to have a heterozygous novel missense variant in *KCNK3* (encoding potassium channel, subfamily K, member 3).[143] *KCNK3* encodes a pH-sensitive potassium channel in the two-pore domain superfamily,[144] which is sensitive to hypoxia and plays a role in the regulation of resting membrane potential and pulmonary vascular tone,[145,146] thus potentially representing a novel target for PAH treatment.

Pharmacogenetics or pharmacogenomics is a tool to better understand the pathways involved in PH, as well as to improve personalization of drug therapy. Because of genetic heterogeneity in treatment effects and outcomes across the patients, pharmacogenetics that will study polymorphisms that modulate the response to treatment will enable physicians to deliver cost-effective, tailored treatments for all PAH patients in the future. A patient's clinical response to disease-specific therapy is complex, involving the severity of the patient's disease, other comorbidities, appropriateness of the prescribed therapy, and patient compliance.[147] However, this is a relatively new area, so there has been little pharmacogenomic research on PAH. However, there are ongoing clinical trials on the pharmacogenetics and pharmacogenomics in PAH, such as the Pharmacogenomics in Pulmonary Arterial Hypertension Trial (ClinicalTrials.gov Identifier: NCT00593905) with the goal to determine, clinically, in PAH patients whether associations exist between the efficacy and toxicity of sitaxsentan, bosentan, and ambrisentan and several gene polymorphisms in several key disease- and therapy-specific genes. Another ongoing clinical trial is the PILGRIM Trial (ClinicalTrials.gov Identifier: NCT01054105) between *BMPR2* mutations and hemodynamic response by iloprost inhalation. Still, previous studies have demonstrated that the presence of mutations in *BMPR2* and *ACVRL1* genes are associated with a less-favorable clinical response to therapeutic therapies, and an overall poorer prognosis.[147]

CONCLUSIONS

Genetics is likely a contributor to complex traits, such as the risks for diseases and altered responses to drug treatments. Advances in technologies have begun to allow dissection of genetic factors contributing to various pulmonary diseases, including COPD, asthma, and PH. A variety of drug treatments have been developed to control the disease progression or exacerbation as well as relieve symptoms. Pharmacogenetic and pharmacogenomic studies aim to associate genetic variants, either from candidate genes/pathways or whole-genome unbiased scans, to the variability of therapeutic response. During the past decade, progress has been made to identify genetic variants linked to the therapeutic variation in patients with these pulmonary diseases. Ongoing clinical trials will provide critical knowledge of the clinical applications of pharmacogenomic approaches in managing pulmonary diseases. Future incorporation of other genomic features such as various epigenetic systems (e.g., DNA methylation) in pharmacogenomic discovery[148] has the promise to elucidate the complete genomic contributors to therapeutic response, thus facilitating the ultimate goal of personalized care of patients with pulmonary diseases.

THOUGHTS FOR FURTHER CONSIDERATION

1. *Pharmacogenomic loci beyond genetic variants*: The current pharmacogenomic findings have been focused on genetic variants, especially those in the form of SNPs. Since genetic variation can only explain part of the variability in the phenotypes observed, including gene expression and therapeutic response, it may be necessary to incorporate other genomic features beyond genetic variation. For example, given the critical roles of epigenetic factors, such as cytosine modifications (primarily DNA methylation) and histone modifications in gene regulation, future pharmacogenomic studies may need to consider and integrate these factors as well.

2. *Gene × gene and gene × environment interactions*: The current pharmacogenomic studies have targeted the effects of individual genes. Since genetic epistasis, that is, the interaction between SNP and SNP, may affect

complex traits, future pharmacogenomic applications would benefit more from considering these potential effects. Similarly, future pharmacogenomic approaches to considering both gene and environmental factors (such as food intake) will be necessary to provide a more comprehensive picture of the determinants for therapeutic response variation.

STUDY QUESTIONS

1. Why is drug response a complex trait or phenotype?
2. What are the general strategies for pharmacogenomic discovery?
3. Which gene is currently the most studied for pharmacogenomics of COPD?
4. Why are there overlapping pharmacogenomic loci between COPD and asthma?
5. Are there any ongoing clinical trials for pharmacogenetics and pharmacogenomics in pulmonary hypertension?

Answers

1. Drug response is likely affected by multiple factors including both genetic and nongenetic factors (e.g., environment, food intake).
2. A candidate gene–based approach may be used if there is reasonable prior knowledge. In contrast, an unbiased approach may be used for a genome-wide scan for associated loci.
3. To date, most studies have been focused on the roles of common genetic variants in the ADRB2 gene (encoding β2-adrenergic receptor) on bronchodilator responsiveness (BDR).
4. These two disorders share certain pathogenesis pathways. Many of the therapies for asthma are also used for treating COPD.
5. Yes, for example, there are the Pharmacogenomics in Pulmonary Arterial Hypertension Trial with the goal to determine, clinically, in PAH patients whether associations exist between the efficacy and toxicity of sitaxsentan, bosentan, and ambrisentan and several gene polymorphisms in several key disease- and therapy-specific genes, and the PILGRIM Trial between BMPR2 mutations and hemodynamic response by iloprost inhalation.

REFERENCES

1. Lander ES, Linton LM, Birren B, et al. Initial sequencing and analysis of the human genome. *Nature* 2001;409:860–921.
2. Venter JC, Adams MD, Myers EW, et al. The sequence of the human genome. *Science* 2001;291:1304–51.
3. Frazer KA, Murray SS, Schork NJ, Topol EJ. Human genetic variation and its contribution to complex traits. *Nature Reviews* 2009;10:241–51.
4. Hindorff LA, Sethupathy P, Junkins HA, et al. Potential etiologic and functional implications of genome-wide association loci for human diseases and traits. *Proceedings of the National Academy of Sciences of the United States of America* 2009;106:9362–7.

5. Holbrook A, Schulman S, Witt DM, et al. Evidence-based management of anticoagulant therapy: Antithrombotic Therapy and Prevention of Thrombosis, 9th ed: American College of Chest Physicians Evidence-Based Clinical Practice Guidelines. *Chest* 2012;141:e152S–84S.

6. Gardiner SJ, Begg EJ. Pharmacogenetics, drug-metabolizing enzymes, and clinical practice. *Pharmacological Reviews* 2006;58:521–90.

7. Reich DE, Cargill M, Bolk S, et al. Linkage disequilibrium in the human genome. *Nature* 2001;411:199–204.

8. Barnes PJ. Chronic obstructive pulmonary disease. *The New England Journal of Medicine* 2000;343:269–80.

9. Rabe KF, Hurd S, Anzueto A, et al. Global strategy for the diagnosis, management, and prevention of chronic obstructive pulmonary disease: GOLD executive summary. *American Journal of Respiratory and Critical Care Medicine* 2007;176:532–55.

10. Coxson HO, Dirksen A, Edwards LD, et al. The presence and progression of emphysema in COPD as determined by CT scanning and biomarker expression: A prospective analysis from the ECLIPSE study. *The Lancet* 2013;1:129–36.

11. Kim V, Han MK, Vance GB, et al. The chronic bronchitic phenotype of COPD: An analysis of the COPDGene Study. *Chest* 2011;140:626–33.

12. Kim V, Criner GJ. Chronic bronchitis and chronic obstructive pulmonary disease. *American Journal of Respiratory and Critical Care Medicine* 2013;187:228–37.

13. Corhay JL, Vincken W, Schlesser M, Bossuyt P, Imschoot J. Chronic bronchitis in COPD patients is associated with increased risk of exacerbations: A cross-sectional multicentre study. *International Journal of Clinical Practice* 2013;67:1294–301.

14. van der Molen T, Miravitlles M, Kocks JW. COPD management: Role of symptom assessment in routine clinical practice. *International Journal of Chronic Obstructive Pulmonary Disease* 2013;8:461–71.

15. Vos T, Flaxman AD, Naghavi M, et al. Years lived with disability (YLDs) for 1160 sequelae of 289 diseases and injuries 1990–2010: A systematic analysis for the Global Burden of Disease Study 2010. *The Lancet* 2012;380:2163–96.

16. Lozano R, Naghavi M, Foreman K, et al. Global and regional mortality from 235 causes of death for 20 age groups in 1990 and 2010: A systematic analysis for the Global Burden of Disease Study 2010. *The Lancet* 2012;380:2095–128.

17. Decramer M, Janssens W, Miravitlles M. Chronic obstructive pulmonary disease. *The Lancet* 2012;379:1341–51.

18. WHO. *The global burden of disease: 2004 update.* WHO, Geneva, Switzerland; 2008.

19. Iribarren C, Tekawa IS, Sidney S, Friedman GD. Effect of cigar smoking on the risk of cardiovascular disease, chronic obstructive pulmonary disease, and cancer in men. *The New England Journal of Medicine* 1999;340:1773–80.

20. Salvi S. Tobacco smoking and environmental risk factors for chronic obstructive pulmonary disease. *Clinics in Chest Medicine* 2014;35:17–27.

21. Salvi SS, Barnes PJ. Chronic obstructive pulmonary disease in non-smokers. *The Lancet* 2009;374:733–43.

22. Mayer AS, Newman LS. Genetic and environmental modulation of chronic obstructive pulmonary disease. *Respiration Physiology* 2001;128:3–11.

23. Foreman MG, Campos M, Celedon JC. Genes and chronic obstructive pulmonary disease. *The Medical Clinics of North America* 2012;96:699–711.

24. Brode SK, Ling SC, Chapman KR. Alpha-1 antitrypsin deficiency: A commonly overlooked cause of lung disease. *CMAJ* 2012;184:1365–71.

25. Salpeter SR. Bronchodilators in COPD: Impact of beta-agonists and anticholinergics on severe exacerbations and mortality. *International Journal of Chronic Obstructive Pulmonary Disease* 2007;2:11–18.

26. Singh S, Loke YK, Furberg CD. Inhaled anticholinergics and risk of major adverse cardiovascular events in patients with chronic obstructive pulmonary disease: A systematic review and meta-analysis. *JAMA* 2008;300:1439–50.

27. Ram FS. Use of theophylline in chronic obstructive pulmonary disease: Examining the evidence. *Current Opinion in Pulmonary Medicine* 2006;12:132–9.

28. Sadowska AM, Klebe B, Germonpre P, De Backer WA. Glucocorticosteroids as antioxidants in treatment of asthma and COPD. New application for an old medication? *Steroids* 2007;72:1–6.

29. Gordon J, Panos RJ. Inhaled albuterol/salbutamol and ipratropium bromide and their combination in the treatment of chronic obstructive pulmonary disease. *Expert Opinion on Drug Metabolism & Toxicology* 2010;6:381–92.

30. Dalonzo GE, Jr. Levalbuterol in the treatment of patients with asthma and chronic obstructive lung disease. *The Journal of the American Osteopathic Association* 2004;104:288–93.

31. Weber RW, Nelson HS. Pirbuterol hydrochloride: Evaluation of beta adrenergic agonist activity in reversible obstructive pulmonary disease and congestive heart failure. *Pharmacotherapy* 1984;4:1–10.

32. Chung KF. Salmeterol/fluticasone combination in the treatment of COPD. *International Journal of Chronic Obstructive Pulmonary Disease* 2006;1:235–42.

33. Keating GM, McCormack PL. Salmeterol/fluticasone propionate: A review of its use in the treatment of chronic obstructive pulmonary disease. *Drugs* 2007;67:2383–405.

34. Steiropoulos P, Tzouvelekis A, Bouros D. Formoterol in the management of chronic obstructive pulmonary disease. *International Journal of Chronic Obstructive Pulmonary Disease* 2008;3:205–15.

35. King P. Role of arformoterol in the management of COPD. *International Journal of Chronic Obstructive Pulmonary Disease* 2008;3:385–91.

36. Yawn BP, Raphiou I, Hurley JS, Dalal AA. The role of fluticasone propionate/salmeterol combination therapy in preventing exacerbations of COPD. *International Journal of Chronic Obstructive Pulmonary Disease* 2010;5:165–78.

37. Hoogsteden HC, Verhoeven GT, Lambrecht BN, Prins JB. Airway inflammation in asthma and chronic obstructive pulmonary disease with special emphasis on the antigen-presenting dendritic cell: Influence of treatment with fluticasone propionate. *Clinical and Experimental Allergy* 1999;29(Suppl 2):116–24.

38. Scott LJ. Budesonide/formoterol Turbuhaler(R): A review of its use in chronic obstructive pulmonary disease. *Drugs* 2012;72:395–414.

39. Calverley PM, Kuna P, Monso E, et al. Beclomethasone/formoterol in the management of COPD: A randomised controlled trial. *Respiratory Medicine* 2010;104:1858–68.

40. Doherty DE, Tashkin DP, Kerwin E, et al. Effects of mometasone furoate/formoterol fumarate fixed-dose combination formulation on chronic obstructive pulmonary disease (COPD): Results from a 52-week phase III trial in subjects with moderate-to-very severe COPD. *International Journal of Chronic Obstructive Pulmonary Disease* 2012;7:57–71.

41. Salloum A, Elbaage TY, Soubani AO. Prednisone for chronic obstructive pulmonary disease. *The New England Journal of Medicine* 2003;349:1288–90; author reply 1288–90.

42. Li H, He G, Chu H, Zhao L, Yu H. A step-wise application of methylprednisolone versus dexamethasone in the treatment of acute exacerbations of COPD. *Respirology* 2003;8:199–204.

43. Emerman CL, Connors AF, Lukens TW, May ME, Effron D. A randomized controlled trial of methylprednisolone in the emergency treatment of acute exacerbations of COPD. *Chest* 1989;95:563–7.

44. Nelson HS. Advair: Combination treatment with fluticasone propionate/salmeterol in the treatment of asthma. *The Journal of Allergy and Clinical Immunology* 2001; 107:398–416.

45. Cigarette smoking and health. American Thoracic Society. *American Journal of Respiratory and Critical Care Medicine* 1996;153:861–5.
46. Wood AM, Stockley RA. The genetics of chronic obstructive pulmonary disease. *Respiratory Research* 2006;7:130.
47. Provinciali M, Cardelli M, Marchegiani F. Inflammation, chronic obstructive pulmonary disease and aging. *Current Opinion in Pulmonary Medicine* 2011;17(Suppl 1):S3–10.
48. Celli BR, MacNee W. Standards for the diagnosis and treatment of patients with COPD: A summary of the ATS/ERS position paper. *The European Respiratory Journal* 2004; 23:932–46.
49. Wouters EF. Local and systemic inflammation in chronic obstructive pulmonary disease. *Proceedings of the American Thoracic Society* 2005;2:26–33.
50. Churg A, Wang RD, Tai H, Wang X, Xie C, Wright JL. Tumor necrosis factor-alpha drives 70% of cigarette smoke-induced emphysema in the mouse. *American Journal of Respiratory and Critical Care Medicine* 2004;170:492–8.
51. Sevenoaks MJ, Stockley RA. Chronic obstructive pulmonary disease, inflammation and co-morbidity—A common inflammatory phenotype? *Respiratory Research* 2006;7:70.
52. Huang SL, Su CH, Chang SC. Tumor necrosis factor-alpha gene polymorphism in chronic bronchitis. *American Journal of Respiratory and Critical Care Medicine* 1997;156:1436–9.
53. Sakao S, Tatsumi K, Igari H, Shino Y, Shirasawa H, Kuriyama T. Association of tumor necrosis factor alpha gene promoter polymorphism with the presence of chronic obstructive pulmonary disease. *American Journal of Respiratory and Critical Care Medicine* 2001;163:420–2.
54. Sakao S, Tatsumi K, Igari H, et al. Association of tumor necrosis factor-alpha gene promoter polymorphism with low attenuation areas on high-resolution CT in patients with COPD. *Chest* 2002;122:416–20.
55. Hersh CP, Demeo DL, Lange C, et al. Attempted replication of reported chronic obstructive pulmonary disease candidate gene associations. *American Journal of Respiratory Cell and Molecular Biology* 2005;33:71–8.
56. Sandford AJ, Chagani T, Weir TD, Connett JE, Anthonisen NR, Pare PD. Susceptibility genes for rapid decline of lung function in the lung health study. *American Journal of Respiratory and Critical Care Medicine* 2001;163:469–73.
57. Keppel KG, Pearcy JN, Heron MP. Is there progress toward eliminating racial/ethnic disparities in the leading causes of death? *Public Health Reports* 2010; 125:689–97.
58. Cho MH, Castaldi PJ, Wan ES, et al. A genome-wide association study of COPD identifies a susceptibility locus on chromosome 19q13. *Human Molecular Genetics* 2012; 21:947–57.
59. Pillai SG, Ge D, Zhu G, et al. A genome-wide association study in chronic obstructive pulmonary disease (COPD): Identification of two major susceptibility loci. *PLoS Genetics* 2009;5:e1000421.
60. Cho MH, McDonald ML, Zhou X, et al. Risk loci for chronic obstructive pulmonary disease: A genome-wide association study and meta-analysis. *The Lancet* 2014; 2:214–25.
61. Palmer LJ, Celedon JC, Chapman HA, Speizer FE, Weiss ST, Silverman EK. Genome-wide linkage analysis of bronchodilator responsiveness and post-bronchodilator spirometric phenotypes in chronic obstructive pulmonary disease. *Human Molecular Genetics* 2003;12:1199–210.
62. Albert P, Agusti A, Edwards L, et al. Bronchodilator responsiveness as a phenotypic characteristic of established chronic obstructive pulmonary disease. *Thorax* 2012; 67:701–8.

63. Hizawa N. Pharmacogenetics of beta2-agonists. *Allergology International* 2011;60: 239–46.
64. Konno S, Makita H, Hasegawa M, et al. Beta2-adrenergic receptor polymorphisms as a determinant of preferential bronchodilator responses to beta2-agonist and anti-cholinergic agents in Japanese patients with chronic obstructive pulmonary disease. *Pharmacogenetics and Genomics* 2011;21:687–93.
65. Hizawa N, Makita H, Nasuhara Y, et al. Beta2-adrenergic receptor genetic polymorphisms and short-term bronchodilator responses in patients with COPD. *Chest* 2007;132:1485–92.
66. Mokry M, Joppa P, Slaba E, et al. Beta2-adrenergic receptor haplotype and broncho-dilator response to salbutamol in patients with acute exacerbations of COPD. *Medical Science Monitor* 2008;14:CR392–8.
67. Cazzola M, Matera MG, Lotvall J. Ultra long-acting beta 2-agonists in development for asthma and chronic obstructive pulmonary disease. *Expert Opinion on Investigational Drugs* 2005;14:775–83.
68. Yelensky R, Li Y, Lewitzky S, et al. A pharmacogenetic study of ADRB2 polymor-phisms and indacaterol response in COPD patients. *The Pharmacogenomics Journal* 2012;12:484–8.
69. Wenzel SE. Asthma phenotypes: The evolution from clinical to molecular approaches. *Nature Medicine* 2012;18:716–25.
70. Fanta CH. Asthma. *The New England Journal of Medicine* 2009;360:1002–14.
71. Woodruff PG, Modrek B, Choy DF, et al. T-helper type 2-driven inflammation defines major subphenotypes of asthma. *American Journal of Respiratory and Critical Care Medicine* 2009;180:388–95.
72. Izuhara K, Arima K, Yasunaga S. IL-4 and IL-13: Their pathological roles in allergic diseases and their potential in developing new therapies. *Current Drug Targets* 2002; 1:263–9.
73. Fish SC, Donaldson DD, Goldman SJ, Williams CM, Kasaian MT. IgE generation and mast cell effector function in mice deficient in IL-4 and IL-13. *The Journal of Immunology* 2005;174:7716–24.
74. Grunig G, Warnock M, Wakil AE, et al. Requirement for IL-13 independently of IL-4 in experimental asthma. *Science* 1998;282:2261–3.
75. Chatila TA. Interleukin-4 receptor signaling pathways in asthma pathogenesis. *Trends in Molecular Medicine* 2004;10:493–9.
76. Akinbami LJ, Moorman JE, Bailey C, et al. *Trends in asthma prevalence, health care use, and mortality in the United States, 2001–2010. NCHS data brief, no. 94.* National Center for Health Statistics, Hyattsville, ML; 2012.
77. Janssens T, Ritz T. Perceived triggers of asthma: Key to symptom perception and man-agement. *Clinical and Experimental Allergy* 2013;43:1000–8.
78. Ritz T, Steptoe A, Bobb C, Harris AH, Edwards M. The asthma trigger inventory: Validation of a questionnaire for perceived triggers of asthma. *Psychosomatic Medicine* 2006;68:956–65.
79. Ritz T, Kullowatz A, Bobb C, et al. Psychological triggers and hyperventilation symp-toms in asthma. *Annals of Allergy, Asthma and Immunology* 2008;100:426–32.
80. Ege MJ, Mayer M, Normand AC, et al. Exposure to environmental microorganisms and childhood asthma. *The New England Journal of Medicine* 2011;364:701–9.
81. Karvonen AM, Hyvarinen A, Gehring U, et al. Exposure to microbial agents in house dust and wheezing, atopic dermatitis and atopic sensitization in early child-hood: A birth cohort study in rural areas. *Clinical and Experimental Allergy* 2012; 42:1246–56.
82. Koh YY, Kim CK. The development of asthma in patients with allergic rhinitis. *Current Opinion in Allergy and Clinical Immunology* 2003;3:159–64.

83. Demehri S, Morimoto M, Holtzman MJ, Kopan R. Skin-derived TSLP triggers progression from epidermal-barrier defects to asthma. *PLoS Biology* 2009;7:e1000067.
84. Wechsler ME, Kelley JM, Boyd IO, et al. Active albuterol or placebo, sham acupuncture, or no intervention in asthma. *The New England Journal of Medicine* 2011; 365:119–26.
85. Jat KR, Khairwa A. Levalbuterol versus albuterol for acute asthma: A systematic review and meta-analysis. *Pulmonary Pharmacology & Therapeutics* 2013;26:239–48.
86. Beumer HM. Pirbuterol versus orciprenaline aerosols in the treatment of bronchial asthma. *International Journal of Clinical Pharmacology, Therapy, and Toxicology* 1983;21:147–66.
87. Cates CJ, Oleszczuk M, Stovold E, Wieland LS. Safety of regular formoterol or salmeterol in children with asthma: An overview of Cochrane reviews. *The Cochrane Database of Systematic Reviews* 2012;10:CD010005.
88. Montuschi P, Peters-Golden ML. Leukotriene modifiers for asthma treatment. *Clinical and Experimental Allergy* 2010;40:1732–41.
89. Dimov VV, Casale TB. Immunomodulators for asthma. *Allergy, Asthma & Immunology Research* 2010;2:228–34.
90. Misawa M, Takenouchi K, Shirakawa Y, Yanaura S. Effects of mast cell stabilizers on a new bronchial asthma model using compound 48/80 in dogs. *Japanese Journal of Pharmacology* 1987;44:197–205.
91. Melani AS. Flunisolide for the treatment of asthma. *Expert Review of Clinical Pharmacology* 2014;7:251–8.
92. Berger WE, Bensch GW, Weinstein SF, et al. Bronchodilation with mometasone furoate/formoterol fumarate administered by metered-dose inhaler with and without a spacer in children with persistent asthma. *Pediatric Pulmonology* 201;49:441–50.
93. Price D, Hillyer EV. Fluticasone propionate/formoterol fumarate in fixed-dose combination for the treatment of asthma. *Expert Review of Respiratory Medicine* 2014;8:275–91.
94. Profita M, Riccobono L, Bonanno A, et al. Effect of nebulized beclomethasone on airway inflammation and clinical status of children with allergic asthma and rhinitis: A randomized, double-blind, placebo-controlled study. *International Archives of Allergy and Immunology* 2013;161:53–64.
95. Pilcher J, Patel M, Smith A, et al. Combination budesonide/formoterol inhaler as maintenance and reliever therapy in Maori with asthma. *Respirology* 2014;19:842–51.
96. Mometasone/formoterol (dulera) for asthma. *The Medical Letter on Drugs and Therapeutics* 2010;52:83–4.
97. Ober C, Yao TC. The genetics of asthma and allergic disease: A 21st century perspective. *Immunological Reviews* 2011;242:10–30.
98. Grotenboer NS, Ketelaar ME, Koppelman GH, Nawijn MC. Decoding asthma: Translating genetic variation in IL33 and IL1RL1 into disease pathophysiology. *The Journal of Allergy and Clinical Immunology* 2013;131:856–65.
99. Watson B, Gauvreau GM. Thymic stromal lymphopoietin: A central regulator of allergic asthma. *Expert Opinion on Therapeutic Targets* 2014;18:771–85.
100. Birben E, Sahiner UM, Karaaslan C, et al. The genetic variants of thymic stromal lymphopoietin protein in children with asthma and allergic rhinitis. *International Archives of Allergy and Immunology* 2014;163:185–92.
101. Savenije OE, Mahachie John JM, Granell R, et al. Association of IL33-IL-1 receptor-like 1 (IL1RL1) pathway polymorphisms with wheezing phenotypes and asthma in childhood. *The Journal of Allergy and Clinical Immunology* 2014;134:170–7.
102. Wu H, Romieu I, Sienra-Monge JJ, Li H, del Rio-Navarro BE, London SJ. Genetic variation in ORM1-like 3 (ORMDL3) and gasdermin-like (GSDML) and childhood asthma. *Allergy* 2009;64:629–35.

103. Leung TF, Sy HY, Ng MC, et al. Asthma and atopy are associated with chromosome 17q21 markers in Chinese children. *Allergy* 2009;64:621–8.
104. Tse SM, Tantisira K, Weiss ST. The pharmacogenetics and pharmacogenomics of asthma therapy. *The Pharmacogenomics Journal* 2011;11:383–92.
105. Israel E. Genetics and the variability of treatment response in asthma. *The Journal of Allergy and Clinical Immunology* 2005;115:S532–8.
106. Duan QL, Tantisira KG. Pharmacogenetics of asthma therapy. *Current Pharmaceutical Design* 2009;15:3742–53.
107. Finkelstein Y, Bournissen FG, Hutson JR, Shannon M. Polymorphism of the ADRB2 gene and response to inhaled beta-agonists in children with asthma: A meta-analysis. *The Journal of Asthma* 2009;46:900–5.
108. Carroll CL, Stoltz P, Schramm CM, Zucker AR. Beta2-adrenergic receptor polymorphisms affect response to treatment in children with severe asthma exacerbations. *Chest* 2009;135:1186–92.
109. Bleecker ER, Yancey SW, Baitinger LA, et al. Salmeterol response is not affected by beta2-adrenergic receptor genotype in subjects with persistent asthma. *The Journal of Allergy and Clinical Immunology* 2006;118:809–16.
110. Bleecker ER, Nelson HS, Kraft M, et al. Beta2-receptor polymorphisms in patients receiving salmeterol with or without fluticasone propionate. *American Journal of Respiratory and Critical Care Medicine* 2010;181:676–87.
111. Cabana MD, Lara M, Shannon J. Racial and ethnic disparities in the quality of asthma care. *Chest* 2007;132:810S–7S.
112. Moore PE, Ryckman KK, Williams SM, Patel N, Summar ML, Sheller JR. Genetic variants of GSNOR and ADRB2 influence response to albuterol in African-American children with severe asthma. *Pediatric Pulmonology* 2009;44:649–54.
113. Badesch DB, Champion HC, Sanchez MA, et al. Diagnosis and assessment of pulmonary arterial hypertension. *Journal of the American College of Cardiology* 2009; 54:S55–66.
114. Galie N, Hoeper MM, Humbert M, et al. Guidelines for the diagnosis and treatment of pulmonary hypertension: The Task Force for the Diagnosis and Treatment of Pulmonary Hypertension of the European Society of Cardiology (ESC) and the European Respiratory Society (ERS), endorsed by the International Society of Heart and Lung Transplantation (ISHLT). *European Heart Journal* 2009;30:2493–537.
115. Peacock AJ, Murphy NF, McMurray JJ, Caballero L, Stewart S. An epidemiological study of pulmonary arterial hypertension. *The European Respiratory Journal* 2007; 30:104–9.
116. Badesch DB, Raskob GE, Elliott CG, et al. Pulmonary arterial hypertension: Baseline characteristics from the REVEAL Registry. *Chest* 2010;137:376–87.
117. Sitbon O, Lascoux-Combe C, Delfraissy JF, et al. Prevalence of HIV-related pulmonary arterial hypertension in the current antiretroviral therapy era. *American Journal of Respiratory and Critical Care Medicine* 2008;177:108–13.
118. Fonseca GH, Souza R, Salemi VM, Jardim CV, Gualandro SF. Pulmonary hypertension diagnosed by right heart catheterisation in sickle cell disease. *The European Respiratory Journal* 2012;39:112–18.
119. Machado RF, Gladwin MT. Pulmonary hypertension in hemolytic disorders: Pulmonary vascular disease: The global perspective. *Chest* 2010;137:30S–8S.
120. Austin ED, Loyd JE. Heritable forms of pulmonary arterial hypertension. *Seminars in Respiratory and Critical Care Medicine* 2013;34:568–80.
121. Zhang X, Zhang W, Ma SF, et al. Hypoxic response contributes to altered gene expression and precapillary pulmonary hypertension in patients with sickle cell disease. *Circulation* 2014;129:1650–58.

122. Yao W, Mu W, Zeifman A, et al. Fenfluramine-induced gene dysregulation in human pulmonary artery smooth muscle and endothelial cells. *Pulmonary Circulation* 2011; 1:405–18.

123. Souza R, Humbert M, Sztrymf B, et al. Pulmonary arterial hypertension associated with fenfluramine exposure: Report of 109 cases. *The European Respiratory Journal* 2008;31:343–8.

124. Barst RJ, Abenhaim L. Fatal pulmonary arterial hypertension associated with phenyl-propanolamine exposure. *Heart* 2004;90:e42.

125. Seferian A, Chaumais MC, Savale L, et al. Drugs induced pulmonary arterial hypertension. *Presse Medicale* 2013;42:e303–10.

126. Yakel DL, Jr., Eisenberg MJ. Pulmonary artery hypertension in chronic intravenous cocaine users. *American Heart Journal* 1995;130:398–9.

127. Abenhaim L, Moride Y, Brenot F, et al. Appetite-suppressant drugs and the risk of primary pulmonary hypertension. International Primary Pulmonary Hypertension Study Group. *The New England Journal of Medicine* 1996;335:609–16.

128. Naeije R, Vanderpool R. Pulmonary hypertension and chronic mountain sickness. *High Altitude Medicine & Biology* 2013;14:117–25.

129. Xu XQ, Jing ZC. High-altitude pulmonary hypertension. *European Respiratory Review* 2009;18:13–17.

130. Seferian A, Simonneau G. Therapies for pulmonary arterial hypertension: Where are we today, where do we go tomorrow? *European Respiratory Review* 2013;22:217–26.

131. Moncada S, Gryglewski R, Bunting S, Vane JR. An enzyme isolated from arteries transforms prostaglandin endoperoxides to an unstable substance that inhibits platelet aggregation. *Nature* 1976;263:663–5.

132. Rubin LJ, Badesch DB, Barst RJ, et al. Bosentan therapy for pulmonary arterial hypertension. *The New England Journal of Medicine* 2002;346:896–903.

133. Lee YH, Song GG. Meta-analysis of randomized controlled trials of bosentan for treatment of pulmonary arterial hypertension. *The Korean Journal of Internal Medicine* 2013;28:701–7.

134. Pulido T, Adzerikho I, Channick RN, et al. Macitentan and morbidity and mortality in pulmonary arterial hypertension. *The New England Journal of Medicine* 2013;369:809–18.

135. Clarke M, Walter C, Agarwal R, Kanwar M, Benza RL. Macitentan (Opsumit) for the treatment of pulmonary arterial hypertension. *Expert Review of Clinical Pharmacology* 2014;7:415–21.

136. Hong IS, Coe HV, Catanzaro LM. Macitentan for the treatment of pulmonary arterial hypertension. *The Annals of Pharmacotherapy* 2014;48:538–47.

137. Montani D, Chaumais MC, Savale L, et al. Phosphodiesterase type 5 inhibitors in pulmonary arterial hypertension. *Advances in Therapy* 2009;26:813–25.

138. Wilkins MR, Wharton J, Grimminger F, Ghofrani HA. Phosphodiesterase inhibitors for the treatment of pulmonary hypertension. *The European Respiratory Journal* 2008;32:198–209.

139. Lang M, Kojonazarov B, Tian X, et al. The soluble guanylate cyclase stimulator riociguat ameliorates pulmonary hypertension induced by hypoxia and SU5416 in rats. *PLoS One* 2012;7:e43433.

140. Soubrier F, Chung WK, Machado R, et al. Genetics and genomics of pulmonary arterial hypertension. *Journal of the American College of Cardiology* 2013;62:D13–21.

141. Harrison RE, Flanagan JA, Sankelo M, et al. Molecular and functional analysis identifies ALK-1 as the predominant cause of pulmonary hypertension related to hereditary haemorrhagic telangiectasia. *Journal of Medical Genetics* 2003;40:865–71.

142. Machado RD, Eickelberg O, Elliott CG, et al. Genetics and genomics of pulmonary arterial hypertension. *Journal of the American College of Cardiology* 2009;54:S32–42.

143. Ma L, Roman-Campos D, Austin ED, et al. A novel channelopathy in pulmonary arterial hypertension. *The New England Journal of Medicine* 2013;369:351–61.
144. Patel AJ, Honore E, Lesage F, Fink M, Romey G, Lazdunski M. Inhalational anesthetics activate two-pore-domain background K+ channels. *Nature Neuroscience* 1999; 2:422–6.
145. Hartness ME, Lewis A, Searle GJ, O'Kelly I, Peers C, Kemp PJ. Combined antisense and pharmacological approaches implicate hTASK as an airway O(2) sensing K(+) channel. *The Journal of Biological Chemistry* 2001;276:26499–508.
146. Olschewski A, Li Y, Tang B, et al. Impact of TASK-1 in human pulmonary artery smooth muscle cells. *Circulation Research* 2006;98:1072–80.
147. Smith BP, Best DH, Elliott CG. Genetics and pharmacogenomics in pulmonary arterial hypertension. *Heart Failure Clinics* 2012;8:319–30.
148. Eadon MT, Wheeler HE, Stark AL, et al. Genetic and epigenetic variants contributing to clofarabine cytotoxicity. *Human Molecular Genetics* 2013;22:4007–20.

10 Pharmacogenomics and Alternative Medicine
Merge of Future and Tradition

Wei Zhang, Xiao-Ping Chen, and Hong-Hao Zhou

CONTENTS

KEY CONCEPTS

- Although current experimental evidence and population studies cannot provide sufficiently clear recommendations on personalized alternative medicine (CAM) based on genetic characteristics, experience has already been accumulated in the interactions between SNPs and vitamins, minerals, trace elements, and natural herbs.
- Pharmacogenomic mechanisms are proved to involve drug response and affect herb–drug interactions via regulating pharmacokinetic and pharmacodynamics gene expressions, which will probably increase the risk of treatment failure or adverse reactions alone or in combination with other drugs.
- Reduced folate carrier (RFC-1, encoded by the gene *SLC19A1*) is a folic acid transporter in vivo, and it is involved in folic acid transport across the placenta, blood–brain barrier, renal, and other physiological processes. RFC-1 gene polymorphism can affect the neural system development of fetus by regulating the transportation of folic acid in vivo.
- Subjects with one or two copies of the variant alleles for *VKORC1* had a significant reduction in *S*-warfarin EC50 (concentration of *S*-warfarin that produces 50% inhibition of prothrombin complex activity) when warfarin was coadministered with cranberry juice extract.
- The rs1142345 (A>G) SNP in the thiopurine *S*-methyltransferase (*TPMT*) gene was significantly associated with the hypoglycemic effect of the drug.

INTRODUCTION

Alternative medicine refers to all the complementary therapies other than conventional Western medicine in the United States, but it is also called complementary medicine in the United Kingdom, defined as all supporting therapies of Western medicine. Currently, alternative medicine is generally called complementary and alternative medicine (CAM). CAM includes traditional medicine and folk medicine, some of which is empirical, aiming to enhance human innate immunity and prevent diseases, but is not well validated by modern sciences and biotechnologies. CAM also includes traditional herbal medicine, and therapy associated with the use of food, vitamins, minerals, trace elements, meditation, hypnosis, homeopathy, massage, aroma, acupuncture, and more.[1,2] In the 21st century, CAM, as a new health-care model, has been growing fast in Europe, Japan, and South Korea, and CAM-associated pharmacogenomic research has made rapid progress as anticipated. In this chapter, we will summarize current status and future perspectives for

CAM-related pharmacogenomic research based on knowledge from vitamin, minerals, and trace elements, as well as other aspects of traditional herbal medicine.

VITAMINS

Vitamins are a collective name for a group of trace organic substances, which are necessary and important for humans and animals to maintain normal physiological function, metabolism, development, and growth. Although all vitamins vary in their chemical structures and biochemical effects, they have some shared features as summarized: (1) most vitamins in food are precursors of their own; (2) vitamins are not a component of the organ, tissue, or cell, and do not produce any energy. The major role of vitamins is to regulate metabolism; (3) the human body can neither synthesize most vitamins nor produce sufficient amounts of certain vitamins. Most vitamins are obtained from food,[3] but the amount of vitamins that humans require are small.[4] Daily amount is often measured as milligrams (mg) or micrograms (μg). However, if vitamins in the body are deficient or lacking, human health status can be affected, or even threatened, in some cases.

Human genetic polymorphisms that affect the fate of vitamins in the body may affect vitamin effects or disease treatment (see Tables 10.1 and 10.2).

CASE REPORT

A girl was born full term, and her antenatal and postnatal histories were unremarkable. Intermittent limb tremor presented at three years, followed by seizures six months later. The epileptic seizures occurred every two months. During seizures,

TABLE 10.1
Effects of Genetic Polymorphisms on Lipid-Soluble Vitamins

Vitamin	Gene	Polymorphism	Effects of Genetic Polymorphisms on Vitamins	Disease
D	CYP3A4		$1\alpha\ 25(OH)_2D_3$: Enhances the transcription of CYP3A4-deficient 25-hydroxylase and impaired metabolism of vitamin D	Rickets, osteomalacia, osteoporosis
	CYP2R1	rs10500804 rs11023380 rs2060793		
E	CYP4F2	rs2108622	Increased circulating α-tocopherol	Prostate cancer
	SCARB1	rs11057830	Plasma concentrations of α-tocopherol affected	
K	APOE E3/4 E4/4		APOE4 allele (E3/4 + E4/4): Phylloquinone concentrations; higher bioavailability of vitamin K to bone	Easy bleeding, anemia
	VKORC1	Rs9923231	Carriers of the G allele at rs9923231 decreased INR value.	

TABLE 10.2

Effects of Genetic Polymorphisms on Aqueous-Soluble Vitamins

Vitamin	Gene	Polymorphisms	Effects of Genetic Polymorphisms on Vitamins	Disease
B_{12}	TCN2 C776G	rs1801198	G allele: Decreased transcription and concentrations of cellular and plasma transcobalamin.	Pernicious anemia
	FUT2	rs602662	A allele: Increased vitamin B_{12}	
C	Haptoglobin Hp2-2		Serum vitamin C concentrations were associated with haptoglobin type, showing lowest values in serum from Hp2-2 subjects.	Capillary fragility; gums bleeding
	Vitamin C transporter SLC23A1	rs6139591 rs1776964 rs33972313	A higher rate of L-ascorbic acid oxidation in Hp2-2 carriers. The rs6139591 TT genotype and a lower than median dietary vitamin C intake had a higher risk of acute coronary syndrome. Reduction in circulating concentrations of L-ascorbic acid per minor allele.	
Folic acid	MTR A2756G	rs1805087	G allele: Increased Hcy concentration.	Cardiovascular disease
	MTHFR 677C→T	rs1801133	T/T: Elevated plasma homocysteine concentrations.	

the girl experienced sudden loss of consciousness, followed by myoclonus of right limbs, lasting for about 30 minutes. Electroencephalogram revealed spikes and slow waves in the left occipital and left posterior temporal. Paroxysmal limbs trembling and seizures were presented from the age of 3. Intracranial calcification was noted by CT. At the age of 5, mental regression, lower-extremity weakness, and sleeping problems were observed. Her plasma folate decreased to 4.49 nmol/L (normal value >6.8 nmol/L). Plasma total homocysteine elevated to 28.11 μmol/L (normal value <15 μmol/L). Folate and 5-methyltetrahydrofolate in cerebrospinal fluid (CSF) were significantly decreased to an undetectable level.

Folinic calcium (15 mg per day, p.o.) was initiated at the age of 5. One week later, she showed hematological normalization. Blood HGB, MCV, folate, and total homocysteine returned to normal. One month after treatment, intermittent limb tremor and epileptic seizures disappeared. Sleeping difficulties and irritability disappeared one year after treatment. Progressive improvement in mental development was observed. However, leg weakness was observed when going up and down stairs. *SLC46A1* gene encoding proton-coupled folate transporter (PCFT) was the only gene responsible for hereditary folate malabsorption. On *SLC46A1* gene, a novel mutation, c.1A>T (M1L), and a reported mutation c.194-195insG (p.Cys66LeufsX99) were identified, supported the diagnosis of hereditary folate malabsorption. Each parent carried one of the two mutations. Folinic calcium supplement resulted in rapid clinical improvement. She is currently six years old with normal development and routine blood features.[5,6]

Vitamin B$_{12}$

Vitamin B$_{12}$ is also called cobalamin, existing in the human body mainly in the form of adenosine cobalamin and methyl-cobalamin, both of which participate in two types of important biochemical reaction processes. As a methyl malonyl coenzyme A mutase, adenosylcobalamin transforms methyl malonyl coenzyme A into succinyl coenzyme A; and as the coenzyme of methionine synthetase, methyl-cobalamin turns the methylation of homocysteine to methionine. Vitamin B$_{12}$ in the human body comes mainly from food through microbial biosynthesis in the digestive tract, and then through the absorption, transportation, and cell uptake by haptocorrin (HC); intrinsic factor (IF); and cobalamin transfer protein II (TCII). Vitamin B$_{12}$ is involved in erythrocytes formation, DNA synthesis, and other in vivo critical processes, such as maintenance of the nerve sheath. Vitamin B$_{12}$ deficiency leads to body immune defects, megaloblastic anemia, coronary heart disease, and gastrointestinal and nervous system diseases. Vitamin B$_{12}$ deficiency can also lead to hyperhomocysteinemia, which is an independent risk factor for cardiovascular diseases, and is also associated with type 2 diabetes, end-stage renal disease, nerviduct defect, and Alzheimer disease.

Beef, pork, fish, and chicken are rich in vitamin B$_{12}$. Vegetarians have a higher risk of vitamin B$_{12}$ deficiency compared to nonvegetarians. Atrophic gastritis can also cause vitamin B$_{12}$ malabsorption. In recent years, through the candidate genes and genome-wide association study (GWAS), it has been confirmed that a number of gene polymorphisms are significantly associated with levels and metabolism of vitamin B$_{12}$ and corresponding diseases.

Fucosyltransferase (FUT2) gene polymorphism is related to the plasma level of vitamin B$_{12}$. The involvement of sites such as rs602662 (G772A) and rs492602 has been confirmed. Notably, rs602662, which has been studied repeatedly by GWAS, is significantly associated with changes in vitamin B$_{12}$ plasma levels. In 1146 Asian Indian people, a research (which used a single base extension method for *FUT2* genotyping and regression analysis, taking into account age, sex, dietary habits, and associated diseases such as coronary heart disease, high blood pressure, and diabetes) has shown that people homozygous for the rs602662-A/A had an average plasma vitamin B$_{12}$ level of 175.3 pmol/L, while people with the G/G genotype had a mean vitamin B$_{12}$ level of 149.5 pmol/L. The reduction ratio is approximately 0.15. For people with the G/A genotype, the level is 152.7 pmol/L. The results showed a characteristic gene–dose effect ($p = 4.0 \times 10^{-5}$). The research also found that in vegetarians, 772G/G had a 20% lower vitamin B$_{12}$ level in plasma than 772A/A (140.7 vs. 174.2 pmol/L). This research has shown that besides vegetarianism, a significant reduction in plasma vitamin B$_{12}$ is also most likely caused by *FUT2* gene polymorphism.[3]

The form of methyl-cobalamin of vitamin B$_{12}$, as a coenzyme of methionine synthase, can transfer the methyl group from 5-methyl-tetrahydrofolate. Vitamin B$_{12}$ deficiency leads to blockade of re-methylation of homocysteine into methionine and raises the level of homocysteine in blood circulation, probably causing cardiovascular disease, abnormal pregnancy, and other clinical illness. In the blood, vitamin B$_{12}$ mainly binds to haptocorrin and transcobalamin. Haptocorrin binds 80% of the plasma vitamin B$_{12}$ and is not involved in cell vitamin uptake, whereas transcobalamin binds 20% of plasma vitamin B$_{12}$ and is critical to the vitamin uptake.

Cobalamin transfer protein–vitamin B_{12} complex is recognized by specific receptors on the cell membrane and transported into the cell, and unbound vitamin B_{12} and haptocorrin–vitamin B_{12} complexes cannot be recognized and absorbed. Therefore, cobalamin transfer protein gene polymorphism may affect the functionality of vitamin B_{12} complex cellular uptake process. The commonly seen mutation in cobalamin is 776C>G (proline is replaced by arginine). Carriers of the 776G/G variant account for 20% of the general population; wild-type homozygotes (776C/C) and mutant 776C/G heterozygotes account for 30% and 50% of the general population, respectively. Research has found that 776C>G mutation not only affects the affinity between cobalamin and vitamin B_{12}, but also affects the transportation capacity of the cobalamin–vitamin B_{12} complex. Miller and colleagues have found that carriers of 776G/G variant had significantly lower levels of haptocorrin–vitamin B_{12} complex, lower percentage of binding of total vitamin B12 and cobalamin, and significantly higher concentrations of plasma methylmalonic acid (MMA) compared with the C/C genotype.[7] These results have shown that 776C>G gene polymorphism can change the way the cells take vitamin B_{12}, exacerbating the deficiency of vitamin B_{12}. Kristina et al. have found that in 359 young women, 776G/G homozygotes had significantly lower levels of plasma cobalamin–vitamin B_{12} complex, compared with the wild-type 776C/C individuals (74 vs. 87 pmol/L, $p = 0.02$). The 776 C>G mutation affects alterations in the homocysteine levels in the body by changing the plasma cobalamin–vitamin B_{12} complex concentration, thus further affecting the occurrence of cardiovascular and other diseases.[7]

Vitamin C

Vitamin C is a potent antioxidant in the human body, and it is used to relieve ascorbate peroxidase substrate oxidative stress. Many important biosynthetic processes also require involvement of vitamin C. For example, through anti-oxidation and improvement of endothelial function, vitamin C plays an important role in cardioprotection and collagen synthesis. Vitamin C deficiency leads to the reduction of the collagen in atherosclerotic plaques, easily leading to plaque rupture, and in severe cases leading to blood clots and even death. The normal function of sodium-dependent vitamin C transporters (SVCT, encoded by gene *SLC23A2*) 1 and 2 help maintain the homeostasis of vitamin C in vivo. SVCT1 is mainly distributed in the intestine and kidney, controlling the intake and discharge of vitamin C; SVCT2 is mainly distributed in metabolically highly active tissue, ensuring the intracellular inverse concentration gradient accumulation of ascorbic acid in the aorta and other specific tissues. Ascorbic acid is the key factor for artery wall and plaque cap collagen synthesis, and it reduces endothelial dysfunction and inflammation, protecting and stabilizing the vascular plaques. Human *SVCT2* gene polymorphism is associated with premature birth and a variety of tumors. The two polymorphic loci rs6139591 and rs2681116 of *SVCT2* are associated with the changes in vitamin C intake and the concentrations of circulating vitamin C. A case–cohort study of 57,053 subjects with a 6.4-year follow-up has found that women carrying rs6139591 T/T gene absorbed less vitamin C from food, having a 5.39-fold increased risk of suffering from acute coronary syndrome (ACS), compared with carriers of the rs6139591 C/C. Women with rs1776964 T/T genotype absorbed more vitamin C from food compared with carriers of the rs1776964 C/C genotype, with an odds ratio of 3.45. The results have shown that the *SVCT2* genetic polymorphisms are associated with ACS in females. Supplementing vitamin C may be an effective way of preventing the disease.[8,9]

Vitamin D

Vitamin D is a lipid-soluble vitamin, belonging to a steroid family. Among all the required vitamins in humans, vitamin D is very special. It is a hormone precursor and the human body can synthesize vitamin D_3 under sunny conditions. 7-Dehydrocholesterol is converted to vitamin D_3 after exposure to ultraviolet light. Animal skin cells contain 7-dehydrocholesterol. So exposure to sunlight is an easy way to get vitamin D_3 synthesized. However, the activity of vitamin D_3 is not high. Vitamin D_3 must be converted to calcitriol (1,25-dihydroxy cholecalciferol) in the liver and kidney. Calcitriol is the most active form, which can regulate the absorption and metabolism of calcium in the small intestine, kidney, and bone.

Dysfunction of vitamin D–related endocrine systems can cause thyroid autoimmune disease. Notably, $1,25(OH)_2$vitamin D_3 reduces HLAII molecule expression through immunomodulation in thyroid cells, and it also inhibits lymphocyte proliferation and inflammatory cytokine secretion. Therefore, in autoimmune thyroid disease cases, plasma $1,25(OH)_2$vitamin D_3 decreases significantly. Vitamin D–binding protein (DBP), the main transporter of $1,25(OH)_2$vitamin D_3 system, mediates endocytosis of $1,25(OH)_2$vitamin D_3. The genetic polymorphism of *DBP* microsatellite sequences can significantly affect the function of vitamin D through regulating the affinity between DBP and $1,25(OH)_2$vitamin D_3. Michael et al. have found that a genetic polymorphism in intron 8 of *DBP* gene, variable tandem repeats (TAAA)n, is significantly associated with Graves disease. In DBP knockout mice, vitamin D metabolism was significantly affected.[10]

DBP plays an important role in in vivo transport and metabolism of vitamin D (in particular $1,25(OH)_2$vitamin D_3). In healthy male subjects and male patients with osteoporosis-related vertebral fractures ($n = 170$), DBP (TAAA)n-Alu*10 and *11 alleles had a protective effect on osteoporosis (OR = 0.39, $p < 0.0005$; OR = 0.09, $p < 0.007$). In other words, when people carry 19–20 repeats (genotypes 9/10, 9/11, and 10/10), the concentrations of circulating DBP and free vitamin D are high, suggesting a higher bone density and a lower risk of osteoporosis.[11]

Vitamin E

Vitamin E, also known as tocopherol, is an important antioxidant in the human body. The main features of vitamin E include preventing polyunsaturated fatty acids and phospholipids from being oxidized, thus maintaining cell membrane integrity. Vitamin E also protects vitamin A from oxidative damage and strengthens the function of vitamin A. Vitamin E reduces blood lipid peroxide, prevents excessive platelet aggregation, increases erythrocyte membrane stability, promotes the synthesis of red blood cells, and maintains cellular respiration.

Vitamin E is transported by triglycerides (TG)-rich lipoprotein in vivo. Apolipoprotein A5 (*APOA5*) gene polymorphism can significantly affect the occurrence of hypertriglyceridemia; therefore, the transport of vitamin E in vivo can be regulated by APOA5. In 169 patients with type 2 diabetes, researchers found that the vitamin E levels in patients with *APOA5*-1131T/C genotype was about 13% higher than that in the normal population, and the T/C genotype probability in the population with higher vitamin E level was about 2.6 times of the probability in the normal population.[12]

Plasminogen activator inhibitor type 1 (PAI-1) is an independent risk factor for cardiovascular diseases. Its expression is increased in type 2 diabetic patients. PAI-1 gene 4G/5G polymorphism can regulate the expression of PAI-1 protein, thus affecting cardiovascular diseases. Vitamin E can effectively lower PAI-1 levels. Testa et al. gave 500 IU of vitamin E per day to 93 type 2 diabetic patients for 10 consecutive weeks and then detected the 4G/5G polymorphism and PAI-1 level. They found that the PAI-1 levels started to drop in the 10th week in 4G/5G and 4G/4G patients while the levels started to drop in the 5th week in 5G/5G patients and 4G/5G patients ($p < 0.01$). The results have shown that the 4G alleles of *PAI-1* affects the pharmacological function of vitamin E. Vitamin E has a faster onset in patients with 5G/5G genotype, indicating that vitamin E may have a better effect on cardiovascular disease prevention in these patients.[13]

Vitamin K

Vitamin K is vital for blood coagulation and bone growth. Its key features include the following: (1) Blood clotting: It is a blood coagulation factor of gamma-carboxylase cofactor. Synthesis of coagulation factors VII, IX, and X is also dependent on vitamin K. (2) Bone metabolism: It is involved in the synthesis of vitamin K–dependent protein, which can regulate the synthesis of calcium phosphate in the bone. Plasma vitamin K levels and bone density of elderly people are positively correlated.

Outcomes of treatment of excessive anticoagulation using vitamin K in clinical practice also show significant differences among individuals. Vitamin K epoxide reductase complex subunit 1 (VKORC1) may be used to partially explain this phenomenon. In excessive anticoagulation patients (international normalized ratio [INR] ≥ 4), 2.5–5 mg vitamin K was given to the patients based on their INR values. The INR value was rechecked at 3, 6, 24, and 72 hours after the treatment. Researchers noticed that patients with *VKORC1*-1639G/A mutation had significantly lower INR 3 hours after taking vitamin K, compared with carriers of the A/A genotype ($p < 0.001$). These results have shown that *VKORC1*-1639G>A gene polymorphism is significantly associated with an acute procoagulant effect of vitamin K.[14]

Folic Acid

Folic acid, as an important carbon carrier, plays a critical role in the nucleotide synthesis, re-methylation of homocysteine, and other important physiological functions, especially in the rapid cell division and growth processes (such as infant development and pregnancy). Folic acid can promote immature cells in the bone marrow to form morphologically normal red blood cells, thus avoiding anemia.

Meta-analysis has shown that the 677T/T genotype in the gene *MTHFR* encoding methylenetetrahydrofolate reductase (MTHFR) was associated with 14–21% higher risk of developing cardiovascular diseases compared with carriage of 677 C allele.[15] Maria et al. have found that folic acid interacts with *MTHFR* 677C/T polymorphism, affecting the incidence of colorectal cancer ($p = 0.037$): if folic acid level is low and with *MTHFR* 677T/T genotype, patients would be at higher risk for colorectal cancer (OR = 2.4), indicating that the body's folic acid levels are more important for patients with the *MTHFR* 677T/T genotype.[16]

Reduced folate carrier (RFC-1, encoded by the gene *SLC19A1*) is a folic acid transporter in vivo, and it is involved in folic acid transport across the placenta, blood–brain barrier, renal, and other physiological processes. Folic acid deficiency can cause intracellular DNA methylation abnormalities, ultimately leading to congenital diseases, cancer, cardiovascular diseases, and neuropsychiatric diseases. *RFC-1* gene 80A/G polymorphism and increased plasma homocysteine levels are associated with decreased folic acid levels, likely causing congenital disorders. A meta-analysis involving 930 mothers of Down syndrome (DS) children and 1240 mothers of normal controls has shown that *RFC-1* gene 80G/G genotype has a 1.27-fold higher risk of DS. The single G allele is associated with a 1.14-fold increased risk of DS.[17] This result shows that *RFC-1* gene polymorphism can affect the neural system development of the fetus by regulating the transportation of folic acid in vivo.

MINERALS AND TRACE ELEMENTS

Minerals are a general name for a variety of elements that constitute human tissues and maintain normal physiological functions. Minerals are also essential nutrients for the human body. Carbon, oxygen, hydrogen, nitrogen, and other elements are mainly in the form of organics, and the remaining 60 elements are often referred to as minerals (inorganics) in the human body.

Calcium, magnesium, potassium, sodium, phosphorus, sulfur, and chlorine account for approximately 60–80% of total minerals in the body. Therefore, these seven elements are called macroelements. Iron, copper, iodine, zinc, selenium, manganese, molybdenum, cobalt, chromium, tin, vanadium, silicon, nickel, and fluorine account for less than 0.005% of total minerals in the body; therefore, these 14 elements are called microelements or trace elements. Although minerals cannot provide energy, they play an important role in human physiology. Human bodies are unable to synthesize these minerals and must get them from the outside environment. Minerals are the staples that make up the organs and tissues. For example, calcium, phosphorus, and magnesium are the main ingredients of bones and teeth. Minerals are also necessary for normal osmotic pressure and acid–base balance of the body. Synthesis of some specific biological molecules, such as hemoglobin and thyroid hormone, requires iodine. The interactions between human gene polymorphism and minerals/trace elements may affect the normal physiological functions of the human body and the occurrence of diseases.

CASE REPORT

Twins born at 36 weeks' gestation were treated with vitamin D (400 IU from fortified milk formula + 1500 IU supplementation because of prematurity). One presented with vomiting and failure to thrive at six months of age. He had severe hypercalcemia (3.68 mmol/L; normal value: < 2.6) with low PTH serum level (1 ng/L, normal value: 10–55), slightly elevated serum level of 25(OH)D at 140 nmol/L, normal serum level of 1,25(OH)2D at 96 pmol/L (normal value: 60–120), and hypercalciuria (Ca/creatinine ratio: 4.3) associated with nephrocalcinosis. Normalization of calcemia was observed within three weeks after switching from milk to an infant formula containing low calcium and no vitamin D content (Locasol; Nutricia-France, SA, St. Ouen, France). Vitamin D supplement was withdrawn together with appropriate hydration and furosemide therapy. Later, the child tolerated vitamin D fortified milk at 600 IU/d with normal serum levels of 25(OH)D. However, the summer serum level of 1,25(OH)2D subsequently rose to 360 pmol/L, despite sun protection. Polymerase chain reaction (PCR) amplification of exons 1–11 of *CYP24A1* was performed in this twin showing no PCR product for exons 9 and 11. This was later confirmed by quantitative PCR using specific TaqMan (Assays Life Technologies, St. Aubin, France) probes for intron 8 and 10, respectively. The deletion was shown to be inherited from heterozygous nonconsanguineous parents who displayed only one allele for intron 10.

The patient is now 18 years old, and his nephrocalcinosis appears unchanged on ultrasound examination. During the summer months, his calcium levels still rise to the upper limit of normal but with elevated serum levels of 1,25(OH)2D and low PTH levels. His unaffected twin, who received the same high amount of vitamin D per day did not exhibit IIH and did not have the *CYP24A1* mutation.[18]

SELENIUM

Selenium has antioxidative activity, which protects the human body from free radicals and carcinogens. Selenium may also relieve inflammation, enhance immunity to fight against infections, promote heart health, and enhance the role of vitamin E. Selenium is essential for the male reproductive system and metabolism. Selenium intake deficiency may be associated with cancer, premature aging, cataracts, hypertension, recurrent infections, and so on.

Alzheimer disease (AD) is a complex genetic disease, characterized by a variety of cognitive disorders. Oxidative stress may play a key role in its pathogenesis. Cytosolic glutathione peroxidase (GPx1) widely exists in many tissues that have high oxidative stress. The *GPx1* gene Pro*198*Leu can lead to reduced enzyme activity. In one research involving elderly AD patients and healthy controls, researchers have found that *GPx1* Pro198Leu polymorphism itself was not associated with the pathogenesis of AD. However, the selenium levels in the plasma and erythrocyte of the AD patients homozygous for the Pro198Pro were significantly lower than that in healthy subjects homozygous for the Pro198Pro (31.44 vs. 54.87 μg/L, $p = 0.002$; 40.25 vs. 87.75 μg/L, $p = 0.0004$), the selenium levels in erythrocytes were positively correlated with *GPx1* activity in AD patients and healthy controls, both of which were homozygous for Pro198Pro genotype ($r = 0.59$, $p < 0.005$; $r = 0.72$, $p < 0.0001$), but this correlation in *GPx1* mutant population did not exist.[19] These results indicate that *GPx1* Pro198Leu polymorphism is correlated, to some extent, with the selenium level in the AD patients, suggesting that the effect of selenium, as a health supplement, can be affected by *GPx1* genotype.

Prostate cancer risk is associated with low selenium level. Selenoprotein P (SEPP1), the most abundant plasma selenoprotein, is responsible for the transportation of dozens of selenocysteine residues. *SEPP1* gene polymorphism may affect selenium function and prostate cancer pathogenesis. In a study including 1352 prostate cancer patients and 1382 healthy controls, *SEPP1* gene rs13168440 polymorphic loci were significantly correlated with plasma selenium levels. The study also showed that patients with the second allele and high plasma selenium levels had lower prostate cancer risk ($P_{interaction} = 0.01$). This correlation in the population with the main allele did not exist.[20] In other research targeting selenoprotein P gene 1 (*SEP15*), whose protein is highly expressed in the prostate, researchers compared four common polymorphisms in *SEP15* gene in 1286 prostate cancer patients and 1267 controls, and found that these polymorphisms were not associated with the prostate cancer, but rs561104 locus and plasma selenium levels had significant interaction with the prostate cancer mortality ($P_{interaction} = 0.02$); that is the patient who did not carry risky rs561104 genotype had lower prostate cancer mortality if he or she had higher plasma selenium levels.[20,21] These studies illustrate that prostate cancer patients with a specific genotype may reduce the risk of death by supplementing selenium.

In another research which included 567 prostate cancer patients and 764 healthy controls found that manganese superoxide dismutase (*MnSOD*) gene codon 16 valine (V) to alanine (A) gene polymorphism was significantly associated with the in vivo antioxidant levels. ($P_{interaction} \leq 0.05$): A/A genotype patients who had high plasma selenium levels had lower risk of prostate cancer (OR = 0.3), and this A/A-type plus high selenium level also had a protective effect on progressive prostate cancer (OR = 0.2), but selenium had less of a protective effect on V/A and V/V-type patients (OR = 0.6 and 0.7, respectively). For A/A-type patients, selenium, lycopene, and vitamin E had a tenfold protective effect compared with other genotype patients.[22]

Calcium

Calcium, an essential mineral nutrient for humans, is the trigger for many biochemical and physiological processes, such as muscle contraction, hormone release, spike transmission, blood clotting, heart rate regulation, milk secretion, and so on. Calcium is closely associated with the function of the body's immunity and nervous, endocrine, digestive, circulatory, motor, and reproductive systems.

Parathyroid hormone (PTH) and vitamin D are two major calcium regulatory endogenous hormones in vivo. Primary hyperparathyroidism (PHPT) generates excessive PTH, accelerating bone turnover, thus resulting in hypercalcemia and low bone density. Research has found that vitamin D receptor (*VDR*) gene polymorphism rs7975232A/C is associated with lumbar spine bone density, and that carriers of the A allele have significantly lower bone density than noncarriers (1.01, or 1.04 vs. 1.21 g/cm^2, p = 0.003), indicating that rs7975232C/C genotype is more conducive to the lumbar spine bone calcium deposition (Table 10.3).[23] Irene et al. have proven that compared to those with *VDR* b/b genotype males, premenopausal female subjects carrying *VDR*b allele had higher Z scores for cervical spine bone density, and higher lumbar spine bone density as well, with 3% higher plasma calcium concentration (9.5 vs. 9.23 mg/dL, p = 0.008).[24] Research on pigs also found that in gene complement factor 2 (C2), which plays an important role in humoral immune response, showed a strong correlation of 1963A>G mutation with plasma calcium levels (p = 5.9 × 10^{-5}). This result provides a good reference for human subject research.[25]

TABLE 10.3
Macroelements and Gene Polymorphisms

Macroelements	Gene	Polymorphisms	Effects	Disease
Calcium	VDR	rs7975232	C/C: More conducive to the lumbar spine bone calcium deposition.	PHPT
	CASR	A986S and R990G	Carriers of the variant R: Lower blood calcium levels.	
	VDR	Bsml polymorphism	Carriers of the B allele: Higher levels of calcium.	Bone disease and MS
	C2	1963A<G	Showed a strong correlation with P.	Humoral immune abnormalities
	VDR	Fok*I* polymorphism	The positive association between 1,25(OH)D and eGFR was steeper in Fok*I*CT and CC polymorphisms than Fok*I*TT polymorphism.	CKD in patients with type 2 diabetes
Magnesium	*MUC1*, *ATP2B1*	Multiple SNPs	These genes were associated with serum magnesium levels.	Hypomagnesemia
Potassium	*CYP17A1*	rs11191548	Reduced renin activity and plasma potassium levels.	Hypertension
	SLC12A3	G264A	Carriers of the variant A: Excreted more potassium.	
Phosphorus	VDR Fok*I*	rs10735810	1,25(OH)D were negatively associated with P levels.	CKD in type 2 diabetes
Sodium	*ADD1*	rs4961	rs4961 is related to systolic blood pressure and blood sodium concentration.	hypertension
	SCNN1B	G442V	Greater sodium retention.	

Iron

Iron is the most abundant trace element in the human body. It is essential for normal body functions, such as oxygen transportation. Normal iron level is important. Higher level leads to a Fenton reaction characterized by excessive hydroxyl radicals and tissue damage; if it is too low, it leads to anemia. Absorption and storage of iron are critical in maintaining normal iron levels. Carlos et al. studied SNP sites of 10 candidate genes and their relationship to iron levels in humans, and found that calcium channel gene *CACNA2D3* intron rs1375515 locus is significantly associated with mean corpuscular volume, hemoglobin, and ferritin levels (Table 10.4). The G allele is related to the reduction of above levels and might be a risk factor for iron-deficient anemia. rs1375515 or its linked loci may regulate body iron stores by influencing the function of calcium channels, and it may also affect the therapeutic effect of iron supplementation.[26]

Nonalcoholic fatty liver disease (NAFLD) can affect iron transportation in the liver, causing iron accumulation in the liver, and leading to hepatocellular carcinoma and severe liver injury. Transmembrane protease serine 6 (*TMPRSS6*) gene can affect iron metabolism by regulating the transcription of hepatic hepcidin. In 216 NAFLD patients, a study found *TMPRSS6* gene p.Ala736Val polymorphic loci Val homozygotes are not only associated with hepatic iron accumulation and decreased ferritin levels (223 vs. 308 ng/mL, $p = 0.01$), but also associated with oxidative stress–induced hepatocyte ballooning degeneration. In the iron overload HFE-negative population, *TMPRSS6* gene p.Ala736Val polymorphic loci have more influence on the hepatic iron accumulation, and 736Val genotype and hepatic iron accumulation are negatively correlated (OR = 0.59), indicating that the locus can significantly affect the secondary iron accumulation in the liver of NAFLD patients.[27]

Magnesium

Magnesium is the second most abundant intracellular cation. It acts as a cofactor in many important physiological processes, such as nucleic acid synthesis and enzymatic reactions. About 60% of magnesium ions is present in the bone, 20% in muscle, and another 20% in the soft tissue. Plasma magnesium concentrations have been linked to a variety of chronic diseases, such as diabetes, hypertension, and osteoporosis. In a study from the European CHARGE syndrome involving 15,336 human GWAS, researchers studied the effect of 2.5 million gene polymorphisms on plasma magnesium levels. Using microarrays, at $p < 5 \times 10^{-8}$ or $p < 4 \times 10^{-7}$ (related) sites, researchers further found in 8463 subjects that six gene regions were correlated with plasma magnesium levels at GWAS level: *MUC1*, *ATP2B1*, *DCDC5*, *TRPM6*, *SHROOM3*, and *MDS1*. Multiple SNPs in these genes are associated with hypomagnesemia. Magnesium ion transporters CNNM2, CNNM3, and CNNM4 are significantly correlated with magnesium concentrations in vivo. The large-scale research provided important targets for further studies on human blood magnesium homeostasis and its regulation mechanism.[28]

TABLE 10.4

Trace Elements and Gene Polymorphisms

Trace Elements	Gene	Polymorphisms	Effects	Disease
Selenium	GPx1	P198L	P198L is correlated to the selenium level.	AD
	SEPP1	rs13168440	Selenium levels were negatively associated with PCa risk only among men with the minor allele.	
	SEP15	rs561104	rs561104 and plasma selenium levels had significant interaction with PCa mortality.	PCa
	MnSOD	V to A	A/A: high plasma selenium level.	
Iron	*CACNA2D3*	rs1375515	The G allele: reduced hemoglobin and ferritin.	Iron deficiency anemia
	TMPRSS6	A736V	Homozygosity for the p.736Val allele was associated with lower hepatic iron stores and ferritin levels.	Severe liver damage and hepatocellular carcinoma
	HFE	H67D (homologous to human H63D)	Brain iron management proteins' expressions were altered in the H67D mice.	Neurodegenerative diseases
	A1AT	E342 K E264 V	Plasma concentrations of A1AT was negatively related to ferritin.	Iron balance disorders
Chromium	XRCC1	R399Q	XRCC1 gene Arg399Gln mutations may become protective biomarkers for hexavalent chromium-induced DNA damage.	Occupational chromium exposure disease
Fluorine	CTR	Taq polymorphisms	The interactive effect of F burden and CTR genotype was significant.	The F bone injury
Copper	*ATP7B*	rs2147363	The ATP7B gene plays a key role in controlling body copper balance.	AD

POTASSIUM

Potassium is an essential macroelement in human nutrition, playing an important role in fluid and electrolyte balance. The secretion of potassium ions in the kidney is subjected to aldosterone, and, in turn, the body potassium storage capacity is regulated. Excessive aldosterone will cause a huge loss of potassium and magnesium ions through the urine, resulting in lower plasma potassium and higher plasma sodium, causing hypertension, whereas with higher potassium and lower sodium, hypotension may occur. CYP17A1 is a member of the cytochrome P450 superfamily and it plays an important role in the synthesis of cortisol and mineralocorticoid. In the GWAS of European populations, researchers have found that the *rs11191548* SNP locus, which is close to *CYP17A1,* is associated with hypertension and systolic blood pressure. In a research involving 1101 primary hypertension patients and 1109 healthy controls, Li et al. found that *rs11191548* SNP was associated with hypertension onset, and that the C allele was significantly correlated with reduced potassium levels in the hypertension patients who used non-renin–angiotensin–aldosterone system antagonists (−0.093 vs. −0.067, $p = 0.003$), indicating that the *CYP17A1* gene polymorphism may reduce renin activity and plasma potassium levels by changing CYP17A1 activity.[29]

CHROMIUM

Chromium is an essential trace element for the body's glucose and lipid metabolism. Trivalent chromium ions are beneficial to humans, but hexavalent chromium ions are toxic. Chromium and other metabolism-controlling substances, such as hormones, insulin, and a variety of enzymes and cell genetic materials (DNA and RNA), play a role in physiological functions. The physiological functions of chromium include regulation of glucose and lipid metabolism, stabilization of nucleic acids (DNA and RNA), and more. Constant exposure to chromium can cause lung cancer and other occupational diseases. Zheng et al. have studied the relationship between hexavalent chromium ions and nine major DNA repair gene mutations. Using Olive Tail Moments (OTMs), they evaluated the extent of DNA damage and found that at *XRCC1* gene G399A polymorphism, G/G genotype has significantly higher OTMs, compared with G/A and A/A type (0.93 vs. 0.73 or 0.5, $p = 0.048$), suggesting that carriers of the A allele are less susceptible to the damaging effects of hexavalent chromium than noncarriers (OR = 0.39). *XRCC1* gene G399A mutations may become protective biomarkers for hexavalent chromium-induced DNA damage.[30]

HERBAL MEDICINES

Herbal medicines have long been used for a variety of ailments in Asian countries and have become more popular worldwide over the last two decades. It has been reported by the WHO that approximately 70% of the global population currently uses medicinal herbs as CAM. All herbal medicines are actually a combination of potentially biologically active compounds that possess various inherent pharmacological activities. Because the metabolism of these compounds usually occurs with the same mechanisms as those of the drug, there is considerable potential for the interactions between herbal components and drugs. Genetic variants in drug-metabolizing enzymes, transporters, and drug targets may affect their activity for different substrates, and thus may influence drug–herb interactions (DHIs) (Table 10.5). The concentrations of the herbal extract and the drug may determine the degree of DHI, and

TABLE 10.5

Gene Polymorphisms and Herb–Drug Interactions

Herbal Medicine	Drug Affected	Gene	Polymorphism	Pharmacogenomics in HDIs Effects
Baicalin	Rosuvastatin	OATP1B1	*1b/*1b; *1b/*15; *15/*15	After baicalin treatment, the AUC_{0-72} and $AUC_{0-\infty}$ of rosuvastatin were decreased according to OATP1B1 haplotype *1b/*1b ($47.0 \pm 11.0\%$ and $41.9 \pm 7.19\%$), *1b/*15 ($21.0 \pm 20.6\%$ and $23.9 \pm 8.66\%$), *15/*15 ($9.20 \pm 11.6\%$ and $1.76 \pm 4.89\%$), respectively.
Berberine Evodiamine	Fluoxetine, sertraline	5-HTT	S, XS11, XL17, XL18 alleles and A>G (rs25531)	When tested against the S, XS11, LG, LA, XL17, and XL18 alleles, 100 mm berberine increased 5-HTT promoter activities by 67%, 128.7%, 106.9%, 100.4%, 26.2%, and 82%, respectively, 2 mm evodiamine increased 5-HTT promoter activities by 216.7%, 81.6%, 305.6%, 181.5%, 175.3%, and 102.2%, respectively.
Dictamnus dasycarpus Pseudoepicoccum cocos Rhus verniciflua stokes	Digoxin, daunorubicin	MDR1	2677G/T/A (rs2032582)	Digoxin effluxes were significantly decreased in MDR1 gene variants of 2677T/893Ser with the treatment of *P. cocos*, and of 2677G/893Ala and 2677T/893Ser with the treatment of *D. dasycarpus*.
Radix Astragali (RA)	Fexofenadine	ABCB1	C3435T	$T_{1/2}$ of fexofenadine in ABCB1 3435T mutation allele carriers was longer compared to ABCB1 3435CC carriers (4.43 ± 1.44 h vs. 2.54 ± 0.21 h), while RA extract pretreatment lengthened $T_{1/2}$ in ABCB1 3435CC carriers and abolished such genotype-related differences.
GFJ	Ebastine	MDR1	C3435T	The grapefruit juice–induced inhibition of its transport/formation (mean-fold decrease ± SD: 1.5 ± 0.8, 1.1 ± 0.9, and 0.9 ± 0.4) for CC, CT, and TT carriers, respectively.

(Continued)

TABLE 10.5 (Continued)

Gene Polymorphisms and Herb–Drug Interactions

Herbal Medicine	Drug Affected	Gene	Polymorphism	Pharmacogenomics in HDIs Effects
SJW	Talinolol	MDR1	1236C>T, 2677G>T/A, 3435C>T	Subjects harboring the ABCB1 haplotype comprising 1236C>T, 2677G>T/A, and 3435C>T polymorphisms had lower intestinal MDR1 mRNA levels and showed an attenuated inductive response to St. John's wort as assessed by talinolol disposition.
SJW	Repaglinide	SLCO1B1	c.521T>C	SLCO1B1 c.521TT genotype subjects presented a trend for lower mean concentrations and AUCs of insulin than c.521TC and CC genotypes subjects, but this trend did not reach statistical significance.
Baicalin	Bupropion	CYP2B6	*1/*1, *1/*6, *6/*6	The AUC ratio of hydroxybupropion to bupropion tended to be more lower in *6/*6 compared with *1/*1 genotype patients (5.3 vs. 8.0) after baicalin treatment.
Echinacea	Dextromethorphan	CYP2D6	Extensive metabolizers Poor metabolizer	Echinacea dosing reduced the oral clearance of dextromethorphan by 28% and increased AUC by 42% in the CYP2D6 poor metabolizers, while extensive metabolizers were not affected.
GB	Omeprazole	CYP2C19	EMs (*1/*1, *1/*2 and *1/*3); PMs (*2/*2 and *2/*3)	After $G.$ $biloba$, the mean decreases in omeprazole $AUC_{0-\infty}$ were 41.5%, 27.2%, and 40.4% in the homozygous EMs, heterozygous EMs and PMs, respectively. Whereas the decreases in omeprazole sulfone $AUC_{0-\infty}$ were 41.2%, 36.0%, and 36.0%, respectively.
GFJ	Lansoprazole	CYP2C19	*1/*1, *1/*2, *1/*3, *2/*2, *2/*3	GFJ treatment significantly increased total AUC of lansoprazole (26,661 ± 7407 vs. 34,487 ± 10,850 ng•h/mL) in *2/*2 and *2/*3 subjects, whereas the total AUC ratio of lansoprazole sulfone/lansoprazole was significantly decreased (0.07 ± 0.05 vs. 0.04 ± 0.05) in *1/*1 genotype carriers.

($Continued$)

TABLE 10.5 (Continued)
Gene Polymorphisms and Herb–Drug Interactions

Herbal Medicine	Drug Affected	Gene	Polymorphism	Pharmacogenomics in HDIs Effects
GFJ	Lansoprazole	CYP2C19	homEMs (*1/*1) hetEMs (*1/*2, *1/*3); PMs (*2/*2, *2/*3)	The mean plasma concentrations of lansoprazole were not increased by GFJ, whereas GFJ slightly prolonged the t_{max} in the three different CYP2C19 genotype groups.
Liu Wei Di Huang Wan (LDW)	Omeprazole, dextromethorphan hydrobromide, midazolam	CYP2C19,CYP2D6, CYP3A4	CYP2C19*1/*1 CYP2C19*2/*2	LDW is unlikely to cause pharmacokinetic interaction when it is combined with other medications predominantly metabolized by CYP2C19, CYP2D6, and CYP3A4 enzymes.
Silymarin	Losartan	CYP2C9	*1/*1, *1/*3	The metabolic ratio of losartan (ratio of $AUC_{0-\infty}$ of E-3174 to $AUC_{0-\infty}$ of losartan) after a 14-day treatment with silymarin decreased significantly higher in individuals with the CYP2C9*1/*1 genotype ($48.78 \pm 25.85\%$) than the CYP2C9*1/*3 genotype ($20.09 \pm 16.87\%$).
SJW	Nifedipine	PXR CYP3A4	PXR haplotypes H1 and H2	Administration of St. John's wort induces higher metabolic activity of CYP3A4 in H1/H1 than in H1/H2 and H2/H2 subjects, with the $AUC_{0-\infty}$ of nifedipine decreased by 42.4% (H1/H2), 47.9% (H2/H2), and 29.0% (H1/H1), whereas that of dehydronifedipine increased by 20.2%, 33.0%, and 106.7%, respectively.
	Gliclazide	CYP2C9	*1/*1; *1/*2 or *2/*2; *1/*3	Treatment with St. John's wort significantly increases the apparent clearance of gliclazide by 50%. For CYP2C9*2 allele carriers, the increase is slightly lower.

(Continued)

TABLE 10.5 (Continued)
Gene Polymorphisms and Herb–Drug Interactions

Herbal Medicine	Drug Affected	Gene	Polymorphism	Pharmacogenomics in HDIs Effects
	Voriconazole	CYP2C19	*1/*1; *1/*2 *2/*2	The AUC of voriconazole was increased by 22% the first day and decreased by 59% after 15 days, with a 144% increase (carriers of 1 or 2 deficient CYP2C19*2 alleles were smaller than wild-type carries) in its oral clearance (CL/F) after St. John's wort administration
	Omeprazole	CYP3A4 CYP2C19	*1/*1; *2/*2; *2/*3	St. John's wort can induce CYP3A4 and increase higher CYP2C19 activity in wild type than poor metabolizers (*2/*2 or *2/*3), leading to increased metabolites of omeprazole in genotype-dependent manner
	Mephenytoin and caffeine	CYP2C19	*1/*1; *2/*2; *2/*3	St. John's wort treatment significantly increased CYP2C19 activity in *1/*1 subjects, with mephenytoin metabolites excretion raised by 151.5% ± 91.9%, which is in contrast to *2/*2 and *2/*3 individuals
Tianqi Jiangtang	Unknown	TPMT	rs1142345	The effective ratio of subjects with homozygotes (AA) of the wild-type allele of TPMTrs1142345 was 2.8 times higher than that of subjects with TPMT heterozygotes (AG)
Yin Zhi Huang (YZH)	Omeprazole	CYP3A4 CYP2C19	CYP2C19 (*1/*1; *1/*2; *1/*3; *2/*2)	YZH induces CYP3A4 and CYP2C19 metabolism of omeprazole with the decrease of the $AUC_{0-\infty}$ ratio of omeprazole/5-hydroxyomeprazole in CYP2C19*1/*1 and CYP2C19*1/*2 or *3 greater than in CYP2C19*2/*2
Cranberry juice and Garlic	Warfarin	VKORC1	VKORC1 1173T>C	Subjects with CT and TT genotypes coadministered with cranberry juice extract significantly reduced S-warfarin EC_{50} by 22% and 11%, respectively, in subjects with CT and TT genotypes. In contrast, an increased EC_{50} (by 22%) was observed when subjects with the VKORC1 wild-type genotype was coadministered with garlic

ABCB1, ATP-binding cassette, sub-family B1; GB, G. biloba; 5-HTT, Serotonin transporter; GFJ, Grapefruit juice; MDR1, multidrug transporter; OATP1B1, solute carrier organic anion transporter family B1; PXR, nuclear receptor subfamily; SJW, St John's wort; TPMT, thiopurine S-methyltransferase; VKORC1, vitamin K epoxide reductase complex, subunit 1.

therefore polymorphisms in drug-metabolizing enzymes and drug transporters that alter the systemic exposure to the substrate drugs or active components of herbs may affect the risk of interactions.

CASE REPORT

A case report described an interaction of St. John's wort (*Hypericum perforatum*) with clozapine via inducing P450s, especially CYP1A2 and CYP3A4. A 41-year-old woman with disorganized schizophrenia was stable on a fixed daily dose of 500 mg, with stable plasma level (0.46–0.57 mg/L) of clozapine for the last six months. However, it was revealed that she had started using St. John's wort (three tablets daily, 300 mg per tablet) shortly before the first measurement of a low plasma clozapine concentration (0.19 mg/L). A month after discontinuation of St. John's wort, the clozapine concentration was 0.32 mg/L, and after one month thereafter it was 0.41 mg/L. Also, the psychiatric condition of the patient was improved.[31] As the author discussed, St. John's wort may pharmacokinetically interact with clozapine through inducing CYP3A4, CYP1A2, CYP2C9, and CYP2C19, all of which are responsible for clozapine metabolism. In addition, P-glycoprotein can also be induced by this herb. This eventually leads to decreased plasma clozapine levels and diminished effectiveness. In another case report, *Ginkgo biloba* can negatively influence the effect of antiretroviral drug efavirenz (EFV) in an HIV-infected male patient. This may be due to the induction of *G. biloba* on CYP2B6 and CYP3A4, both of which are the drug-metabolizing enzymes for EFV.[32] We can believe that both the genetic polymorphisms and medicines or herbal products can cause the variations of enzymes' activities in individuals.

ST. JOHN'S WORT

St. John's wort is an herb most commonly used for depression and conditions such as anxiety, tiredness, loss of appetite, and trouble sleeping. There is some strong evidence that it is effective for mild-to-moderate depression. Currently, there have been 95 drugs (296 brand and generic names) known to have a major interaction with St. John's wort. After administration of St. John's wort, the $AUC_{0-\infty}$ of nifedipine and dehydronifedipine decreased by 42.4% and 20.2% in PXR H1/H2 genotype; 47.9 and 33.0% in H2/H2 genotypes; whereas for the H1/H1 the $AUC_{0-\infty}$ of nifedipine decreased 29.0%, but the $AUC_{0-\infty}$ of dehydronifedipine increased by 106.7%.[33] St. John's wort treatment significantly increased phenytoin clearance in CYP2C19 extensive metabolizers (EMs) but not in poor metabolizers (PMs) and decreased the plasma concentrations of omeprazole in a genotype-dependent manner.[34,35] Subjects harboring the ABCB1 haplotype comprising 1236C>T, 2677G>T/A, and 3435C>T polymorphisms had lower intestinal MDR1 mRNA levels and showed an attenuated inductive response to St. John's wort as assessed by talinolol disposition.[36] The AUC of voriconazole was decreased by 59% with St. John's wort treatment, with a 144% increase in oral clearance of voriconazole. The apparent oral clearance of voriconazole and the absolute increase in apparent oral clearance were smaller in *CYP2C19*2* carriers than those with *CYP2C19*1/*1* genotype.[37]

GINKGO

Ginkgo is often used for memory disorders, including Alzheimer disease. It is also used for conditions that seem to be due to reduced blood flow in the brain, especially in older people. These conditions include memory loss, headache, ringing in the ears, vertigo, difficulty concentrating, mood disturbances, and hearing disorders. Ginkgo enhanced omeprazole hydroxylation in a CYP2C19 genotype-dependent manner. The decrease was greater in *CYP2C19* PMs (*2, *3) than EMs.[38]

BAICALIN

Baicalin is a flavone glucuronide purified from the medicinal plant Radix scutellariae through uridine diphosphate glucuronation. Nowadays, baicalin has begun to be used in bilirubin-lowering therapy, both prescribed and over the counter, in China. The mean changes in AUC ratio of bupropion was lower for subjects with *CYP2B6*6/*6* genotype compared with those with *CYP2B6*1/*1* genotype following baicalin use, indicating induction of CYP2B6-catalyzed bupropion hydroxylation by baicalin. And administration of baicalin decreased the $AUC_{0-\infty}$ of rosuvastatin by about 42%, 24%, and 1.8% in carriers of the *SLCO1B1* *1b/*1b, *1b/*15, and *15/*15, respectively.[39,40]

GARLIC

Garlic is widely used worldwide for its pungent flavor as a seasoning or condiment. There is some scientific evidence that garlic can lower high cholesterol after a few months of treatment. Garlic seems to also lower blood pressure in people with high blood pressure and possibly slow "hardening of the arteries." There is also some evidence that eating garlic might reduce the chance of developing some cancers such as cancer of the colon, and possibly of stomach and prostate. Coadministration of garlic did not significantly alter warfarin pharmacokinetics or pharmacodynamics. However, subjects with the *VKORC1* wild-type genotype showed an increase in the S-warfarin EC50 when warfarin was administered with garlic.[41]

GRAPEFRUIT JUICE

Grapefruit juice, a potent inhibitor of CYP3A4, can affect the metabolism of a variety of drugs, increasing their bioavailability. In some cases, it can lead to a fatal interaction with drugs such as astemizole or terfenadine. Grapefruit juice significantly increased total AUC of lansoprazole in *CYP2C19* PMs, and the total AUC of lansoprazole sulfonic/lansoprazole was significantly decreased in *CYP2C19* EMs (*1/*1).[42] Homozygous wild-type genotype of ABCB1 3435C>T, but not the other genotypes, showed a significant decrease in the active metabolite carebastine urinary excretion after grapefruit juice.[43]

CRANBERRY JUICE

Cranberry juice contains phytochemicals, which may help prevent cancer and cardiovascular diseases. Cranberry juice is high in oxalate and has been suggested to increase the risk for developing kidney stones, although more recent studies have indicated that it may lower the risk. Cranberry significantly increased the area under the INR–time curve by 30% when administered with warfarin, without altering pharmacokinetics or plasma protein binding of S- or R-warfarin. Subjects with one or two copies of the variant alleles for *VKORC1* had a significant reduction in S-warfarin EC50 (concentration of S-warfarin that produces 50% inhibition of prothrombin complex activity) when warfarin was coadministered with cranberry juice extract.[41]

TIANQI JIANGTANG

Tianqi Jiangtang is an herbal prescription widely used for diabetes treatment in China. Tianqi Jiangtang consists of 10 Chinese herbal medicines, namely, Radix Astragali, Radix Trichosanthis, Fructus Ligustri Lucidi, Dendrobii Caulis, Radix Ginseng, Cortex Lycii Radicis bone, Rhizoma Coptidis, Asiatic Cornelian cherry fruit, Ecliptae Herba, and Chinese gall. Many of these herbal medicines are correlated with diabetes-related parameters. For example, Rhizoma Coptidis and astragalin in Radix Astragali reduce glucose, similar to Diformin. Berberine in Rhizoma Coptidis improves some glycemic parameters. Ginsenoside Re in Radix Ginseng has significant antihyperglycemic effects. The iridosides of *Cornus officinalis* in Asiatic Cornelian cherry fruit prevent diabetic vascular complications. Only one major effective component of Tianqi Jiangtang has been identified, namely, berberine hydrochloride ($C_{20}H_{18}ClNO_4$), which has been successfully employed in antidiabetes treatments.

A total of 194 impaired glucose tolerance (IGT) subjects treated with Tianqi Jiangtang for 12 months were genotyped for 184 mutations in 34 genes involved in drug metabolism or transportation. The rs1142345 (A>G) SNP in the thiopurine S-methyltransferase (TPMT) gene was significantly associated with the hypoglycemic effect of the drug ($p = 0.001$, FDR $p = 0.043$). The "G" allele frequencies of rs1142345 in the healthy (subjects reverted from IGT to normal glucose tolerance), maintenance (subjects still had IGT), and deterioration (subjects progressed from IGT to T2D) groups were 0.094, 0.214, and 0.542, respectively. rs1142345 was also significantly associated with the hypoglycemic effect of the drug between the healthy and maintenance groups ($p = 0.027$, OR = 4.828) and between the healthy and deterioration groups ($p = 0.001$, OR = 7.811). Therefore, rs1142345 was associated with the clinical effect of traditional hypoglycemic herbs. This is the first study to utilize the ADME gene chip in the pharmacogenetic study of traditional herbs.[44]

ARTICHOKE

Artichoke leaf extracts (ALE) possess hypocholesterolemic and antioxidant properties. Clinical trial indicated that ALE could increase HDL cholesterol, while decreasing total cholesterol and LDL cholesterol.[45] And its vasoprotective role works by upregulating endothelial-type nitric oxide synthase gene and downregulating inducible nitric oxide synthase expressions.[46,47] ALE compounds are also reported to modulate the genetic toxicity of the alkylating agent ethyl methanesulfonate.[48]

POMEGRANATE JUICE

Pomegranate juice (PJ) was previously reported to have anti-atherogenic and anti-oxidative properties. This antioxidant effect may be due to PJ phenolic, which can stimulate *PON2* expression via PPAR and AP-1 pathway activation.[49,50] It may also reduce cellular cholesterol accumulation and foam cell formation, and even lessen chemoprevention and chemotherapy effects of cancers.[51] Studies have been conducted by researches to find whether PJ could involve the pharmacokinetic pathways to influence drug responses. Faria et al. found that male mice that consumed PJ decreased hepatic CYP content by 43%, mainly CYP1A2 and CYP3A.[52] Studies by Hidaka et al. showed that components of pomegranate inhibited enteric CYP3A, therefore inhibiting CYP3A-mediated metabolism of carbamazepine and further altered the carbamazepine pharmacokinetics (the AUC of carbamazepine increased by 1.5-fold).[53] PJ is also reported to influence the pharmacokinetics of nitrendipine. The increase in AUC of nitrendipine in coadministered and pretreated groups was 1.8- and 4.99-fold, and in C_{max} was 1.4- and 4.1-fold, respectively. This effect may due to the chronic inhibition by PJ on intestinal P-gp-mediated efflux and CYP3A-mediated metabolism of nitrendipine.[54]

CINNAMON

The Chinese herb medicine cinnamon is known to have antidiabetic effects. It can improve both insulin resistance and glucose metabolism. The antidiabetic effect of cinnamon may work by activating AMP-activated protein kinase pathways, upregulating mitochondrial uncoupling proteins-1 (UCP-1), and enhancing the production and translocation of glucose transporter 4 (GLUT4).[55,56] Currently, there is a lack of genetic variation study on the differences in cinnamon treatment response. Studies have reported that functional polymorphisms rs5435 (C→T) of the *GLUT4* gene is associated with diabetes, with the odds ratio for the C/T + T/T genotype being 1.26 when taking the C/C genotype as reference.[57] Whether the cinnamon treatment effect is affected by this glucose transporter gene polymorphism still remains to be validated.

BUTTERBUR

Butterbur or *Petasites hybridus* is a herbal remedy for add-on therapy of asthma because of its anti-allergic and anti-inflammatory effects.[58] Its preventive treatment of migraine has also been widely reported to be effective.[59] A bioactive compound, *S*-petasin, isolated from butterbur, is reported to inhibit adrenocorticotropin or cAMP production and decrease the activities of cytochrome P450 side-chain cleavage enzyme (P450scc).[60] In vitro study discovered that butterbur extracts can inhibit COX-2 and PGE2 release by direct interaction with the enzyme and by preventing p42/44 MAP kinase activation.[61]

Together with the above-mentioned herbs, artichoke, PJ, and cinnamon are well-known CAM for diseases. Pharmacogenomic mechanisms are proved to involve drug response and affect herb–drug interactions via regulating pharmacokinetic and pharmacodynamics gene expressions. In these four herbs, however, there is still a lack of pharmacogenomic study. They will probably increase the risk of treatment failure or adverse reactions alone or in combination with other drugs.

CONCLUSIONS AND FUTURE PERSPECTIVES

Although the current experimental evidence and population studies cannot provide sufficiently clear recommendations on personalized CAM based on genetic characteristics, experience has already been accumulated in the interactions between SNPs and vitamins, minerals, trace elements, and natural herbs. These CAM-based pharmacogenomic research results are critical to repeated tests in different population or races. Existing evidence for gene–CAM interaction is not particularly strong, and many studies are limited by a short period of observation, small sample size, or flawed experimental design. Future studies need to improve based on the past results on experimental design, sufficient certainty, randomized samples, longer observation period, and more strict inclusion criteria. We hope that in the coming years, CAM-related pharmacogenetics and pharmacogenomics can achieve more solid, repeatable, and high-quality results, accelerating individualized CAM.

STUDY QUESTIONS

1. Which of the choices belong to CAM?
   a. Prescription drugs
   b. Traditional medicine
   c. Chinese medicine
   d. Folk medicine
2. Vitamins are reported to be affected by genetic polymorphisms including_____?
   a. Vitamin C
   b. Vitamin D
   c. Vitamin E
   d. Vitamin K
3. The elements below are macroelements in the human body except for___?
   a. Calcium
   b. Oxygen
   c. Carbon
   d. Potassium
4. Herbal medicines, responses, such as ____, are influenced by genetic variations and consequently differences in herb–drug interactions.
   a. Cranberry juice
   b. Magnesium
   c. Ginkgo
   d. St. John's wort
5. Which of the following processes are influenced by pharmacogenomics?
   a. Absorption
   b. Distribution
   c. Metabolism
   d. Excretion

Answer Key
1. bcd
2. abcd
3. ad
4. acd
5. abcd

REFERENCES

1. Teut, M. and K. Linde. Scientific case research in complementary and alternative medicine—A review. *Complement Ther Med*, 2013;21(4):388–95.
2. Weeks, L.C. and T. Strudsholm. A scoping review of research on complementary and alternative medicine (CAM) and the mass media: Looking back, moving forward. *BMC Complement Altern Med*, 2008;8:43.
3. Tanwar, V.S., et al. Common variant in FUT2 gene is associated with levels of vitamin B(12) in Indian population. *Gene*, 2013;515(1):224–8.
4. Bender. *Nutritional biochemistry of the vitamins*. Cambridge, UK: Cambridge University Press; 2003.
5. Diop-Bove, N., et al. A novel deletion mutation in the proton-coupled folate transporter (PCFT; SLC46A1) in a Nicaraguan child with hereditary folate malabsorption. *Gene*, 2013;527(2):673–4.
6. Wang, Q., et al. The first Chinese case report of hereditary folate malabsorption with a novel mutation on SLC46A1. *Brain Dev*, 2015;37(1):163–7.
7. Von Castel-Dunwoody, K.M., et al. Transcobalamin 776C→G polymorphism negatively affects vitamin B-12 metabolism. *Am J Clin Nutr*, 2005;81(6):1436–41.
8. Cahill, L.E. and A. El-Sohemy. Vitamin C transporter gene polymorphisms, dietary vitamin C and serum ascorbic acid. *J Nutrigenet Nutrigenomics*, 2009;2(6):292–301.
9. Dalgard, C., et al. Variation in the sodium-dependent vitamin C transporter 2 gene is associated with risk of acute coronary syndrome among women. *PLoS One*, 2013;8(8):e70421.
10. Pani, M.A., et al. Vitamin D 1alpha-hydroxylase (CYP1alpha) polymorphism in Graves' disease, Hashimoto's thyroiditis and type 1 diabetes mellitus. *Eur J Endocrinol*, 2002;146(6):777–81.
11. Al-oanzi, Z.H., et al. Vitamin D-binding protein gene microsatellite polymorphism influences BMD and risk of fractures in men. *Osteoporos Int*, 2008;19(7):951–60.
12. Girona, J., et al. The apolipoprotein A5 gene −1131T→C polymorphism affects vitamin E plasma concentrations in type 2 diabetic patients. *Clin Chem Lab Med*, 2008;46(4):453–7.
13. Testa, R., et al. Effect of 4G/5G PAI-1 polymorphism on the response of PAI-1 activity to vitamin E supplementation in type 2 diabetic patients. *Diabetes Nutr Metab*, 2004;17(4):217–21.
14. Zuchinali, P., et al. Influence of VKORC1 gene polymorphisms on the effect of oral vitamin K supplementation in over-anticoagulated patients. *J Thromb Thrombolysis*, 2014;37:338–44.
15. McNulty, H., et al. Nutrition throughout life: Folate. *Int J Vitam Nutr Res*, 2012;82(5):348–54.
16. Torre, M.L., et al. MTHFR C677T polymorphism, folate status and colon cancer risk in acromegalic patients. *Pituitary*, 2014;17:257–66.
17. Coppede, F., V. Lorenzoni, and L. Migliore. The reduced folate carrier (RFC-1) 80A>G polymorphism and maternal risk of having a child with Down syndrome: A meta-analysis. *Nutrients*, 2013;5(7):2551–63.

18. Castanet, M., E. Mallet, and M.L. Kottler. Lightwood syndrome revisited with a novel mutation in CYP24 and vitamin D supplement recommendations. *J Pediatr*, 2013; 163(4):1208–10.

19. Cardoso, B.R., et al. Glutathione peroxidase 1 Pro198Leu polymorphism in Brazilian Alzheimer's disease patients: Relations to the enzyme activity and to selenium status. *J Nutrigenet Nutrigenomics*, 2012;5(2):72–80.

20. Penney, K.L., et al. Selenoprotein P genetic variants and mrna expression, circulating selenium, and prostate cancer risk and survival. *Prostate*, 2013;73(7):700–5.

21. Penney, K.L., et al. A large prospective study of SEP15 genetic variation, interaction with plasma selenium levels, and prostate cancer risk and survival. *Cancer Prev Res*, 2010;3(5):604–10.

22. Li, H., et al. Manganese superoxide dismutase polymorphism, prediagnostic antioxidant status, and risk of clinical significant prostate cancer. *Cancer Res*, 2005;65(6): 2498–504.

23. Christensen, M.H., et al. 1,25-Dihydroxyvitamin D and the vitamin D receptor gene polymorphism Apa1 influence bone mineral density in primary hyperparathyroidism. *PLoS One*, 2013;8(2):e56019.

24. Lambrinoudaki, I., et al. Vitamin D receptor Bsm1 polymorphism, calcium metabolism and bone mineral density in patients with multiple sclerosis: A pilot study. *Neurol Sci*, 2013;34(8):1433–39.

25. Lee, J.B., et al. A missense mutation (c.1963A<G) of the complementary component 2 (C2) gene is associated with serum Ca(+)(+) concentrations in pigs. *Mol Biol Rep*, 2012; 39(10):9291–97.

26. Baeza-Richer, C., et al. Identification of a novel quantitative trait nucleotype related to iron status in a calcium channel gene. *Dis Markers*, 2013;34(2):121–9.

27. Valenti, L., et al. The A736V TMPRSS6 polymorphism influences hepatic iron overload in nonalcoholic fatty liver disease. *PLoS One*, 2012;7(11):e48804.

28. Meyer, T.E., et al. Genome-wide association studies of serum magnesium, potassium, and sodium concentrations identify six loci influencing serum magnesium levels. *PLoS Genet*, 2010;6(8):pii:e1001045.

29. Li, X., et al. Common polymorphism rs11191548 near the CYP17A1 gene is associated with hypertension and systolic blood pressure in the Han Chinese population. *Am J Hypertens*, 2013;26(4):465–72.

30. Zhang, X., et al. XRCC1 Arg399Gln was associated with repair capacity for DNA damage induced by occupational chromium exposure. *BMC Res Notes*, 2012;5:263.

31. Van Strater, A.C. and J.P. Bogers. Interaction of St John's wort (*Hypericum perforatum*) with clozapine. *Int Clin Psychopharmacol*, 2012;27(2):121–4.

32. Naccarato, M., D. Yoong, and K. Gough. A potential drug-herbal interaction between *Ginkgo biloba* and efavirenz. *J Int Assoc Physicians AIDS Care*, 2012;11(2):98–100.

33. Wang, X.D., et al. Impact of the haplotypes of the human pregnane X receptor gene on the basal and St John's wort-induced activity of cytochrome P450 3A4 enzyme. *Br J Clin Pharmacol*, 2009;67(2):255–61.

34. Wang, L.S., et al. The influence of St John's wort on CYP2C19 activity with respect to genotype. *J Clin Pharmacol*, 2004;44(6):577–81.

35. Wang, L.S., et al. St John's wort induces both cytochrome P450 3A4-catalyzed sulfoxidation and 2C19-dependent hydroxylation of omeprazole. *Clin Pharmacol Ther*, 2004; 75(3):191–7.

36. Schwarz, U.I., et al. Induction of intestinal P-glycoprotein by St John's wort reduces the oral bioavailability of talinolol. *Clin Pharmacol Ther*, 2007;81(5):669–78.

37. Rengelshausen, J., et al. Opposite effects of short-term and long-term St John's wort intake on voriconazole pharmacokinetics. *Clin Pharmacol Ther*, 2005;78(1):25–33.

38. Yin, O.Q., et al. Pharmacogenetics and herb-drug interactions: Experience with *Ginkgo biloba* and omeprazole. *Pharmacogenetics*, 2004;14(12):841–50.
39. Fan, L., et al. Induction of cytochrome P450 2B6 activity by the herbal medicine baicalin as measured by bupropion hydroxylation. *Eur J Clin Pharmacol*, 2009; 65(4):403–9.
40. Fan, L., et al. The effect of herbal medicine baicalin on pharmacokinetics of rosuvastatin, substrate of organic anion-transporting polypeptide 1B1. *Clin Pharmacol Ther*, 2008;83(3):471–6.
41. Mohammed, A.M., et al. Pharmacodynamic interaction of warfarin with cranberry but not with garlic in healthy subjects. *Br J Pharmacol*, 2008;154(8):1691–700.
42. Uno, T., et al. Lack of significant effect of grapefruit juice on the pharmacokinetics of lansoprazole and its metabolites in subjects with different CYP2C19 genotypes. *J Clin Pharmacol*, 2005;45(6):690–4.
43. Gervasini, G., et al. The effect of CYP2J2, CYP3A4, CYP3A5 and the MDR1 polymorphisms and gender on the urinary excretion of the metabolites of the H-receptor antihistamine ebastine: A pilot study. *Br J Clin Pharmacol*, 2006;62(2):177–86.
44. Li, X., et al. The rs1142345 in TPMT affects the therapeutic effect of traditional hypoglycemic herbs in prediabetes. *Evid Based Complement Alternat Med*, 2013; 2013:327629.
45. Rondanelli, M., et al. Beneficial effects of artichoke leaf extract supplementation on increasing HDL-cholesterol in subjects with primary mild hypercholesterolaemia: A double-blind, randomized, placebo-controlled trial. *Int J Food Sci Nutr*, 2013; 64(1):7–15.
46. Li, H., et al. Flavonoids from artichoke (*Cynara scolymus* L.) up-regulate endothelial-type nitric-oxide synthase gene expression in human endothelial cells. *J Pharmacol Exp Ther*, 2004;310(3):926–32.
47. Xia, N., et al. Artichoke, cynarin and cyanidin downregulate the expression of inducible nitric oxide synthase in human coronary smooth muscle cells. *Molecules*, 2014; 19(3):3654–68.
48. Jacociunas, L.V., et al. Protective activity of *Cynara scolymus* L. leaf extract against chemically induced complex genomic alterations in CHO cells. *Phytomedicine*, 2013; 20(12):1131–34.
49. Shiner, M., B. Fuhrman, and M. Aviram. Macrophage paraoxonase 2 (PON2) expression is up-regulated by pomegranate juice phenolic anti-oxidants via PPAR gamma and AP-1 pathway activation. *Atherosclerosis*, 2007;195(2):313–21.
50. Rosenblat, M., N. Volkova, and M. Aviram. Pomegranate juice (PJ) consumption antioxidative properties on mouse macrophages, but not PJ beneficial effects on macrophage cholesterol and triglyceride metabolism, are mediated via PJ-induced stimulation of macrophage PON2. *Atherosclerosis*, 2010;212(1):86–92.
51. Malik, A., et al. Pomegranate fruit juice for chemoprevention and chemotherapy of prostate cancer. *Proc Natl Acad Sci U S A*, 2005;102(41):14813–18.
52. Faria, A., et al. Pomegranate juice effects on cytochrome P450S expression: In vivo studies. *J Med Food*, 2007;10(4):643–9.
53. Hidaka, M., et al. Effects of pomegranate juice on human cytochrome p450 3A (CYP3A) and carbamazepine pharmacokinetics in rats. *Drug Metab Dispos*, 2005; 33(5):644–8.
54. Voruganti, S., et al. Effect of pomegranate juice on intestinal transport and pharmacokinetics of nitrendipine in rats. *Phytother Res*, 2012;26(8):1240–45.
55. Shen, Y., et al. Verification of the antidiabetic effects of cinnamon (*Cinnamomum zeylanicum*) using insulin-uncontrolled type 1 diabetic rats and cultured adipocytes. *Biosci Biotechnol Biochem*, 2010;74(12):2418–25.

56. Cao, H., D.J. Graves, and R.A. Anderson. Cinnamon extract regulates glucose transporter and insulin-signaling gene expression in mouse adipocytes. *Phytomedicine*, 2010;17(13):1027–32.
57. Cao, H., M.M. Polansky, and R.A. Anderson. Cinnamon extract and polyphenols affect the expression of tristetraprolin, insulin receptor, and glucose transporter 4 in mouse 3T3-L1 adipocytes. *Arch Biochem Biophys*, 2007;459(2):214–22.
58. Lee, K.P., et al. Anti-allergic and anti-inflammatory effects of bakkenolide B isolated from *Petasites japonicus* leaves. *J Ethnopharmacol*, 2013;148(3):890–4.
59. Lipton, R.B., et al. *Petasites hybridus* root (butterbur) is an effective preventive treatment for migraine. *Neurology*, 2004;63(12):2240–44.
60. Chang, L.L., et al. Effects of *S*-petasin on cyclic AMP production and enzyme activity of P450scc in rat zona fasciculata-reticularis cells. *Eur J Pharmacol*, 2004;489(1–2): 29–37.
61. Fiebich, B.L., et al. *Petasites hybridus* extracts in vitro inhibit COX-2 and PGE2 release by direct interaction with the enzyme and by preventing p42/44 MAP kinase activation in rat primary microglial cells. *Planta Med*, 2005;71(1):12–19.

11 Pharmacoeconogenomics
A Good Marriage of Pharmacoeconomics and Pharmacogenomics

Hong-Guang Xie

CONTENTS

KEY CONCEPTS

- The cost of healthcare is rapidly increasing worldwide, thereby becoming a common barrier to the widespread use of an expensive new drug, medical innovation, or biological technology in clinical settings.
- Economic considerations are of paramount importance for optimal selection of the drug and dosage in patient care.
- Better safety, higher efficacy, and lower cost have been widely recognized to be the gold standard for optimal drug therapy.

- Pharmacogenomics (PGx) promises to improve healthcare outcomes of the patient by increasing drug safety and efficacy in patient care, but it cannot be assumed to be cost-effective if relevant economic evaluation is not available in terms of potential additional costs and benefits as a result of genotyping. Therefore, a PGx-based intervention needs to answer a key question of whether or not it provides the patient with improved healthcare outcomes at a reasonable additional cost when compared with conventional strategies.
- The term *pharmacoeconogenomics*, coined in this chapter, refers to the study of economic evaluation of the PGx intervention in terms of the fact that the PGx component has been incorporated into the research field of pharmacoeconomics.
- Not all patients would benefit cost-effectively from only genotype-guided medications.
- Marked cost-effectiveness would more likely be anticipated for the drug whose therapeutic window is narrow, whose treatment response is highly variable among individual patients, and whose indications are severe chronic diseases that are difficult to treat or to measure.
- Information on pharmacoeconogenomics should be detailed in drug labeling and be updated over time.

INTRODUCTION

The marketed drug or medication is a class of special goods that is approved to treat, prevent, or diagnose diseases. Currently, disease-burden and healthcare spending have risen at meteoric rates throughout the world because of scarce health resources; upward-spiraling costs of technological innovation; rapidly escalating costs of drug research and development (R&D); gradually declining R&D productivity; increased demands for evidence of safety and efficacy when a new drug is approved; widening gaps in the ratio of R&D successes to failures; relatively shorter patent exclusivity periods; and fast-growing global populations, in particular, the aging populations (who often need polypharmacy). Therefore, evaluation of the costs and benefits of healthcare has become increasingly important in the management and distribution of limited resources. In other words, economic considerations are of paramount importance for optimal choice of the drug and dosage in patient care. A large number of clinical research studies have well documented that the cost has become a common barrier to the widespread use of an expensive new drug, medical innovation, or biological technology in clinical settings.[1] Therefore, better safety, higher efficacy, and lower costs have been widely accepted to be the gold standard of optimal drug therapy.[2-6] In short, pharmacoeconomics comes of age.[4]

Optimal drug therapy is desired to be cost-effective (less costly and more effective). Pharmacogenomics (PGx) promises to improve healthcare outcomes of the patient by increasing drug efficacy and minimizing adverse drug reactions. PGx testing performance is a critical factor that could affect cost-effectiveness of drug therapy in terms of potential additional costs and benefits as a result of genotyping.[7] Although PGx-based personalized medicine has the potential to increase drug efficacy and safety, which results in substantial economic savings by decreasing costs associated with therapy failure or drug toxicity to be avoided by genotyping, PGx cannot be assumed to be cost-effective if any economic evaluation is not available for a certain PGx intervention. To determine whether PGx technologies or interventions are implemented in a cost-effective manner, all costs and benefits resulting from a PGx intervention should be properly evaluated according to the basic principles and practice of pharmacoeconomics.[5,7–10] In this chapter, a novel term—*pharmacoeconogenomics*—is coined to better describe the marriage of pharmacoeconomics and PGx in patient care. In other words, pharmacoeconogenomics is the study of economic evaluation of PGx interventions. Through this chapter, the readers can better follow the basic principles of pharmacoeconomics and some well-defined examples in the PGx research field, and further understand how the PGx component is incorporated into pharmacoeconomics, and why not all patients benefit more from a cost-effective perspective, using only genotype-guided medications.

PRINCIPLES OF PHARMACOECONOGENOMICS

The term *cost* is defined as the value of resource input relative to "charge," which is the amount charged to the payer. The direct costs refer to all costs where funds are paid out as a consequence of the intervention, whereas indirect costs refer to all other costs not categorized as direct costs. On the other hand, healthcare outcomes or outputs of these costs are measured in different ways in pharmacoeconomic evaluation, such as cost per quality-adjusted life years (QALY) gained, quality of life-adjusted life expectancy, the number and severity of medical events, the number of deaths averted, or the costs saved by prevention of hospitalization of a patient with adverse drug reaction events after use of an alternative medication or another intervention, and so on. Therefore, measurement and assessment of the outcomes produced by the resource inputs is the key for a pharmacoeconomic study—in particular, PGx-based strategies.[11]

PGx is the study of how genetic differences affect variation in response to medication, and thus testing of a patient's genetic make-up has the potential to predict intended or undesired responses to the medication in patient care. How such tests can best be performed and applied in a cost-effective way is the key to guide an economic evaluation of PGx testing in clinical settings. Therefore, the major principles and methodologies of pharmacoeconomic analyses are also applied to relevant PGx studies.

COST-EFFECTIVENESS ANALYSIS

Cost-effectiveness analysis (CEA) is a widely used tool to measure and assess the value of new healthcare interventions relative to empiric or current conventional therapy with standard dosing, and also a quantitative method to systematically compare the costs and healthcare outcomes of competing healthcare technologies or interventions. By definition, CEA can use various outcomes, including clinical events or life years saved. In fact, CEA is a specific type of economic evaluation that measures costs in relation to all types of the outcomes gained, such as life years saved by an alternative or intervention. In the case of the PGx-based CEA study, costs and effects of the PGx assay are compared with that of conventional clinical practice without using PGx information for decision making of drug therapy, because this approach can answer the following basic question[9]: Does a PGx-based intervention provide a patient with improved healthcare outcomes and quality of life at a reasonable additional cost when compared with the conventional treatment strategies (standard care without PGx testing)? In general, PGx-based treatment strategies would be likely to be cost-effective when the following criteria can be met to a greater extent: (1) the genetic variant of interest is relatively prevalent in the patient population; (2) there is a strong or established genotype–phenotype association; (3) patients may face severe clinical outcomes or life-threatening status and even death if the treatment strategy is not appropriate; (4) severe adverse drug reactions would be minimized or avoided by genotype-guided personalized medicine; and (5) genetic testing is highly specific, sensitive, relatively cheap, and easy to use.[7,9,12,13] In short, PGx interventions would be cost-effective when severe clinical and/or economic consequences could be avoided after use of PGx testing. However, PGx-based therapies are not always cost-effective in systematic review articles of eligible CEA studies about PGx interventions for patient care.[7,14] More importantly, PGx interventions are often applied to the drug with a narrow therapeutic window, or a drug with high variation in drug response, not to all marketed drugs.[9,15]

COST-UTILITY ANALYSIS

Cost-utility analysis (CUA) measures the benefit in patient-oriented (or clinical) terms (such as cost per QALY saved by the intervention), and thus different interventions used for patients can be directly compared by standardizing the denominator.[10] By definition, CUA uses QALY only as its outcome. For example, an alternative or new intervention can be compared with conventional clinical practice in an incremental analysis. In other words, the new or alternative interventions would be believed to be cost-effective if they produce health benefits at a certain cost comparable to or less than that of other currently accepted drug therapies (or so-called conventional treatment strategies) in clinical settings. Clearly, as a specific type of CEA, CUA has been more accepted in healthcare than other types of economic evaluation,[14] due to more emphases on use of the QALY in pharmacoeconogenomic studies.

The incremental cost-effectiveness (ICE)[9,10,16] ratio is defined as

$$ICE = (C2 - C1)/(E2 - E1)$$

where C2 and E2 denote the cost and effectiveness of the new intervention being evaluated, and C1 and E1 refer to the cost and effectiveness of the standard care (or current medical practice), respectively. For a CUA study of PGx, the ICE ratio also represents the difference in costs between the two alternatives divided by the difference in effectiveness between the same two alternatives,[8] such as cost per QALY gained, where the quality values for the alternatives (also known as utility values) are either estimated as part of the study or retrieved from the existing literature, and one

QALY is equivalent to one year of perfect health. In general, cost-effectiveness may vary by the strategy or assumptions,[7] and an alternative intervention is considered cost-effective when the ICE ratio is typically less than USD 50,000 per QALY gained, a generally accepted cutoff value.[7,9,10,17]

COST-BENEFIT ANALYSIS

Cost-benefit analysis (CBA) is used to value and measure all costs and all benefits in economic terms, and to compute a net monetary gain/loss or a cost-benefit ratio. CBA intends to ignore clinical procedures that may improve the quality or length of life to be gained, without any direct measurable savings that can be anticipated. In short, for CBA, the costs are compared with healthcare outcomes in dollar terms only. Because it is somehow difficult to value the outcomes in dollar terms, CBA studies are performed less frequently than other types of economic evaluation for healthcare services, and are also less frequently accepted by healthcare policy makers.[9,18]

COST-MINIMIZATION ANALYSIS

Cost-minimization analysis (CMA), a specific type of economic evaluation, is used to compare the costs without quantitative comparison to healthcare outcomes, and then to find the least costly intervention among those shown or assumed to be of equal benefit. In fact, it is useful only if the healthcare outcomes that need to be compared are assumed to be the same.

PRACTICE OF PHARMACOECONOGENOMICS IN CLINICAL SETTINGS

The widespread use of PGx testing in clinical practice is either limited by conflicting results or indefinite clinical utility,[9,14,15] in particular, for frequently prescribed, long-term medications. Developing and validating a clinically significant and practical PGx testing in a cost-effective manner is challenging. For a genetic variant that occurs in a patient with an unusual drug response (side effects[16,19–21] or impaired efficacy) or whose functional alteration is known, bringing PGx testing to the bedside may be cost-effective to some extent.[21] In most (if not all) situations, patients would benefit more from genotyping of some variant genes throughout their whole life at a one-time cost, in particular, for those with high-gene penetrance or those associated with clinical phenotype. The factors that could be used to predict more cost-effectiveness for PGx-based drug therapies are summarized in Table 11.1. Obviously, PGx testing cannot provide more additional benefits or ICE ratio for all drugs, diseases, genes, and patients[7]; and the ICE ratio of PGx-guided drug therapy needs to be evaluated on a case-by-case basis. Some well-defined examples are briefly summarized below to better understand why pharmacoeconogenomics is more important for some medications, rather than for all medications in patient care.

TABLE 11.1

Factors for Prediction of Relative Cost-Effectiveness for PGx-Based Interventions

	Feature Favoring ICE Ratio	Gene–Drug Interaction
Drugs	With a narrow therapeutic window	*CYP2C9* (warfarin, phenytoin)
	The presence of marked interpatient or ethnic variation in drug disposition and/or response	*CYP2C19* (clopidogrel)
	Clinically severe or expensive drug side effects to be avoided	*TPMT* (6-MP, AZT)
	Developed for genetically identified subpopulations	HER2 (trastuzumab) BCR-ABL (imatinib)
	Difficult to measure and evaluate drug response	
	Long-term medications or expensive drug costs	
	Identification of nonresponders or patients who require higher doses of medication	
Diseases/ patients	Carriage of some inherited (germline) mutations with known functional alteration	*CYP2C9*3* (phenytoin)
	Experiencing an unusual drug response (severe drug toxicity or therapy failure)	*CYP2D6* (codeine)
	Difficult to treat	
	Long-term, severe, or expensive clinical outcomes that may be improved or avoided	
	Difficult to predict or monitor disease prognosis or treatment outcomes	
	Suffering from chronic, severe, or even life-threatening diseases	
	The presence of acquired somatic mutations from tumor	EGFR (gefitinib)
	Identification of patients who would never be clinically detected under the status quo	
Biomarkers/ genes	A strong or established genotype–phenotype association	*CYP2D6* (codeine)
	Relatively common polymorphisms in the populations	*CYP2C19*2* (omeprazole)
	More virulent viral genotyping	HCV (interferon, ribavirin)
	A significantly clinical or economic benefit gained from naturally occurring, less frequent variant alleles	*CYP2C9*3* (warfarin)
	A relatively high positive predictive value of genotyping	*HLA* B*5701 (abacavir) HER2 (trastuzumab)
	A rapid and relatively inexpensive genotyping assay or validated genetic diagnosis	
	Identification of candidates for a novel therapy that is not available without the test	HER2 (trastuzumab)

Note: AZT, azathioprine; BCR-ABL, breakpoint cluster region-abelson; CYP, cytochrome P450; EGFR, epidermal growth factor receptor; HCV, hepatitis C virus; HER2, human epidermal growth factor receptor 2; HLA, human leukocyte antigen; ICE: incremental cost-effectiveness; 6-MP, 6-mercaptopurine; TPMT, thiopurine *S*-methyltransferase.

TPMT and the Thiopurine Drugs

TPMT testing before initiation of drug therapy is one of the first clinical examples of use of PGx testing in patient care. The thiopurine drugs are frequently prescribed for additional indications other than acute lymphoblastic leukemia (ALL), such as rheumatoid arthritis, inflammatory bowel disease, and other autoimmune conditions, and also for immune suppression following organ transplants. Accumulated evidence has well documented that 6-mercaptopurine (6-MP, launched in the United States in 1953) dramatically improves the cure rate of ALL children,[22] but that severe myelosuppression (or leukopenia) or fatal sepsis is the most severe, even life-threatening adverse drug reaction associated with the use of purine drugs, such as 6-MP[23] and azathioprine (AZT, used as an immunosuppressant in transplant patients since the 1950s).[24] Isolated in 1980, thiopurine *S*-methyltransferase (TPMT) has been identified as the key catabolic enzyme responsible for the inactivation of thiopurine drugs, including 6-MP,[25] and thus TPMT-inherited deficiency (due to germline mutations) is associated with severe hematopoietic toxicity (such as myelosuppression) when treated with standard doses of 6-MP.[26-29] Clinical studies have observed that 0.3% of white subjects could be at increased risk of suffering thiopurine-induced myelosuppression due to the presence of two deficient copies of the *TPMT* gene, which are termed as a poor metabolizer phenotype,[27,30,31] and that approximately 10% of patients who have intermediate levels of TPMT activity would also be at increased risk of myelosuppression from standard doses of thiopurine drugs,[30] although their severe toxicity would be less likely. Therefore, it is necessary to identify or predict whether an individual patient could be an intermediate or poor metabolizer phenotype of TPMT before starting thiopurine drugs, and lower dosage or alternative drugs would be prescribed for patients with intermediate TPMT activity levels or deficient metabolizers.

In terms of the fact that the consequences for a patient who is not tested for TPMT before use of thiopurine drugs could be life-threatening, phenotyping or genotyping assay of TPMT prior to thiopurine treatment may have a favorable cost-effectiveness ratio in ALL patients.[12] Phenotyping assay of TPMT in humans is performed as described elsewhere,[32] using red blood cells as the source of the enzyme TPMT, and phenotypic monitoring of TPMT through red blood cell counting may reduce the drive to implement *TPMT* genotyping testing. However, the drawbacks of the phenotyping assay include labor-intensive, time-consuming, required expensive instrumentation; and it is impractical in patients receiving transfusions. In contrast, the PCR-based RFLP genotyping of *TPMT* is used to determine the presence of *TPMT*-deficient alleles, which has been validated to predict the TPMT activity levels and to further predict the risk of severe neutropenia for 6-MP and AZT.[27]

TPMT testing before initiation of drug therapy is one of the best examples of PGx that is not only clinically useful but also cost-effective. In Europe (United Kingdom, Ireland, Germany, and the Netherlands), the actual level of implication was estimated to be 12% according to a multicenter survey.[1] When stratified by the biomarker, TPMT is one of the most common biomarkers evaluated in a total of 34 eligible articles about economic evaluation of PGx strategies published from 1999 to 2009.[14] In a CEA study, screening for TPMT before treating rheumatoid arthritis patients with AZT indicated that *TPMT* genotyping was relatively cost-effective.[33]

HER2 AND TRASTUZUMAB

Human epidermal growth factor receptor-2 (HER2) was identified as a potential monoclonal antibody target in the early 1980s, and thereafter the humanized IgG1 monoclonal antibody trastuzumab moved into clinical trials in 1992 and was approved by the US FDA as Herceptin in 1998 and by the EMEA's centralized procedure in 2000, respectively. Herceptin, specifically developed to target the receptor HER2 and to cause cell death, is an anticancer target therapy for use in patients with breast cancer who are HER2-positive. Genetic testing helps identify overexpression of HER2. There are two different commercial testing kits currently available in the market: one is immunohistochemistry (IHC) testing (which detects overexpression of HER2 protein/antigen in breast cancer tissue) and the other is cytogenetic fluorescent in situ hybridization (FISH) testing (which detects amplification of the gene encoding the HER2), both of which have been approved by the US FDA for detection of HER2 overexpression. The use of the HER2 testing has been required in the drug labeling of trastuzumab treatment. In Europe (United Kingdom, Ireland, Germany, and the Netherlands), HER2 testing was performed in approximately 84% of breast cancer women who received trastuzumab treatment, indicating a larger consensus on the clinical utility of HER2 testing in that clinical setting.[1] Because HER2 testing service can identify who would be good responders to trastuzumab treatment, this testing can reduce inefficacious use of trastuzumab in some breast cancer patients before they receive that biologic. According to a multicenter survey in Europe,[1] the clinical utility is perceived with a higher level of implementation of HER2 testing, and the benefits of HER2 testing clearly outweigh its costs, as measured with cost per life years gained, improved life expectancy, or the number of deaths averted. In addition, HER2 is thought to be a biomarker that has both clinical validity and utility due to significantly improved patient care.[14]

*HLA B*5701* AND ABACAVIR HYPERSENSITIVITY

Abacavir, an HIV reverse transcriptase inhibitor, is often used in combination with other antivirals in the treatment of HIV infection. Hypersensitivity reactions to abacavir occur in approximately 5% of HIV-infected patients taking the drug, which may result in potentially life-threatening hypotension if drug therapy is not ceased.[34] Clinical research studies have also indicated that patients with the human leukocyte antigen *HLA B*5701* are at greater risk of a hypersensitivity reaction to abacavir.[19,20] In terms of the higher prevalence of the *HLA B*5701* (9–18%), an established genotype–phenotype association, the low cost of *HLA B*5701* testing (e.g. $62), and the high cost to treat abacavir-induced hypersensitivity (e.g., $3730), *HLA B*5701* testing before drug therapy would be most likely in a cost-effective fashion.[21] Another study also indicated that the ICE ratio of *HLA* genotyping was cost savings of up to $32,500 per hypersensitivity reaction avoided.[16] Therefore, *HLA* genotyping is considered cost-effective as anticipated,[21] according to a reasonable cutoff value of less than US$50,000 per QALY gained.[7,9,10,17] Furthermore, *HLAB*5701* is considered to have both clinical validity and clinical utility, which lead to improved patient outcomes.[14]

CYP2C9 AND THE ANTICOAGULANT WARFARIN

Warfarin anticoagulation therapy is increasing rapidly in the elderly population because of the increasing prevalence of atrial fibrillation and longer life spans. In 2007, more than 20 million new prescriptions were written for this drug in the United States, according to the National Institutes of Health. Initiation and maintenance therapy of warfarin is difficult because of wide intersubject variability in its dose requirements and response, and thus warfarin dosing is monitored and adjusted to maintain appropriate prothrombin time and international normalized ratio within a target range of 2–3 for each patient. Warfarin is one of the first medications to be relabeled by the US FDA to explain why an individual's genetic make-up could affect response to the drug, and PGx testing for warfarin is one of the first such tests to get US FDA approval.

Although warfarin is a generic medication that is relatively inexpensive and the cost of testing for *CYP2C9* and *VKORC1* also seems to be relatively less expensive (ranging from $250 to $630), the main cost of warfarin dosing is the healthcare spending associated with bleeding complications and death, in particular during the first few months of warfarin initiation therapy to titrate warfarin doses. Because adverse events associated with warfarin therapy are common, and many are preventable through a genotype-guided algorithm, the number of patients who could benefit more from the PGx testing would be large, suggesting the presence of potential cost-effectiveness for genotyping. In 34 eligible pharmacoeconogenomic articles published from 1999 to 2009, *CYP2C9* alone or in combination with *VKORC1* (vitamin K epoxide reductase complex subunit 1) are the most common genes that were evaluated in the clinical settings.[14] *CYP2C9**2 or *3 variant alleles are associated with reduced maintenance dose requirements of warfarin therapy and increased risk of major bleeding events;[35,36] the marginal cost per additional major bleeding event averted was estimated to be $5778 for *CYP2C9* genotype-guided dosing of warfarin,[37] and €4233 for *CYP2C9* genotype-guided dosing of acenocoumarol.[38] Testing of *CYP2C9**2, *3 and *VKORC1* A/A before receiving warfarin was considered to have clinical validity[39] but unclear clinical utility.[40,41] However, in atrial fibrillation patients treated with warfarin, the cost per QALY gained was $60,725 and $170,000 for *CYP2C9* and *VKORC1* genotype-guided dosing, respectively,[42,43] suggesting little cost-effectiveness. Currently, the US Centers for Medicare and Medicaid Services have not found sufficient evidence to cover the cost of genotyping for warfarin dosing.[44] In the future, multicenter, large-scale, prospective, randomized clinical trials will be required to confirm whether genotype-guided warfarin dosing would be less costly and more effective than its conventional regimen.

CYP2C19 AND THE ANTIPLATELET DRUG CLOPIDOGREL

Clopidogrel is a prodrug whose bioactivation is dependent largely on CYP2C19 activity in the liver (Figure 11.1), and thus the *CYP2C19* genotype is the major predictor of adverse cardiovascular events in clopidogrel-treated patients with acute coronary syndrome (ACS) or those undergoing percutaneous coronary intervention (PCI).[45] In 2010, the US FDA consecutively released the black box warning three times for the clopidogrel labeling, alerting clinicians of the role of *CYP2C19**2 and *3 loss-of-function variants in response to that drug in patient care. Although the

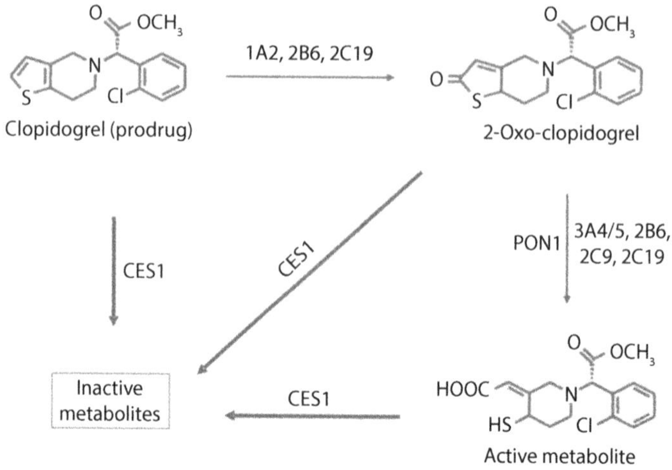

FIGURE 11.1 Clopidogrel and its metabolic pathways in humans.

warning states that clopidogrel poor metabolizers can be identified by genetic tests, the drug labeling does not actually direct clinicians to use any of the tests to guide the personalized medicine of the drug in clinical settings.

A CEA of a *CYP2C19*2* genotype-guided strategy of antiplatelet therapy was performed in ACS patients undergoing PCI, as compared with two "no testing" strategies (empiric clopidogrel or prasugrel).[46] The *CYP2C19* genetic testing was assumed to cost up to $500 per patient,[46,47] generic clopidogrel was estimated to cost $1 every day, and the net whole sale price for prasugrel was estimated as $5.45 per day. Results indicated that a genotype-guided strategy yielded similar outcomes to empiric approaches to treatment but was marginally less costly and more effective.[46] The similar result was replicated with a New Zealand evaluation.[48] Because Eastern Asians, Maoris, and Pacific Islanders have a significantly higher allele frequency of *CYP2C19* loss-of-function variants than populations of European or black African descent,[45,47–49] use of a genetic test to guide clopidogrel therapy in patients with ACS or undergoing PCI would be more cost-effective among Eastern Asians, Maoris, and Pacific Islanders than black and white populations.[47,48]

ECONOMIC EVALUATION OF GENETIC TESTING

Performance of a genetic test should be evaluated with sensitivity, specificity, positive and negative predictive values, and the area under the receiver operating curve.[50] Of them, high specificity and positive predictive value are particularly important for a genetic test that is used to predict response to a medication or to categorize patients into different subgroups. Before a genetic test is integrated into routine clinical practice, two levels of evidence—clinical validity and utility—must be generated. Clinical validity refers to the presence of a definite genotype–phenotype relationship in a certain stratified population, whereas clinical utility is to confirm whether a genetic test improves health outcomes in patient care and whether such a test is cost-effective relative to a standardized strategy (typically conventional care). In other words, if a genetic test is to move from the bench to the bedside, its clinical relevance should be addressed at the bedside. When a genetic test is applied in the clinical setting, it is more likely to be cost-effective if the savings that would be realized by such a test is greater than the cost of performing that test.

As mentioned above, one example of PGx testing is testing for the *HER2* gene. There are several components that are used for economic evaluation of HER2 testing.[50] There are two validated approaches to testing HER2, in which HIC is easier to perform and less expensive (at one-quarter the cost of FISH testing), and FISH has better reliability and validity.[51] There is 20% variation in HER2 testing accuracy and thus, a standardized testing algorithm that includes both ICH and FISH is recommended to improve the positive predictive value for a subgroup of breast cancer patients who would benefit more from trastuzumab treatment (an estimated at least 20–30% of breast cancer patients are HER2-positive). On the other hand, trastuzumab has potential cardiotoxicity when given in combination with chemotherapy, in particular, anthracycline-based regimens,[52] and thus treatment with trastuzumab would require serial monitoring for that adverse event. Because breast cancer is a common disease worldwide, a large number of affected patients would make HER2 testing economically attractive and clinically feasible.[1]

CONCLUSIONS AND FUTURE PERSPECTIVES

In summary, optimal medication for patient care should be of better safety, higher efficacy, and less cost, to a greater extent, as shown in Figure 11.2. Using PGx to individualize drug therapy will have clinical and economic benefits in some (if not all) patients when treated with certain drugs. In other words, PGx will more likely be cost-effective only for a certain combination of the drug, disease (or patient), gene,

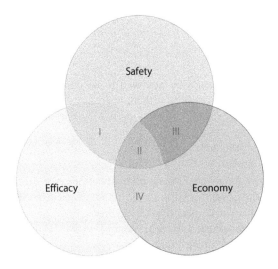

FIGURE 11.2   The optimal drug for patient care should be of improved safety, higher efficacy, and less cost.

and testing assay in nature as summarized in Table 11.1. More ICE ratio is expected for the drug whose therapeutic window is narrow, whose treatment response is highly variable and difficult to measure, and whose indication is a severe chronic disease that is difficult to treat. Because PGx strategies cannot be assumed to be cost-effective, it should be required to perform a comprehensive CEA evaluation of a new PGx-based intervention in advance of its translation to patient care. With rapid advances in genetic technologies, a dramatic decrease in the cost of genetic testing itself has made such tests affordable. To date, most of our research efforts made for PGx studies have been focused on their clinical validity—making the science make sense. Now is the time to increase the research efforts on the clinical utility of economic evaluation of PGx and to further improve the PGx knowledge base for clinical and economic integration and decision making.

ACKNOWLEDGMENTS

This work is supported, at least in part, by the research grants 81473286 by the National Natural Science Foundation of China (NSFC); BL2013001 from the Department of Science and Technology of Jiangsu Province (JSTD); BK2012525 funded by the Jiangsu Natural Science Foundation (JNSF); and HRSS2012-258 from the Ministry of Human Resource and Social Security, China, in addition to support from a research contract #31010300010339 from Nanjing First Hospital, China (all to Dr. Xie).

STUDY QUESTIONS

1. What is the gold standard of current drug therapy in patient care?
2. Which causes are associated with the rapid increase of healthcare spending?
3. Optimal drug therapy is generally required to be
   a. More effective
   b. Less costly
   c. Relatively safe
   d. All of the above
4. Pharmacogenomics cannot be assumed to be cost-effective for a certain pharmacogenomic intervention if there is no
   a. Clinical validity
   b. Clinical utility
   c. Economic evaluation
   d. All of the above
5. Why can't all patients benefit from all medications cost-effectively?
6. The outcome measured for CEA is
   a. Quality-adjusted life year (QALY) gained
   b. Quality of life-adjusted life expectancy
   c. The number of deaths averted
   d. The number of the severe medical events avoided
   e. All of the above
7. What is the outcome measured for CUA?

8. Which analysis has been more accepted in healthcare than other types of economic evaluation?
   a. CEA
   b. CBA
   c. CUA
   d. CMA
   e. None of the above

9. Which alternative intervention is considered cost-effective when the ICE ratio is typically less than per QALY gained, a generally accepted cutoff value?
   a. US$30,000
   b. US$40,000
   c. US$50,000
   d. US$60,000

10. Which economic analysis measures its outcomes in dollar terms only?
    a. CEA
    b. CBA
    c. CMA
    d. CUA
    e. None of the above

11. Which are the most common biomarkers evaluated in a total of 34 eligible articles mentioned in this chapter?
    a. TPMT
    b. HER2
    c. *HLA* B*5701
    d. *VKORC1*, and *CYP2C9**2 or *3
    e. All of the above

12. Why would *CYP2C19* genotyping be more cost-effective among Eastern Asians, Maoris, and Pacific Islanders than black and white populations?

13. Which drugs would be expected to more likely have greater ICE ratio?

Answer Key
   1. Better safety, higher efficacy, and less cost.
   2. See details in the first paragraph of the Introduction section.
   3. d
   4. d
   5. See details in the Conclusions and Future Perspectives section.
   6. e
   7. QALY
   8. CUA
   9. c
   10. b
   11. e
   12. Higher allele frequency of *CYP2C19**2 and *3 among these ethnic populations.
   13. See details in the Conclusions and Future Perspectives section.

REFERENCES

1. Woelderink A, Ibarreta D, Hopkins MM, Rodriguez-Cerezo E. The current clinical practice of pharmacogenetic testing in Europe: TPMT and HER2 as case studies. *Pharmacogenomics J.* 2006;6:3–7.
2. Reidenberg JW. The need to consider the cost factor of drugs in clinical trials. *Clin Pharmacol Ther.* 2003;73:477–478.
3. Waldman SA, Terzic A. Pharmacoeconomics in the era of individualized medicine. *Clin Pharmacol Ther.* 2008; 84:179–182.
4. Walley T, Breckenridge A. Pharmacoeconomics comes of age? *Clin Pharmacol Ther.* 2008;84:279–280.
5. Xie HG, Zhou HH. Pharmacoeconomics: Basic principles and practice. *Chin J Pharmacoepidemiol.* 1995;4:1–4 (in Chinese).
6. Xie HG, Zhou HH. Drugs used for patient care should be more effective and less costly. *Natl Med J China.* 1995;75:3–4 (in Chinese).
7. Phillips KA, van Bebber SL. A systematic review of cost-effectiveness analyses of pharmacogenomic interventions. *Pharmacogenomics.* 2004;5:1139–1149.
8. Phillips KA, Veenstra D, van Bebber S, Sakowski J. An introduction to cost-effectiveness and cost-benefit analysis of pharmacogenomics. *Pharmacogenomics.* 2003;4:231–239.
9. Veenstra DL, Higashi MK, Phillips KA. Assessing the cost-effectiveness of pharmacogenomics. *AAPS Pharm Sci.* 2000;2:E29.
10. Weinstein MC, Siegel JE, Gold MR, Kamlet MS, Russell LB. Recommendations of the panel on cost-effectiveness in health and medicine. *JAMA.* 1996;276:1253–1258.
11. Payne K. Towards an economic evidence base for pharmacogenetics: Consideration of outcomes is key. *Pharmacogenomics.* 2008;9:1–4.
12. van den Akker-van Marle ME, Gurwitz D, Detmar SB, Enzing CM, Hopkins MM, Gutierrez de Mesa E, Ibarreta D. Cost-effectiveness of pharmacogenomics in clinical practice: A case study of thiopurine methyltransferase genotyping in acute lymphoblastic leukemia in Europe. *Pharmacogenomics.* 2006;7:783–792.
13. Flowers CR, Veenstra D. The role of cost-effectiveness analysis in the era of pharmacogenomics. *Pharmacoeconomics.* 2004;22:481–493.
14. Wong WB, Carlson JJ, Thariani R, Veenstra DL. Cost effectiveness of pharmacogenomics: A critical and systematic review. *PharmacoEconomics.* 2010;28:1001–1013.
15. Xie HG, Frueh FW. Pharmacogenomics steps toward personalized medicine. *Per Med.* 2005;2:325–337.
16. Hughes DA, Vilar FJ, Ward CC, Alfirevic A, Park BK, Pirmohamed M. Cost-effectiveness analysis of HLA B*5701 genotyping in preventing abacavir hypersensitivity. *Pharmacogenetics.* 2004;14:335–342.
17. Garber AM, Phelps CE. Economic foundations of cost-effective analysis. *J Health Econ.* 1997;16:1–31.
18. Ofman JJ, Sullivan SD, Neumann PJ, Chiou CF, Henning JM, Wade SW, Hay JW. Examining the value and quality of health economic analyses: Implications of utilizing the QHES. *J Manag Care Pharm.* 2003;9:53–61.
19. Hetherington S, Hughes AR, Mosteller M, Shortino D, Baker KL, Spreen W, Lai E, et al. Genetic variations in HLA-B region and hypersensitivity reactions to abacavir. *Lancet.* 2002;359:1121–1122.
20. Mallal S, Nolan D, Witt C, Masel G, Martin AM, Moore C, Sayer D, et al. Association between presence of HLA-B*5701, HLA-DR7, and HLA-DQ3 and hypersensitivity to HIV-1 reverse-transcriptase inhibitor abacavir. *Lancet.* 2002;359:727–732.
21. Veenstra DL. Bringing genomics to the bedside: A cost-effective pharmacogenomic test? *Pharmacogenetics.* 2004;14:333–334.

22. Coulthard SA, Matheson EC, Hall AG, Hogarth LA. The clinical impact of thiopurine methyltransferase polymorphisms on thiopurine treatment. *Nucleosides Nucleotides Nucleic Acids.* 2004;23:1385–1391.

23. Philips FS, Sternberg SS, Hamilton S, Clarke DA. The toxic effects of 6-mercaptopurine and related compounds. *Ann N Y Acad Sci.* 1954;60:283–296.

24. Fulginiti VA, Scribner R, Groth CG, Putnam CW, Brettschneider L, Gilbert S, Porter KA, Starzl TE. Infections in recipients of liver homografts. *N Engl J Med.* 1968;279:619–626.

25. Weinshilboum RM, Sladek SL. Mercaptopurine pharmacogenetics: Monogenic inheritance of erythrocyte thiopurine methyltransferase activity. *Am J Hum Genet.* 1980;32:651–662.

26. Andersen JB, Szumlanski C, Weinshilboum RM, Schmiegelow K. Pharmacokinetics, dose adjustments, and 6-mercaptopurine/methotrexate drug interactions in two patients with thiopurine methyltransferase deficiency. *Acta Paediatr.* 1998;87:108–111.

27. Krynetski EY, Schuetz JD, Galpin AJ, Pui CH, Relling MV, Evans WE. A single point mutation leading to loss of catalytic activity in human thiopurine S-methyltransferase. *Proc Natl Acad Sci U S A.* 1995;92:949–953.

28. Lennard L, Gibson BE, Nicole T, Lilleyman JS. Congenital thiopurine methyltransferase deficiency and 6-mercaptopurine toxicity during treatment for acute lymphoblastic leukaemia. *Arch Dis Child.* 1993;69:577–579.

29. McLeod HL, Relling MV, Liu Q, Pui CH, Evans WE. Polymorphic thiopurine methyltransferase in erythrocytes is indicative of activity in leukemic blasts from children with acute lymphoblastic leukemia. *Blood.* 1995;85:1897–1902.

30. El Azhary RA. Azathioprine: Current status and future considerations. *Int J Dermatol.* 2003;42:335–341.

31. Tai HL, Krynetski EY, Yates CR, Loennechen T, Fessing MY, Krynetskaia NF, Evans WE. Thiopurine S-methyltransferase deficiency: Two nucleotide transitions define the most prevalent mutant allele associated with loss of catalytic activity in Caucasians. *Am J Hum Genet.* 1996;58:694–702.

32. Ford LT, Cooper SC, Lewis MJ, Berg JD. Reference intervals for thiopurine S-methyltransferase activity in red blood cells using 6-thioguanine as substrate and rapid non-extraction liquid chromatography. *Ann Clin Biochem.* 2004;41:303–308.

33. Marra CA, Esdaile JM, Anis AH. Practical pharmacogenetics: The cost effectiveness of screening for thiopurine S-methyltransferase polymorphisms in patients with rheumatological conditions treated with azathioprine. *J Rheumatol.* 2002;29:2507–2512.

34. Clay PG. The abacavir hypersensitivity reaction: A review. *Clin Ther.* 2002;24:1502–1514.

35. Xie HG, Prasad HC, Kim RB, Stein CM. CYP2C9 allelic variants: Ethnic distribution and functional significance. *Adv Drug Deliv Rev.* 2002;54:1257–1270.

36. Schwarz UI, Ritchie MD, Bradford Y, Li C, Dudek SM, Frye-Anderson A, Kim RB, Roden DM, Stein CM. Genetic determinants of response to warfarin during initial anticoagulation. *N Engl J Med.* 2008;358:999–1008.

37. You JH, Chan FW, Wong RS, Cheng G. The potential clinical and economic outcomes of pharmacogenetics-oriented management of warfarin therapy—A decision analysis. *Thromb Haemost.* 2004;92:590–597.

38. Schalekamp T, Boink GJ, Visser LE, Stricker BH, de Boer A, Klungel OH. CYP2C9 genotyping in acenocoumarol treatment: Is it a cost-effective addition to international normalized ratio monitoring? *Clin Pharmacol Ther.* 2006;79:511–520.

39. Klein TE, Altman RB, Eriksson N, Gage BF, Kimmel SE, Lee MT, Limdi NA, et al. Estimation of the warfarin dose with clinical and pharmacogenetic data. *N Engl J Med.* 2009;360:753–764.

40. Kangelaris KN, Bent S, Nussbaum RL, Garcia DA, Tice JA. Genetic testing before anticoagulation? A systematic review of pharmacogenetic dosing of warfarin. *J Gen Intern Med*. 2009;24:656–664.

41. Garcia DA. Warfarin and pharmacogenomic testing: The case for restraint. *Clin Pharmacol Ther*. 2008;84:303–305.

42. Meckley LM, Gudgeon JM, Anderson JL, Williams MS, Veenstra DL. A policy model to evaluate the benefits, risks and costs of warfarin pharmacogenomic testing. *PharmacoEconomics*. 2010;28:61–74.

43. Eckman MH, Rosand J, Greenberg SM, Gage BF. Cost-effectiveness of using pharmacogenetic information in warfarin dosing for patients with nonvalvular atrial fibrillation. *Ann Intern Med*. 2009;150:73–83.

44. Centers for Medicare and Medicaid Services. Decision memo for pharmacogenomic testing to predict warfarin response (CAG-00400N). Available from: http://www.cms.hhs.gov/site-search/serach-results.html?q=warfarin (accessed April 23, 2014).

45. Xie HG, Zhang YD. Pharmacogenomics and personalized medicine of the antiplatelet drugs. In: Barh D, Dhawan D, Ganguly NK, eds. *Omics for Personalized Medicine*. Springer India, New Delhi, India, 2013, pp. 469–506.

46. Lala A, Berger JS, Sharma G, Hochman JS, Scott Braithwaite R, Ladapo JA. Genetic testing in patients with acute coronary syndrome undergoing percutaneous coronary intervention: A cost-effectiveness analysis. *J Thromb Haemost*. 2013;11:81–91.

47. Thompson CA. Clopidogrel poor metabolizers may need alternative to standard regimen. *Am J Health-Syst Pharm*. 2010;67:779–780.

48. Panattoni L, Brown PM, Te Ao B, Webster M, Gladding P. The cost effectiveness of genetic testing for CYP2C19 variants to guide thienopyridine treatment in patients with acute coronary syndromes: A New Zealand evaluation. *Pharmacoeconomics*. 2012;30:1067–1084.

49. Xie HG, Kim RB, Wood AJJ, Stein CM. Molecular basis of ethnic differences in drug disposition and response. In: Cho AK, Blaschke TF, Insel PA, Loh HH, eds. *Annual Review of Pharmacology and Toxicology*, Vol. 41. Annual Reviews, Palo Alto, CA, 2001, pp. 815–850.

50. Wu AC, Fuhlbrigge AL. Economic evaluation of pharmacogenetic tests. *Clin Pharmacol Ther*. 2008;84:272–274.

51. Hicks DG, Tubbs RR. Assessment of HER2 status in breast cancer by fluorescence in situ hybridization: A technical review with interpretive guidelines. *Hum Pathol*. 2005;36:250–261.

52. Valachis A, Nearchou A, Polyzos NP, Lind P. Cardiotoxicity in breast cancer patients treated with dual HER2 blockade. *Int J Cancer*. 2013;133:2245–2252.

Index

Milton Keynes UK
Ingram Content Group UK Ltd.
UKHW031144141024
449569UK00024B/1079